A Glimpse at Medicine in the Future

Mandana Hasanzad
Editor

A Glimpse at Medicine in the Future

Springer

Editor
Mandana Hasanzad
Medical Genomics Research Center, Tehran Medical Sciences
Islamic Azad University, Tehran
Tehran, Iran

ISBN 978-981-97-4973-7 ISBN 978-981-97-4974-4 (eBook)
https://doi.org/10.1007/978-981-97-4974-4

This Springer imprint is published by the registered company Springer Nature Singapore Pte Ltd.
The registered company address is: 152 Beach Road, #21-01/04 Gateway East, Singapore 189721, Singapore

To my parents,
Source of encouragement, support, and love

Acknowledgments

The production of the present book is truly a collaborative effort. I am greatly indebted to a variety of people. I am sincerely appreciative of their time and dedication.

Additionally, I have great respect for the many people at Springer Nature, whose remarkable efforts I deeply admire.

Contents

Introduction to the Future of Medicine

1

Homa Hemati, George P. Patrinos, Hamid Reza Aghaei Meybodi, Negar Sarhangi, Bagher Larijani, Nazli Khodayari, and Mandana Hasanzad

What Will You Learn in This Chapter?
This chapter will take you on an amazing journey into the evolving landscape of the future of medicine. We will look at the revolutionary advances and transformative technologies that are changing the way we think about healthcare. What you can expect to learn is an introduction to some of these advancements such as genome editing, digital health, personalized medicine, and artificial intelligence.

Rationale and Importance
This brief review of the future of medicine is not only a means of disseminating information but also a catalyst for progress, influencing healthcare practices, public discourse, and policy decisions. Its importance lies in its capacity to shape the narrative around the evolving landscape of medicine, fostering a collective understanding that ultimately benefits healthcare professionals, patients, researchers, and policymakers.

H. Hemati
Faculty of Pharmacy, Department of Pharmacoeconomics and Pharmaceutical Administration, Tehran University of Medical Sciences, Tehran, Iran

Endocrinology and Metabolism Research Center, Endocrinology and Metabolism Clinical Sciences Institute, Tehran University of Medical Sciences, Tehran, Iran

G. P. Patrinos
Department of Pharmacy, School of Health Sciences, University of Patras, Patras, Greece

Department of Genetics and Genomics, College of Medicine and Health Sciences, United Arab Emirates University, Al-Ain, Abu Dhabi, UAE

Zayed Center for Health Sciences, United Arab Emirates University, Al-Ain, Abu Dhabi, UAE

Faculty of Medicine and Health Sciences, Department of Pathology, Clinical Bioinformatics Unit, Erasmus University Medical Center, Rotterdam, The Netherlands

H. R. Aghaei Meybodi · B. Larijani
Endocrinology and Metabolism Research Center, Endocrinology and Metabolism Clinical Sciences Institute, Tehran University of Medical Sciences, Tehran, Iran

N. Sarhangi
Personalized Medicine Research Center, Endocrinology and Metabolism Clinical Sciences Institute, Tehran University of Medical Sciences, Tehran, Iran

Medical Genomics Research Center, Tehran Medical Sciences, Islamic Azad University, Tehran, Iran

N. Khodayari
Division of Pulmonary, Critical Care and Sleep Medicine, Department of Medicine, University of Florida, Gainesville, FL, USA

M. Hasanzad (✉)
Personalized Medicine Research Center, Endocrinology and Metabolism Clinical Sciences Institute, Tehran University of Medical Sciences, Tehran, Iran

M. Hasanzad (ed.), *A Glimpse at Medicine in the Future*,
https://doi.org/10.1007/978-981-97-4974-4_1

1.1 Introduction

The historical progression of medicine has been pivotal in shaping the contemporary healthcare landscape. Initially, medical treatments were exclusive luxuries available to a privileged few. However, notable exceptions, such as the use of digitalis and artemether extracted from natural sources, marked early advancements in treatment efficacy. The twentieth century saw the emergence of consistently effective pharmaceutical agents like aspirin, morphine, insulin, sulphonamides, and penicillin. However, these developments also brought about risks, exemplified by the Elixir Sulfanilamide tragedy in 1937, which prompted the introduction of drug testing and regulation through the Federal Food, Drug, and Cosmetic Act. Advancements in post-marketing surveillance have led to sophisticated systems for monitoring new medicines, enhancing safety protocols, and allowing for orderly drug withdrawals. This has significantly improved the quality of life for individuals suffering from various diseases, leading to an increase in life expectancy globally [1].

Historical estimates indicate a notable increase in life expectancy, rising from less than 40 years in the 1800s to an average of 73.16 years in 2023. Moreover, reductions in child and maternal mortality rates have contributed to this increase in life expectancy. Diseases related to communicable illnesses have also seen a decline, with their Disability-Adjusted Life Years (DALYs) dropping by 50% since 2000 [2].

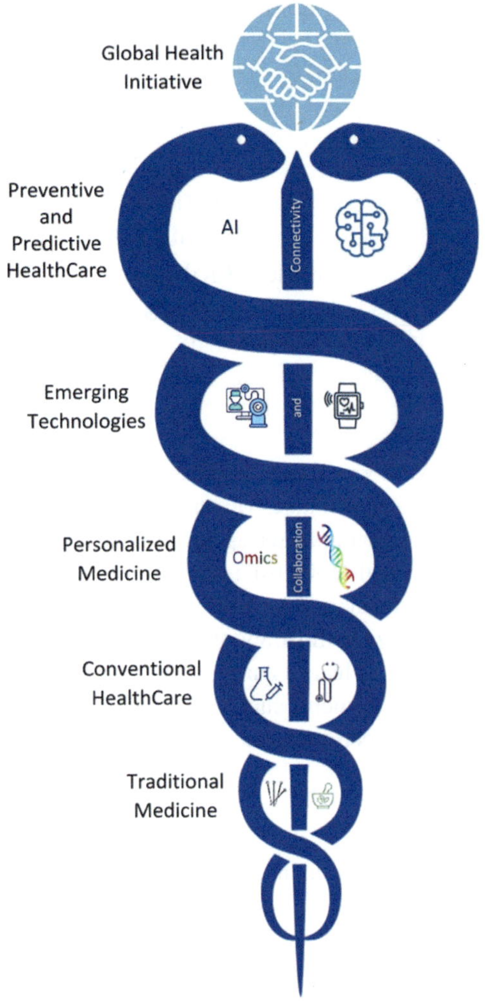

Fig. 1.1 Evolutionary journey in medicine

The progress in medicine has been influenced by numerous factors, including evidence-based medicine, precision medicine, technological advancements, and ethical considerations (Fig. 1.1). The evolution of medical practice has transitioned towards patient-centered care despite the advent of technological and scientific advancements. Additionally, there's a profound cultural shift from a print-based industrial society to a technology-based internet society in the twenty-first century, impacting the medical profession. The integration of AI applications in healthcare, advancements in medical informatics, and the development of various techniques and devices for medical management are notable contributions to the evolution of medicine. These advancements have reshaped medical care, contributing significantly to its progress.

In an era marked by medical breakthroughs, the journey toward comprehensive healthcare faces multifaceted challenges. Although advancements have propelled increases in life expectancy, the pace of healthy life expectancy lags, resulting in a significant portion of old age spent battling disability and illness. Disturbingly, the World Health Organization (WHO) notes an 80% surge in diabetes-related disease burden from 2000 to 2019. Furthermore, cancer incidences and fatalities witnessed a steady rise throughout the decade, ranking second in the global disease burden concerning Disability-Adjusted Life Years (DALYs) in 2019 [3, 4].

Furthermore, key statistics underscore the devastating impact of Noncommunicable diseases (NCDs), claiming 41 million lives annually, comprising a staggering 74% of all global deaths. Premature deaths before the age of 70 from NCDs tally up to 17 million yearly, with a daunting 86% occurring in low- and middle-income nations. Alarming disparities continue as 77% of all NCD fatalities transpire within these socioeconomically challenged regions. Cardiovascular diseases reign supreme, accounting for 17.9 million deaths annually, closely followed by cancers (9.3 million), chronic respiratory diseases (4.1 million), and diabetes (2.0 million, encompassing diabetes-induced kidney disease deaths). Collectively, these ailments contribute to over 80% of all premature NCD deaths. An array of risk factors such as tobacco use, physical inactivity, alcohol abuse, poor diets, and air pollution amplify the threat of succumbing to NCDs.

Noncommunicable diseases, chronic in nature, stem from a confluence of genetic, physiological, environmental, and behavioral factors. Cardiovascular diseases, cancers, chronic respiratory diseases, and diabetes headline the array of NCDs, disproportionately affecting low- and middle-income nations, where more than three-quarters of NCD-related fatalities occur. The vulnerability spans across age brackets, striking not only the elderly but also prematurely claiming 17 million lives before the age of 70, predominantly in resource-constrained nations. Children, adults, and the elderly face risks posed by unhealthy diets, sedentary lifestyles, tobacco exposure, alcohol misuse, and air pollution, precipitating NCDs.

Combatting NCDs hinges on robust management strategies encompassing detection, screening, treatment, and equitable access to palliative care. Investing in essential interventions through primary healthcare systems aids in early detection and prompt treatment, curbing the necessity for costlier interventions down the line. However, the challenges loom larger in countries grappling with inadequate healthcare coverage, impeding universal access to crucial NCD interventions. Effective management becomes imperative in steering toward the Sustainable Development Goal (SDG) targets, anchoring the importance of battling NCDs [5].

These challenges pinpoint the pressing need for innovative prevention, diagnostic, and treatment modalities to confront the looming threat of NCDs. The path ahead demands a concerted global effort, transcending boundaries and disparities, to combat these ailments and forge a healthier, more resilient future for all. In this book, we aim to give the readers insights into the future prospects of eleven areas in medicine to achieve more effective approaches to various diseases.

1.2 Gene Therapy and Gene Editing

Gene therapy and genome editing are two innovative branches of medicine and genetics that hold immense promise for treating various diseases.

Gene therapy, conceptualized over four decades ago, was primarily envisioned to introduce therapeutic genes into specific cells, thereby improving patients' quality of life. Initially proposed by Theodore Friedmann and later championed by George Stamatoyannopoulos, gene therapy involves packaging therapeutic genes into vectors for delivery into patients' cells. There are three main types of gene therapy: ex vivo, in vivo, and in situ. Ex vivo therapy involves engineering cells outside the body, such as blood cells for blood cancers or stem cells for monogenic diseases, followed by reintroduction into the patient. In vivo, therapy delivers vectors directly into the patient's bloodstream or cerebrospinal fluid for diseases affecting organs like the brain or liver. In situ therapy targets specific organs or areas within the body through direct injection or catheter insertion. The application of gene therapy has seen progress in clinical trials. However, its translation into mainstream medical practice remains limited, with only a few therapies receiving marketing authorization despite over 2800 ongoing clinical trials until the last update in August 2018 [6].

1.3 Prospects of Gene Therapy and Genome Editing

The field of gene therapy and genome editing continues to evolve with significant promise, offering a myriad of prospects for treating various genetic diseases. Some of the key prospects include:

1. Precise and targeted treatments: Genome editing techniques like CRISPR/Cas9 enable precise modifications in the DNA, allowing targeted treatments for specific genetic mutations. This precision reduces off-target effects and enhances the effectiveness of therapies.
2. Potential for personalized medicine: Gene therapy and genome editing hold the potential to revolutionize medicine by offering personalized treatments tailored to an individual's genetic makeup. This could lead to more effective and safer therapies, especially for inherited genetic conditions.
3. Advancements in rare disease treatments: These therapies offer hope for patients with rare genetic disorders for which conventional treatments are limited or nonexistent. Techniques like gene addition or targeting/correction aim to address underlying genetic defects directly, providing therapeutic benefits.
4. Broadening therapeutic horizons: Continuous advancements in these technologies expand the scope of treatable diseases. Previously challenging conditions, including various types of cancer, muscular dystrophy, neurodegenerative disorders, and inherited blood disorders, are now potential candidates for gene-based therapies.
5. Overcoming limitations of traditional treatments: For certain diseases, especially those caused by single-gene mutations, gene therapy and genome editing offer the potential for more effective treatments compared to traditional approaches.

1.4 Challenges in Gene Therapy and Genome Editing

Despite the immense promise, several challenges impede the wide-scale implementation of gene therapy and genome editing techniques:

1. Off-target effects and safety concerns: The risk of off-target effects—unintended genetic modifications—remains a significant concern. Ensuring the accuracy and safety of editing techniques without causing unintended mutations or disturbances in the genome is a critical challenge.
2. Immunogenicity and immune responses: Immune reactions against viral vectors or edited cells pose challenges in gene therapy. Patients might develop immune responses that limit the effectiveness of the treatment or cause adverse reactions, necessitating strategies to mitigate these responses.
3. Ethical and regulatory considerations: Ethical implications regarding germline editing and the potential for altering hereditary traits raise significant ethical debates. Furthermore, regulatory frameworks need continuous development and refinement to ensure the safe and ethical application of these technologies.
4. Long-term efficacy and durability: Ensuring that gene therapies and edited cells maintain their therapeutic effects over the long term remains a challenge. Factors such as the persistence of edited cells, stability of genetic modifications, and the need for repeated dosing need careful consideration.
5. Manufacturing and delivery challenges: Standardizing manufacturing processes, ensuring scalability, and devising efficient delivery mechanisms for edited cells or genetic material pose significant technical challenges. Achieving consistency and reproducibility in the production of therapies is crucial for their success.

6. Cost and accessibility: The high cost associated with these therapies and limited accessibility to advanced treatments pose barriers to their widespread adoption. Achieving affordability and accessibility for patients globally remains a challenge.

Addressing these challenges requires interdisciplinary collaboration among scientists, clinicians, ethicists, policymakers, and regulatory bodies. Continued research, technological innovations, and regulatory advancements are essential to unlock the full potential of gene therapy and genome editing while ensuring their safety, efficacy, and ethical use in clinical settings [7].

1.5 Human Genomics Technology

The human genome sequencing stands as a colossal accomplishment in scientific history. Unraveling the 3 billion nucleotides composing the human blueprint within 50 years, akin to the progression from the first flight to a moon landing, signifies a staggering trajectory in scientific advancement. Despite the Human Genome Project's (HGP) lofty cost, it has borne scientific and economic impacts, validating its significance. However, while the completion of the HGP ignited ambitious expectations in transforming disease diagnosis, treatment, and prevention, evaluating its true impact remains critical.

Technological breakthroughs post-HGP have revolutionized genomic medicine. High-density DNA microarrays and next-generation sequencing (NGS) platforms have enabled cost-effective analysis of genetic variations within human genomes, a feat once deemed exorbitant. These innovations, reliant on a high-quality reference genome established by the HGP, allow a comprehensive assessment of both common and rare genetic variants.

The substantial drop in genotyping and sequencing costs within a few years after the HGP's completion resulted in an exponential rise in genotyped individuals and sequenced genomes. Direct-to-consumer genealogy companies alone saw a remarkable surge from less than a million individuals genotyped in 2014 to over 12 million by 2018. This influx has catalyzed research initiatives, including efforts like the UK Biobank and the US All of Us programs, targeting sequencing in millions of individuals.

Genomic medicine's pursuits span various domains, including common and rare inherited diseases, reproductive health, and cancer. Genome-wide association studies (GWASs) have significantly advanced understanding of the genetic underpinnings of common diseases. Yet, challenges mar their success. The genetic component of common diseases comprises numerous variants with subtle effects, challenging risk prediction. Additionally, the hunt for rare variants explaining "missing heritability" has not yielded anticipated results, and the LD structure complicates variant resolution and target identification.

Despite these challenges, several reasons foster optimism in genomic medicine's impact on common diseases. Early GWAS discoveries primarily involved variants with substantial effects; therefore, exploring non-European populations and smaller studies might unveil additional significant signals. Furthermore, GWAS insights have propelled drug discovery and pharmacogenomic studies. Advances in methods for pinpointing causal variants and genes are promising, and polygenic risk scores (PRSs) are proving effective in risk stratification without in-depth biological understanding.

The cancer landscape has been a focal point of genomic medicine. Genomic sequencing and initiatives like The Cancer Genome Atlas (TCGA) have identified numerous driver genes and mutations, paving the way for targeted therapies. While precision therapies guided by genomic information show promise, substantial advancements are needed to deliver on their full potential. Immunotherapy, harnessing the immune system to combat cancer, represents a promising avenue. Strategies involving re-engineered immune cells and personalized vaccines have shown encouraging results, although further validation across different cancer types is imperative. Early cancer detection remains pivotal for improved outcomes. Utilizing circulating tumor DNA (ctDNA) to monitor disease presence and employing complementary strategies based on epigenetic signals and protein biomarkers offer avenues for enhancing the detection and localization of tumors.

While genomic medicine has faced hurdles, its impact on diagnosis, treatment, and prevention of diseases, particularly cancer, holds immense promise. Challenges persist, but ongoing advancements and innovations are crucial in realizing the full potential of decoding the human genome.

Among the initial fervor surrounding the Human Genome Project (HGP), there existed an idealistic expectation that decoding the human genome would serve as a solution for all medical challenges. Yet, while it hasn't lived up to these overblown hopes, the human genome's impact is undeniable. From sequencing a single human genome to over a million within a mere 15 years, our understanding of common disease risk's genetic complexities has grown. The quest for meaningful therapies based on association signals has encountered obstacles, especially in cancer, despite a deeper grasp of molecular lesions.

Acknowledging that initial expectations were overly optimistic, we recognize significant accomplishments and the evolving trajectory of the field. However, aggressive timelines for therapies might overshadow the long-term journey this domain entails. Investing in basic science and foundational technologies beyond the human genome sequence is imperative for maximal impact.

Several ambitious goals for the future of genome sciences are envisioned. Understanding every gene's basic function, scaling the identification of causal variants, developing a molecular atlas of all human cell types, and creating predictive models for sequence variants' impacts represent pivotal challenges. These could potentially expedite progress across various domains of genomic medicine.

The envisioned future involves widespread genome sequencing, integrating genetic information into healthcare records, and expanding sequencing applications beyond human genomes to encompass microbial genomes. Routine sequencing for cancers, prospective parents, and even recurrent sequencing for monitoring conditions is on the horizon.

While genome sequencing might offer marginal benefits to some patients, its enduring impact on understanding human biology remains pivotal. Combined with other advancements like polygenic risk scores, gene therapy, and immunotherapy, our collective genomes will be foundational in disease understanding and therapeutic development.

However, despite the transformative potential of genomics in medicine, tempered expectations are warranted. Significant hurdles persist, necessitating continued long-term investment in basic science and technology. Genomic research, including genomics itself, should be viewed as a societal and species-level investment, with returns that may not fully materialize for decades or even within our lifetimes.

Ethical, legal, and social implications (ELSIs) surrounding genomics are increasingly critical. Concerns regarding privacy, scientific misuse, unethical modifications, insurance industry interests, and embryo selection demand proactive measures to maximize benefits while minimizing harm [8, 9].

1.6 Cancer Therapies

Cancer, a complex and devastating disease, continues to be a global health concern, accounting for a significant portion of worldwide mortality. With an increasing understanding of the underlying pathways involved in cancer progression, treatment strategies have evolved significantly. Traditional modalities like surgery, radiation therapy, and chemotherapy have been augmented by novel approaches such as immunotherapy, targeted drug therapy, gene therapy, and innovative ablation techniques. This chapter provides a comprehensive overview of both conventional and advanced cancer therapies, shedding light on their mechanisms, challenges, and current status in clinical contexts [10].

1.6.1 Conventional Cancer Therapies

Historically, cancer treatments primarily comprised surgery, radiotherapy, and chemotherapy, each exhibiting varying efficacy levels based on the disease's stage. Surgery stands out as particularly effective in the early stages, while radiation and chemotherapy, although effective, exhibit limitations such as damage to healthy cells and the development of drug resistance. Despite their contributions to reducing morbidity and mortality, these modalities often come with adverse effects and challenges related to specificity and dosage selection.

1.6.2 Advanced and Innovative Cancer Therapies

The landscape of cancer therapies has seen significant advancements, leading to the exploration and adoption of cutting-edge approaches. Stem cell therapy, targeting stem cells such as mesenchymal stem cells (MSCs) and cancer stem cells (CSCs), shows promise in tumor treatment. Both adult and pluripotent stem cells present a potential avenue for therapeutic use, albeit with challenges regarding side effects and tumorigenesis [10].

1.6.3 Targeted Therapy

Targeted cancer therapies, often referred to as molecularly targeted drugs or precision medicines, constitute a cutting-edge arsenal in the fight against cancer. These drugs disrupt specific growth molecules crucial for cancer proliferation and metastasis. Unlike conventional chemotherapy, which affects both cancerous and healthy cells, targeted therapies aim to selectively interfere with molecular pathways essential for tumor growth, thus minimizing damage to normal tissues.

The intricate interplay between tumor microenvironments (TMs) and cancerous cells underscores the significance of understanding these dynamic interactions. TMs encompass diverse cell types such as endothelial cells, pericytes, inflammatory cells, and cancer stem cells (CSCs). Researchers are exploring the potential of TM conditions to develop more effective targeting measures for cancer therapy, leveraging the signaling mechanisms and pathways within these microenvironments.

Targeted therapy drugs exhibit diverse mechanisms, from inducing apoptosis and preventing proliferation to interfering with metabolic reprogramming. Notably, these agents have significantly improved survival rates in various cancers. For instance, the addition of erlotinib to standard chemotherapy has enhanced survival rates in advanced pancreatic cancer patients, underscoring the efficacy of targeted therapies [11].

1.7 Targeted Cell Agents: A Multifaceted Approach

Within targeted therapies, agents such as enzyme inhibitors and apoptosis-inducing drugs play pivotal roles in impeding cancer progression. Enzyme inhibitors disrupt cancer cell signals crucial for growth and spread, while apoptosis-inducing drugs, like serine/threonine kinase and protein kinase inhibitors, target mechanisms controlling cell survival. Furthermore, these agents hinder angiogenic factors, curtailing blood supply to tumors and arresting their growth.

The integration of targeted drug delivery into cancer treatment strategies offers promising prospects. Avastin (bevacizumab), in combination with chemotherapy, has prolonged the survival of patients with advanced colorectal carcinoma, exemplifying the potential of targeted drug delivery to extend patient lifespans.

1.8 Immunotherapy: Empowering the Body's Defenses

The concept of immunotherapy traces back to William Coley's pioneering work in the late nineteenth century. Today, immunotherapy has evolved into a multifaceted approach classified as passive or active. Passive immunotherapy aims to augment existing antitumor responses using therapeutic agents like immune checkpoint inhibitors and adoptive cell transfer (ACT), showcasing remarkable clinical efficacy in inducing durable tumor regression [10].

Conversely, active immunotherapy harnesses the immune system's ability to recognize and attack cancer cells through cancer vaccines. The success of therapeutic cancer vaccines hinges on factors such as target antigens, adjuvant utilization, and eliciting antitumor T cell responses within the tumor. microenvironment [12].

1.9 Types of Target Agents

- Monoclonal antibodies: These man-made immune system proteins, like gemtuzumab and ibritumomab tiuxetan, exhibit diverse mechanisms of action, from recruiting host immune functions to targeting essential cancer cell processes or carrying lethal payloads to the target cell.
- Small molecule inhibitors: Smaller in size, these inhibitors, such as gefitinib and sorafenib, interrupt intracellular signaling pathways, inhibiting cell growth, proliferation, and angiogenesis, showing efficacy in various cancers like NSCLC and renal cancer.
- Ablation cancer therapy: Techniques like thermal ablation (RFA, microwave ablation, cryoablation) offer minimally invasive approaches to destroy tumors, eliciting immunogenic responses and modulating the immune system, albeit with limitations in specific tumor locations.
- Gene therapy: Utilizing gene insertion to cure specific disorders, gene therapy has shown promise in cancer treatment, particularly in strategies involving proapoptotic and chemosensitizing genes, antitumor immune responses, and targeted oncogene silencing. Challenges include effective delivery mechanisms and limited efficacy in specific patient subsets.
- Natural antioxidants: Natural antioxidants derived from vitamins, polyphenols, and plant compounds are being explored for their anti-inflammatory and anti-proliferative properties in cancer therapy. Compounds like curcumin, berberine, and quercetin exhibit cytotoxic effects in various tumors and are undergoing clinical trials.

Ongoing clinical trials focus on stem cell therapies, targeted therapies, immunotherapies, and gene therapies due to their promising results and effectiveness in cancer treatment. These trials aim to further refine and optimize treatment approaches for better patient outcomes. The landscape of cancer therapy is continually evolving, with a focus on developing safer and more efficient nanomedicines, targeted drug delivery systems, and innovative treatments. While significant strides have been made, the challenges of drug resistance, toxicity, and achieving comprehensive personalized therapies persist. The future of cancer treatment may lie in combinations of therapies targeting multiple molecular alterations or hallmarks, akin to HIV treatment, offering a more comprehensive approach to combating cancer [13].

1.10 Machine Learning and Artificial Intelligence (AI)

AI has the potential to significantly change medicine in the coming years. As the medical AI community works to overcome the many ethical, technological, and human-centered obstacles necessary for safe and effective translation, the use of medical AI systems in routine clinical practice currently offers a significant but mainly unrealized potential. Additionally, a 2020 study discovered that the US Food and Drug Administration (FDA) is rapidly approving AI, specifically machine learning (ML; a kind of AI), devices [14]. Medical AI is still in its early stages of validation and implementation, despite significant advancements toward large-scale utilization, particularly through prospective research like RCTs and medical picture analysis.

The promise of highly personalized oncology care through AI technologies has been anticipated for years, with recent advances in machine learning (ML) and deep learning algorithms, expansive databases incorporating multi-omics data, and the more accessible computational power paving the way toward realizing this potential. Clinical applications of AI across the cancer continuum, particularly in computer vision-assisted image analysis, have seen success and garnered approval from the U.S. Food and Drug Administration (FDA). Emerging techniques like whole blood multi-cancer detection, virtual biopsies, and natural language processing (NLP) for health trajectory inference from medical notes have shown promise in clinical practice [15]. The broad application spectrum of AI in oncology ranges from basic science drug discovery and translational gene expression to patient-facing tools like chatbots, augmented reality in surgeries, and hospital scheduling systems. AI-driven clinical decision support systems (CDSS) integrated with electronic health records (EHRs) aim to enhance clinical workflows, facilitate clinical trial matching, and identify adverse drug reactions. Despite the considerable progress, challenges persist in AI adoption within oncology, primarily related to data transparency, interpretability issues due to the "black box" nature of AI models, and inherent biases in training data that impact reproducibility and exacerbate healthcare disparities. The future evolution of AI in oncology is expected to encompass complex models using multimodal data elements, potentially culminating in living databases integrating various health aspects for highly personalized treatment and surveillance strategies [16].

AI's historical trajectory in medicine faced skepticism and a decline in adoption following earlier expectations in the 1970s. However, recent advancements fueled by deep learning expanded digital health data availability, and improved computational capabilities have reignited the prospects of AI in healthcare. Machine learning, especially deep learning, operates through neural networks, enabling autonomous feature discovery and extraction without human-defined heuristics, notably impacting medical imaging and natural language processing fields [17, 18].

1.11 Current Landscape and Challenges in AI Adoption

The accelerated growth in computational power, coupled with the availability of vast machine-readable electronic health records (EHRs) and multiomics data, has revolutionized AI algorithms and clinical decision support systems (CDSS) in oncology. Notably, AI-based applications have significantly impacted imaging analysis, genomics, and clinical practices across the cancer care continuum. Ongoing research aims to leverage AI for multicancer early detection, determining tumor origin, transforming cancer screening for rare cancers, and improving surveillance strategies for cancer survivors.

1.12 Advancements and Implications in Clinical Settings

Progress in imaging-based machine learning models holds promise in enhancing diagnostic accuracy, predicting associated morbidity and mortality, and personalizing screening, prevention, treatment approaches, and survivor surveillance. The potential of AI extends to supporting virtual biopsies by classifying pathological and genomic features associated with cancer diagnoses.

1.13 Addressing Challenges and Future Directions

While AI's integration into oncology holds immense promise, overcoming significant challenges is crucial. Addressing data standardization, inherent bias in training datasets, lack of reporting standards for source code and model training conditions, limited prospective clinical validation, and seamless integration into clinical workflows without adding burdens to healthcare providers remain substantial hurdles.

In conclusion, AI's penetration into medical sciences has led to transformative advancements and applications across various domains. While AI holds immense potential to revolutionize healthcare, including oncology, significant challenges, particularly related to data standardization, bias mitigation, and seamless integration, need to be addressed for its widespread and ethical implementation [15, 19, 20].

1.14 Genetic Health

Genetics significantly influences an individual's health, influencing various aspects of their well-being, and it helps to understand how genes contribute to the development and functioning of the human body. In the complex nature of human existence, genetics is the invisible component of human beings, which plays a major role. Genetic health is influencing the direction of healthcare as we stand at the center of scientific discoveries. The mysteries found within our DNA are still being uncovered by research projects, opening the door to novel cures, focused treatments, and a better comprehension of the complexities of human health [21, 22].

1.15 Regenerative Medicine and Bioprinting

The rapid evolution of 3D bioprinting technology has revolutionized regenerative medicine and tissue engineering, primarily focused on addressing the challenges in organ transplantation. Organ scarcity has been a persistent issue, leading to advancements in tissue engineering aiming to replicate natural tissues using biomaterials. The synergy between 3D bioprinting, tissue engineering, microfluidics, and AI has significantly impacted medical interventions, enabling solutions for patients with chronic illnesses, neurodegenerative disorders, and severe accidents.

1.16 Current State and Challenges

The technologies employed, such as inkjet, extrusion, laser-assisted, digital light processing, and stereolithographic 3D bioprinters, have shown promising results in creating tissue constructs. However, challenges remain in scaling up, industrialization, and commercialization of these technologies. Mimicking tissue complexity, developing biomaterials compatible with anatomical features, and ensuring self-repair mechanisms within the implanted biomaterials are among the critical challenges.

1.17 Impact and Future Prospects

Despite challenges, 3D bioprinting has made significant progress in producing transplantable tissues and organs. Moreover, it has contributed to drug testing models and advancements in drug delivery systems. Integration with microfluidics has addressed some limitations, though challenges persist in developing functional vasculature within 3D-bioprinted models.

In conclusion, tissue engineering and regenerative medicine have significantly advanced due to 3D bioprinting technologies, offering potential solutions to organ scarcity and tissue damage. While advancements are promising, unresolved technical hurdles exist, especially in optimizing bioinks and cell sources for wider clinical use [23].

1.18 Personalized Medicine

The landscape of modern biomedical research has been profoundly reshaped by the advent of emerging technologies, such as DNA sequencing, proteomics, imaging protocols, and wireless monitoring devices. The implementation of these data-intensive, high-throughput assays has illuminated an extensive spectrum of inter-individual variation, unraveling the diverse effects, mechanisms, and contributing factors to disease processes. This intricate variability among individuals has sparked a fundamental query regarding the extent to which these differences should influence decisions concerning disease treatment, monitoring, or prevention for a given individual. It is now widely acknowledged that the underlying heterogeneity in many disease processes implies a necessity to customize or "personalize" strategies for treating, monitoring, or preventing diseases based on an individual's unique biochemical, physiological, environmental, and behavioral profile.

Distinct terms such as personalized medicine, individualized medicine, and precision medicine have been employed interchangeably, although some have proposed subtle disparities among them. However, these terminologies generally converge in signifying a tailored approach to healthcare that considers individual diversities. While personalized medicine holds promise, its adoption faces numerous challenges, especially in obtaining approval from regulatory bodies for routine use and garnering acceptance from various healthcare stakeholders. The economic aspect, particularly the comparative performance against traditional therapies, remains a focal point, especially concerning the expense associated with personalized therapies [24].

1.19 Early Examples of Personalized Medicine

Numerous interventions tailored to individual patient profiles, primarily based on genetic profiles, have been exemplars of personalized medicine. While personalized medicine has been practiced not only in disease treatment but also in early disease detection and prevention, its roots in genetically mediated pharmacokinetic aspects of drugs are notably significant. The understanding of drug-metabolizing enzymes and their impact on the body's response to drugs has been pivotal in tailoring drug interventions based on an individual's genetic makeup. One of the classic examples involves Warfarin, a widely used blood-thinning medication. Genetic variation in genes such as *VKORC1* and *CYP2C9* results in variations in an individual's response to Warfarin, leading to differential pharmacokinetic and pharmacodynamic properties across individuals. As a consequence, personalized dosing based on an individual's genotype has been recommended by regulatory bodies to optimize Warfarin therapy [24, 25].

1.20 Emerging and Next-Generation Personalized Medicine Strategies

Recent research and clinical endeavors have propelled personalized medicine into uncharted realms. This section spotlights three key activities in personalized medicine: the utilization of patient-derived cell and organoid avatars for identifying optimal therapies, intense individualized diagnostic and monitoring protocols, and personalized medicine approaches in addressing fertility issues. Also, the expanding number of prospective clinical studies in genome-guided therapeutics, with the PREPARE study standing as the hallmark clinical trial with almost 7000 patients in Europe, will shed more light on pharmacogenomics, which jointly with economic evidence will bring implementation of pharmacogenomics closer to clinical fruition.

1.21 Patient-Derived Cellular Avatars

Advancements in pluripotency induction techniques, notably induced pluripotent stem cell (iPSC) methods, have facilitated the generation of disease-specific cellular models without the need for direct tissue biopsies. These "disease in a dish" models enable the study of molecular pathologies relevant to a patient's condition. Recent technologies such as Clustered Regularly Interspaced Short Palindromic Repeats (CRISPR) have further enhanced these models by enabling the creation of isogenic cells, allowing direct investigation into the effects of specific mutations while controlling for relevant genetic backgrounds.

Organoids, partial organs developed from patient-derived cells, offer deeper insights into molecular pathologies by modeling cell interactions and global tissue function. The integration of patient avatars derived from their cells, along with other patient information, allows for personalized drug screening to identify compounds that address the patient's molecular defects. While this approach has shown promise, concerns exist regarding the relevance of in vitro models to in vivo pathobiology and drug responses, necessitating further research to bridge this gap.

1.22 Intensive Personalized Health Monitoring

The convergence of affordable genotyping, sequencing technologies, and health monitoring devices has empowered individuals and healthcare providers to assess genetically-mediated disease risks and monitor aspects of health continuously or near-continuously. Genetic diagnosis has been instrumental in uncovering potential disease mechanisms and guiding pharmacotherapies. Individuals have leveraged intense health monitoring to identify health status changes linked to genetic susceptibilities.

The utility of genetic assays in diagnosing individuals with idiopathic conditions and continuous monitoring to detect health status changes is increasingly recognized. While establishing personal thresholds for measures rather than relying solely on population thresholds holds promise, evaluating measures collected from individuals lacking established norms remains a challenge.

1.23 Personalized Interventions Involving Fertility and Sterility

Personalized medicine's application extends to fertility treatments, leveraging real-world data, digital medicine, identified genetic variants influencing fertility, and adaptive trial designs to enhance fertility interventions. Emerging strategies such as cryopreservation of oocytes and ovarian tissues, in vitro gametogenesis, and cell reprogramming technologies offer highly personalized approaches for fertility treatments, but with concerns about viability, potential damage, and ethical considerations.

The proliferation of modern biomedical technologies has unraveled the intricate inter-individual variations in disease processes, highlighting the imperative need for personalized medicine. Despite promising prospects, personalized medicine confronts challenges in optimizing individual characterization and crafting and validating personalized medicines. Privacy concerns, manufacturing efficiency, cost considerations, and healthcare professionals' education constitute some of the obstacles that need addressing. Nonetheless, the evolution of personalized medicine continues to be a transformative force in healthcare, promising tailored interventions based on an individual's unique characteristics [25, 26].

1.24 Digital Health

The emergence of "digital health" signifies a profound shift in healthcare, defined as the integration of disruptive technologies that provide accessible, objective data to both patients and caregivers. This transformation aims to establish a more balanced doctor-patient relationship with shared decision-making and democratized care, fundamentally altering medical practice.

Historically, medicine operated with limited tools and relied on experiential knowledge passed down through generations. Even seminal innovations, such as Dr. Laennec's first stethoscope in the early nineteenth century, faced a gradual adoption process within healthcare. The evolving healthcare landscape became dependent on technology, yet medical curricula and care guidelines did not swiftly reflect this evolution. By the 2010s, digitalization in healthcare had become inevitable. The exponential growth of medical knowledge and patient empowerment surged, outpacing stakeholders' readiness. Physicians grappled with mounting responsibilities, leading to burnout, while patients navigated through overwhelming information without adequate guidance. The advent of digital health brought forth an array of technologies, from genome sequencing to smartphone-connected ECGs, presenting opportunities for improved health outcomes but also risking dehumanization of care [27, 28].

1.25 Challenges in Validating Digital Health Solutions

The landscape of digital health boasts significant investment, with billions of dollars poured into the sector annually. However, this expansive environment comprises over 300,000 health apps and a cluttered array of solutions, complicating navigation for stakeholders seeking genuine value. The absence of a reliable mechanism to identify validated digital health solutions poses a considerable challenge. Regulatory oversight remains limited, leaving stakeholders, especially payers, unable to discern quality in this crowded space.

Existing frameworks for overseeing digital health mostly prioritize patient safety. However, healthcare requires a robust, transparent validation process for these products, encompassing technical, clinical, and system validation domains. Efforts are underway to propose standardized, objective validation pathways, essential for all stakeholders' benefit.

1.26 Regulatory Ambiguity and the Need for Transparent Evaluation

The rapid development and promotion of digital health technologies differ significantly from traditional medical product design, safety testing, and clinical efficacy trials. Unlike conventional health technology, digital health products face fewer constraints, fostering rapid creation and distribution. However, the rush to market often sacrifices comprehensive validation, leading to misleading claims and dubious quality. Regulatory ambiguity further complicates matters, especially in areas like clinical decision support functionalities, where the regulatory perspective remains unclear. Efforts from organizations like the FDA and FTC aim to provide clarity but fall short in ensuring products genuinely meet users' needs. This trend indicates a future where a fraction of digital health technologies undergoes regulatory review pre-market, leaving many unproven solutions to proliferate.

1.27 Embracing the Cultural Transformation of Digital Health

Digital health, beyond being a technological revolution, signifies a profound cultural transformation reshaping healthcare delivery and medical practice. It not only necessitates a regulatory framework addressing technology but also demands a cultural shift in the physician-patient relationship and healthcare practices. Embracing patient empowerment and collaborative care models and educating healthcare professionals on technology integration are vital to harnessing the potential of digital health while mitigating its ethical and policy challenges.

This transformation mandates a redefined collaboration between patients and caregivers, emphasizing shared responsibility. It calls for a healthcare landscape where informed decision-making, regulatory transparency, and patient empowerment coexist. Successfully integrating digital health into healthcare systems will redefine medical education, policy-making, and the patient-caregiver dynamic for a digitally transformed era [27–29].

1.28 RNA Therapies

RNA, which was once considered a mere cellular messenger, is now acknowledged as pivotal in almost all biological pathways. The evolution of RNA therapy from a scientific curiosity to a potent therapeutic strategy has been marked by seminal discoveries in understanding RNA's diverse roles and the development of innovative RNA-based drugs.

The understanding of RNA's functional diversity began with the discovery of messenger RNA (mRNA), transfer RNA (tRNA), and ribosomal RNA (rRNA). Sydney Brenner and colleagues identified messenger RNA, which transports genetic information from the DNA in the nucleus to the protein synthesis machinery in the cytoplasm, alongside the identification of tRNA and rRNA. These molecules collectively formed the foundation of RNA's role in gene expression. Furthermore, the concept of antisense RNA emerged, showcasing the ability to suppress target RNA by employing complementary sequences, thereby paving the way for future RNA-based drug development.

Over time, the utilization of mRNA directly for protein production and vaccine development became apparent. Experiments injecting mRNA into mouse skeletal muscle demonstrated protein synthesis without additional delivery materials. This method extended to encapsulating mRNA for influenza nucleoprotein in liposomes and inducing immune responses, suggesting potential applications in RNA-based vaccines. The discovery of RNA interference (RNAi) further revolutionized therapeutic approaches, presenting a mechanism to eliminate disease-causing RNA types. The divergence between antisense oligonucleotide techniques and RNAi's reliance on small interfering RNA (siRNA) highlighted distinct yet complementary strategies in targeting specific RNA sequences.

In parallel with these discoveries, the FDA approval of initial RNA-based drugs marked significant milestones. Fomivirsen, an antisense oligonucleotide, was the first RNA-based drug targeting CMV retinitis, followed by pegaptanib, an RNA aptamer, designed to treat age-related macular degeneration. The emergence of siRNA-based drugs like patisiran and givosiran further exemplified the therapeutic potential of RNA, providing treatments for hereditary transthyretin-mediated amyloidosis and acute hepatic porphyria, respectively.

1.29 Categories of RNA Therapy

RNA therapies encompass diverse categories. Antisense RNAs, including single-stranded molecules targeting specific RNA sequences, have found clinical utility in various diseases. Notable examples such as mipomersen, nusinersen, and eteplirsen modulate mRNA splicing or induce degradation, showing promise in treating familial hypercholesterolemia, spinal muscular atrophy, and Duchenne muscular dystrophy, respectively. The advent of small interfering RNAs (siRNAs) introduced double-stranded RNA molecules leveraging RNAi for targeted mRNA degradation. FDA-approved siRNA drugs like patisiran and givosiran have demonstrated efficacy in suppressing target gene expression, offering treatments for hereditary transthyretin-mediated amyloidosis and acute hepatic porphyria.

RNA aptamers constitute another category, binding specifically to proteins and modifying their function. Notably, pegaptanib's ability to bind VEGF has implications in age-related macular degeneration treatment. Additionally, messenger RNAs, translated into proteins after introduction into cells, present promising avenues for personalized cancer vaccines. This approach capitalizes on identifying cancer-specific mutations and synthesizing mRNAs to elicit immune responses targeting tumor cells, offering a personalized therapeutic strategy.

1.30 Delivery in RNA Therapy

While RNA-based therapies offer immense promise, their effective delivery into target cells remains a significant challenge. The large size and high electric charge of RNA molecules hinder their intracellular penetration and render naked RNA forms inefficient for therapeutic use. To overcome this hurdle, various delivery strategies have been explored. Nanoparticles have emerged as a promising avenue for efficient RNA delivery. Encapsulation of RNA within lipid bilayers or complexes with polymers enhances stability and aids in endocytosis and subsequent escape from endosomes. Notable examples include lipid nanoparticle-encapsulated drugs like patisiran, showcasing improved half-life and efficacy. Moreover, conjugates targeting specific cell surface receptors, such as GalNAc conjugates recognized by hepatocyte receptors, facilitate targeted delivery into specific cells, notably the liver.

1.31 Perspective for the Future of RNA Therapy

The recent strides in RNA-based drug development underscore the immense potential of this therapeutic approach. The ability to target both protein-coding and non-coding RNAs presents a novel opportunity to address previously "undruggable" targets. RNA therapies offer rapid design capabilities, with the ability to manipulate sequences to achieve target specificity. Moreover, the sustained effects observed in siRNA-based drugs like inclisiran highlight the prolonged efficacy of RNA therapies. This extended duration of action is advantageous for patients requiring infrequent treatments. As small molecule-based drugs often lack long-lasting effects, further research into siRNA-based drugs can address this limitation.

However, challenges persist, particularly in the delivery of RNA drugs to tissues beyond the liver. Strategies to efficiently target other tissues for diseases less associated with the liver remain a crucial area for future investigations. Despite these challenges, the substantial progress in RNA therapy, coupled with established small molecule and antibody-based drugs, promises a future where a diverse array of diseases lacking effective treatments today may find resolutions through RNA-based therapeutics [30].

1.32 Ethical Concerns in the Future of Medicine

With technological developments, individualized treatments, and novel therapies, the future of medicine looks bright. However, these potentials bring with them ethical concerns that must be carefully evaluated to ensure responsible and equitable healthcare practices. There are a few medical concerns associated with artificial intelligence, personalized medicine, cancer therapies, genome editing and CRISPR technology, and digital health.

Here are ethical concerns related to artificial intelligence and personalized medicine:

1.33 Ethical Concerns in Artificial Intelligence

The integration of Artificial Intelligence (AI) in healthcare holds immense promise for revolutionizing medical diagnostics, treatments, and patient care. However, this rapid advancement in technology brings forth a multitude of ethical considerations that necessitate in-depth exploration and thoughtful deliberation. Chapter 11 examines the pervasive ethical issues intertwined with AI implementation in healthcare and discusses potential strategies to address these concerns.

1.33.1 Justice and Fairness

1.33.1.1 Bias

In the pursuit of AI-driven healthcare, biases emerge as a critical concern. These biases stem from self-learning algorithms, causing inadvertent preferences or systematic errors in decision-making. Issues like algorithmic bias, overfitting, and biased training data lead to unfair outcomes, especially in cases of sampling or gender bias within healthcare datasets.

1.33.1.2 Fairness

The ethical pillar of fairness intertwines with the opaqueness of AI algorithms, posing challenges in ensuring equitable outcomes. Factors like gender and ethnicity influence decision-making, leading to potential discrimination or unequal treatment, hampering the fair distribution of healthcare services.

1.33.2 Freedom and Autonomy

1.33.2.1 Control

The control of data within AI systems impacts decision-making processes in healthcare. Lack of understanding or transparency regarding AI-driven recommendations can influence patient trust and healthcare providers' ability to exert control over AI-generated decisions.

1.33.2.2 Respecting Human Autonomy

AI's role in decision-making processes can potentially undermine patient autonomy, creating challenges in shared decision-making between patients and healthcare providers. Opacity in AI decisions might impose values on treatment options, limiting patient autonomy.

1.33.2.3 Informed Consent

The importance of informed consent in data usage within AI systems becomes crucial. Consent mechanisms must align with the complexity of AI algorithms, ensuring patient understanding and control over data sharing and usage.

1.33.3 Privacy

1.33.3.1 Data Privacy

Maintaining control over individual health information becomes paramount, raising concerns about data usage, ownership, and confidentiality. Privacy breaches or misuse of sensitive health data pose significant ethical dilemmas.

1.33.3.2 Confidentiality

Ensuring the security and appropriate handling of healthcare data in transit or storage is pivotal to maintaining confidentiality and preventing unauthorized access or breaches.

1.33.4 Transparency

1.33.4.1 Transparency

Understanding the decision-making process of AI systems remains a significant challenge. The opacity of AI algorithms raises concerns regarding their inner workings, affecting trust, especially when decisions impact sensitive medical records.

1.33.4.2 Explainability

The need for explainability in AI decisions becomes evident in building trust and credibility. However, the complexity of AI algorithms makes it challenging to interpret relationships between input data and outcomes.

1.33.5 Patient Safety and Cyber Security

1.33.5.1 Patient Safety

While AI integration promises improvements in patient safety, there's a looming concern about the potential harm caused by erroneous AI recommendations or decisions. The delineation of responsibility for AI-induced errors remains ambiguous.

1.33.5.2 Cyber Security

Protecting healthcare data from breaches or cyber-attacks poses a significant challenge. The vulnerability of patient data during storage, transmission, or sharing raises ethical concerns about privacy and security.

1.33.6 Trust

The pivotal role of trust in the patient-AI interaction demands attention. Issues surrounding data usage, transparency, and biases significantly influence patient trust in AI-enabled healthcare systems.

1.33.7 Beneficence

Ethical considerations in promoting patient welfare and balancing benefits against risks and costs call for careful evaluation and application of AI in healthcare settings.

1.33.8 Responsibility

The delineation of responsibility in AI-induced errors remains a complex ethical issue. It prompts discussions about accountability, especially concerning AI-driven decisions impacting patient care.

1.33.9 Solidarity

Addressing justice and equality concerns, solidarity plays a vital role in allocating healthcare resources and preventing discrimination based on race, gender, or socioeconomic factors.

1.33.10 Sustainability

AI's integration in healthcare must align with sustainable development goals. Balancing conflicting goals, managing uncertainties, and ensuring equitable access to AI-driven healthcare solutions are crucial for sustainable implementation.

1.33.11 Dignity

Respecting human dignity and rights becomes a central ethical consideration in AI-enabled healthcare. Upholding dignity in decision-making processes and ensuring the compatibility of AI systems with human values and freedom is imperative.

1.33.12 Conflicts

The integration of AI in healthcare often leads to conflicts between goals, stakeholders, and decision-makers. Balancing differing interests, avoiding bias, and ensuring responsible decision-making are essential to mitigate conflicts [31].

1.34 Ethical Concerns in Personalized Medicine

Even with the noticeable potential for both immediate and long-term application of precision medicine in routine healthcare, significant ethical and social concerns still need to be thoroughly thought through and resolved to guarantee that patient rights are upheld and that the advantages of this strategy are shared fairly. These include:

1.34.1 Privacy, Confidentiality, and Security

Precision medicine presents various ethical concerns, prominently revolving around patient privacy, confidentiality, and data security. Patient apprehensions regarding privacy breaches, especially concerning genetic data, remain a significant focus. These concerns, shared across diverse ethnic groups, stem from increasing public awareness of privacy issues in an era dominated by online data accessibility. The potential misuse of personal data for commercial gain, insurance eligibility, or employment status poses substantial worries. Patients, particularly those reporting substance abuse, fear that their data could be exploited to investigate unsolved crimes, thus hesitating to participate in precision medicine research programs. While younger individuals are less concerned about privacy breaches, their focus lies more on data accessibility. Moreover, despite a portion of patients expressing willingness to share genomic data widely, this perspective may be subject to participation bias.

1.34.2 Economic Impact

Patients also worry about the economic implications of precision medicine. The direct costs of medical treatments and their coverage by insurance companies, along with concerns about insurability due to genetic conditions, are justifiable concerns. The Genetic Information Nondiscrimination Act (GINA) in the US emerged as a response to address some of these worries related to insurance eligibility.

1.34.3 Potential Harms and Counseling

Ethical concerns surrounding precision medicine extend to potential psychosocial harms, emphasizing the need for genetic counseling. Increased diagnostic scans and the dissemination of information indicating disease risks can lead to heightened psychological distress and anxiety among patients. This distress might adversely impact patients' quality of life, treatment adherence, and engagement in healthy behaviors. Specific worries encompass unexpected paternity revelation and ethical considerations regarding human cloning. Additionally, there's a concern that knowledge of genetic predispositions might lead to neglecting overall health in favor of genetic determinants, indicating a need for education on multifactorial disease causation.

1.34.4 Access Disparities

Ensuring equal access to precision medicine for patients from all backgrounds is crucial. Concerns about racial discrimination and the fear of genetic information being used against individuals have been raised by certain ethnic groups. Historical incidents, such as the Tuskegee Syphilis Study, have contributed to mistrust among these groups.

1.34.5 Informed Consent Challenges

Patients express concerns about the process of informed consent, particularly regarding the usage of genetic information without their explicit consent. Clear communication and comprehensible consent forms are vital to ensuring patients fully understand and consent to their participation in research. Trust in research varies among different ethnic groups, influenced by their familiarity with the healthcare system and previous experiences.

1.34.6 Diagnostic Accuracy and Trust in Technology

Issues regarding the accuracy of genetic testing methods, such as full genome sequencing, contribute to patient anxieties, potentially impacting trust in the healthcare system. Moreover, the increasing reliance on Artificial Intelligence (AI) algorithms in precision medicine raises concerns about the potential deterioration of patient-doctor relationships. Patients worry that overreliance on algorithms might compromise medical experts' judgment and reduce the emphasis on their expertise in decision-making.

1.34.7 Impact on Decision-Making and Faith in Technology

Parents of patients, considered proxies for decision-making, express concerns about how AI algorithms might influence decision-making about their children's health. While some fear confusion and conflicts with medical experts, others display openness to technological advancements in healthcare [32].

1.34.8 Trustworthiness of the Different Types of Genetic Tests

There is a big controversy regarding the trustworthiness of the different types of genetic tests. In other words, there are genetic tests for monogenic genetic diseases and pharmacogenomics that have a very strong evidence basis and are approved by major regulatory bodies, have guidance for disease management, and are therefore trustworthy. However, there are also other genetic tests for polygenic diseases, genome-guided diets, genetic tests for athletic performance, personality genetic tests, and even those aiming to provide ethnic background information. These tests, often referred to as lifestyle genomic tests, have a very weak evidence basis without any guidance for health/disease management and as such should not be broadly available to patients, especially using the direct-to-consumer approach. Also, the issue of using the polygenic risk scores (PRS) for polygenic and multifactorial diseases with a very strong environmental influence is a very controversial one, and genetic tests providing PRS for clinical use should be considered with a very high degree of skepticism.

1.35 Conclusion

The future of medicine has unprecedented promise and transformational potential, driven by breakthroughs in technology, genetics, and artificial intelligence. As we enter a new era, it is clear that precision medicine, tailored treatments, and novel therapeutics will reshape the healthcare environment. The convergence of data analytics, genomics, and wearable technologies is preparing the way for a more patient-centric approach in which treatments are personalized to individual genetic profiles and lifestyle settings. Furthermore, the growing use of AI in diagnostics, drug discovery, and medical management is expected to improve efficiency, accuracy, and accessibility. Collaboration among healthcare professionals, researchers, politicians, and technology specialists will be critical to realizing the full potential of medicine's future. The transition from a reactive to a proactive healthcare approach is ongoing in this dynamic environment, empowering individuals to take control of their health and well-being. The future offers not only longevity but also an improved quality of life through preventive measures, earlier interventions, and more effective therapies. While there will definitely be barriers ahead, the potentials are intriguing, and the track of medical growth speaks to a future in which medicine is predictive, preventive, and customized, rather than merely curative. Adopting these advances while remaining committed to ethical and fair implementation will definitely form a healthier, stronger, and more connected world community.

References

1. Fisher D, Wicks P, Babar Z-U-D. Medicine and the future of health: reflecting on the past to forge ahead. BMC Med. 2016;14(1):169.
2. Saloni Dattani LR, Ritchie H, Ortiz-Ospina E, Roser M. Life expectancy. Oxford: University of Oxford; 2023.
3. Wartman SA. Medicine, machines, and medical education. Acad Med. 2021;96(7):947–50.
4. Han W, Han X, Zhou S, Zhu Q. The development history and research tendency of medical informatics: topic evolution analysis. JMIR Med Inform. 2022;10(1):e31918.
5. World Health Organization. Noncommunicable diseases. Geneva: World Health Organization; 2023.
6. Hasanzad M, Larijani BJ. The pathway from gene therapy to genome editing: a nightmare or dream. Int J Mol Cell Med. 2019;8(2):69–70.
7. Papanikolaou E, Bosio A. The promise and the hope of gene therapy. Front Genome Ed. 2021;3:618346.
8. Shendure J, Findlay GM, Snyder MW. Genomic medicine-progress, pitfalls, and promise. Cell. 2019;177(1):45–57.
9. Houwink EJF. Comment on Future trends in clinical genetic and genomic services by Borle et al. Eur J Hum Genet. 2022;30(5):505–6.
10. Debela DT, Muzazu SG, Heraro KD, Ndalama MT, Mesele BW, Haile DC, et al. New approaches and procedures for cancer treatment: current perspectives. SAGE Open Med. 2021;9:20503121211034366.
11. Shuel SL. Targeted cancer therapies: clinical pearls for primary care. Can Fam Physician. 2022;68(7):515–8.
12. Zhong L, Li Y, Xiong L, Wang W, Wu M, Yuan T, et al. Small molecules in targeted cancer therapy: advances, challenges, and future perspectives. Signal Transduct Target Ther. 2021;6(1):201.
13. Fritah H, Rovelli R, Chiang CL, Kandalaft LE. The current clinical landscape of personalized cancer vaccines. Cancer Treat Rev. 2022;106:102383.
14. Benjamens S, Dhunnoo P, Meskó B. The state of artificial intelligence-based FDA-approved medical devices and algorithms: an online database. NPJ Dig Med. 2020;3(1):118.
15. Shreve JT, Khanani SA, Haddad TC. Artificial intelligence in oncology: current capabilities, future opportunities, and ethical considerations. Proc Am Soc Clin Oncol. 2022;42:1–10.
16. Sebastian AM, Peter D. Artificial intelligence in cancer research: trends, challenges and future directions. Life. 2022;12(12):1991.
17. Bohr A, Memarzadeh K. The rise of artificial intelligence in healthcare applications. In: Artificial intelligence in healthcare. Amsterdam: Elsevier; 2020. p. 25–60. https://doi.org/10.1016/B978-0-12-818438-7.00002-2.
18. Kaul V, Enslin S, Gross SA. History of artificial intelligence in medicine. Gastrointest Endosc. 2020;92(4):807–12.
19. Basu K, Sinha R, Ong A, Basu T. Artificial intelligence: how is it changing medical sciences and its future? Indian J Dermatol. 2020;65(5):365–70.
20. Muehlematter UJ, Daniore P, Vokinger KN. Approval of artificial intelligence and machine learning-based medical devices in the USA and Europe (2015-20): a comparative analysis. Lancet Digit Health. 2021;3(3):e195–203.
21. Hasanzad M, Sarhangi N, Ehsani Chimeh S, Ayati N, Afzali M, Khatami F, et al. Precision medicine journey through omics approach. J Diabetes Metab Disord. 2022;21(1):881–8.

22. Hasanzad M, Sarhangi N, Naghavi A, Ghavimehr E, Khatami F, Ehsani Chimeh S, et al. Genomic medicine on the frontier of precision medicine. J Diabetes Metab Disord. 2022;21(1):853–61.
23. Kantaros A. 3D printing in regenerative medicine: technologies and resources utilized. Int J Mol Sci. 2022;23(23):14621.
24. Hasanzad M. Precision medicine in clinical practice. Cham: Springer; 2022.
25. Goetz LH, Schork NJ. Personalized medicine: motivation, challenges, and progress. Fertil Steril. 2018;109(6):952–63.
26. Badhe Pravin OK, Ashwini B. Personalized medicine: the future of modern medicine. Authorea. 2021.
27. Meskó B, Drobni Z, Bényei É, Gergely B, Győrffy Z. Digital health is a cultural transformation of traditional healthcare. Mhealth. 2017;3:38.
28. Mathews SC, McShea MJ, Hanley CL, Ravitz A, Labrique AB, Cohen AB. Digital health: a path to validation. NPJ Dig Med. 2019;2(1):38.
29. Mesko B, Győrffy Z. The rise of the empowered physician in the digital health era: viewpoint. J Med Internet Res. 2019;21(3):e12490.
30. Kim YK. RNA therapy: current status and future potential. Chonnam Med J. 2020;56(2):87–93.
31. Li F, Ruijs N, Lu Y. Ethics & AI: a systematic review on ethical concerns and related strategies for designing with AI in healthcare. AI. 2023;4(1):28–53.
32. Ahmed L, Constantinidou A, Chatzittofis A. Patients' perspectives related to ethical issues and risks in precision medicine: a systematic review. Front Med. 2023;10:1215663.

Human Genomics Technology

2

Forough Taheri, Naieme Goharifar, Mehri Safari, Akram Goharifar, Arash Salmaninejad, Negar Sarhangi, Mandana Hasanzad, and Mahsa M. Amoli

What Do We Learn in This Chapter?
This chapter aims to comprehensively explore the clinical application of GWAS data and highlights the immense potential of GWAS in shaping personalized medicine, disease prevention, and therapies. The collected data in this chapter is a valuable resource for researchers, clinicians, and healthcare professionals seeking to understand and leverage GWAS data for practical applications in medicine.

We also present a thorough analysis of the present state and impact of next-generation sequencing (NGS) data analysis in clinical settings and explore its potential as a future diagnostic approach in personalized medicine. The focus areas include cancer, neurology, cardiology, pharmacogenetics, and drug therapy.

Rationale and Importance
Genome-wide association studies (GWAS) have emerged as a powerful tool in genetics, providing valuable insights into the genetic architecture of various diseases and traits. The integration of GWAS findings into clinical practice holds

F. Taheri
Endocrinology and Metabolism Research Center, Endocrinology and Metabolism Clinical Sciences Institute, Tehran University of Medical Sciences, Tehran, Iran

N. Goharifar
Faculty of Science and Technology, Department of Biology, ACECR Institute of Higher Education, Isfahan, Iran

M. Safari
Department of Operative Dentistry, Dental School, Islamic Azad University of Isfahan Branch, Isfahan, Iran

A. Goharifar
Department of Prosthodontics, School of Dentistry, Islamic Azad University, Isfahan Branch (Khorasgan), Isfahan, Iran

A. Salmaninejad
Faculty of Medicine, Department of Medical Genetics, Guilan University of Medical Sciences, Guilan, Iran

N. Sarhangi
Personalized Medicine Research Center, Endocrinology and Metabolism Clinical Sciences Institute, Tehran University of Medical Sciences, Tehran, Iran

Medical Genomics Research Center, Tehran Medical Sciences, Islamic Azad University, Tehran, Iran

M. Hasanzad
Personalized Medicine Research Center, Endocrinology and Metabolism Clinical Sciences Institute, Tehran University of Medical Sciences, Tehran, Iran

M. M. Amoli (✉)
Metabolic Disorders Research Centre, Endocrinology and Metabolism Molecular-Cellular Sciences Institute, Tehran University of Medical Sciences, Tehran, Iran
e-mail: amolimm@tums.ac.itr

M. Hasanzad (ed.), *A Glimpse at Medicine in the Future*,
https://doi.org/10.1007/978-981-97-4974-4_2

immense potential for advancing disease prevention, diagnosis, and treatment, ultimately leading to improved patient outcomes and personalized approaches to healthcare.

Next-generation sequencing (NGS) technology has revolutionized genomic research by enabling rapid, cost-effective, and comprehensive analysis of DNA sequences. In oncology, NGS has enhanced our understanding of tumor biology, identified actionable mutations and facilitated the development of targeted therapies. In neurology, NGS has aided the identification of genetic causes of various neurological disorders, paving the way for personalized treatment strategies. In cardiology, NGS has contributed to the identification of genetic variants associated with cardiovascular diseases, enabling risk stratification and guiding therapeutic interventions. Additionally, NGS has been instrumental in pharmacogenetics and drug therapy, allowing for the identification of genetic variants that influence drug response and facilitate personalized treatment selection. However, the full potential of NGS in clinical practice is yet to be realized, and challenges need to be addressed. Despite these challenges, NGS holds tremendous promise for the future of medicine, offering opportunities for precision diagnostics and targeted therapies tailored to individual patients.

2.1 Introduction

Genetic factors play a crucial role in numerous common diseases, and uncovering the specific genetic determinants associated with these conditions has the potential to shed light on disease pathogenesis, identify new targets for therapeutic interventions, and devise effective treatment strategies. Furthermore, genetic factors can serve as valuable biomarkers for diagnosing diseases, stratifying patients, and predicting prognosis or response to treatment. Identifying these factors enables healthcare professionals to implement timely preventive or optimized interventions in patient care [1]. Our coding and noncoding genomic regions contain genetic variants, which encompass qualitative alterations in nucleotide sequencing referred to as single nucleotide polymorphisms (SNPs) [2], as well as quantitative structural changes known as copy-number variants (CNVs) [3]. The HapMap Project extensively characterized over 3.1 million SNPs in humans, and now companies offer genotyping platforms with over 1 million SNPs [4].

Over the past decade, genome-wide association studies (GWAS) have revolutionized the field of disease genetics by examining the genetic variants of numerous individuals to identify associations between genotypes and phenotypes (Fig. 2.1) [5, 6]. This approach has led to the discovery of a vast number of loci that significantly enhance our understanding of genetic factors involved in human diseases [7]. However, the accessibility of high-throughput genotyping services offered by contract-based companies has presented a challenge for future GWAS endeavors, as clinical research groups can now conduct these studies more readily. GWAS is a valuable tool for investigating complex human traits by comparing genomic variations between two groups of individuals, e.g., cancer patients and unaffected controls [8]. The objective is to identify specific regions of the genome, known as quantitative trait loci (QTL), that harbor genetic variants influencing phenotypes and linking them to individual variations and outcomes such as disease recurrence or recovery. This research approach holds significant clinical implications as it can lead to the discovery of novel therapeutic targets and accurate predictions of treatment response [9]. Undoubtedly, personalized medicine is a major objective for the 21st century, aiming to maximize benefits and minimize adverse effects by tailoring treatments to each patient [10]. With the completion of the Human Genome Project, advancements in technology, and more cost-effective high-throughput DNA sequencing techniques, the identification of genetic variants has become more efficient. Genetic variation, encompassing both coding and noncoding regions of our DNA, contributes to our unique characteristics and can also influence our individual susceptibility to diseases [11].

The primary goal of genetic research is to translate biological knowledge into medical

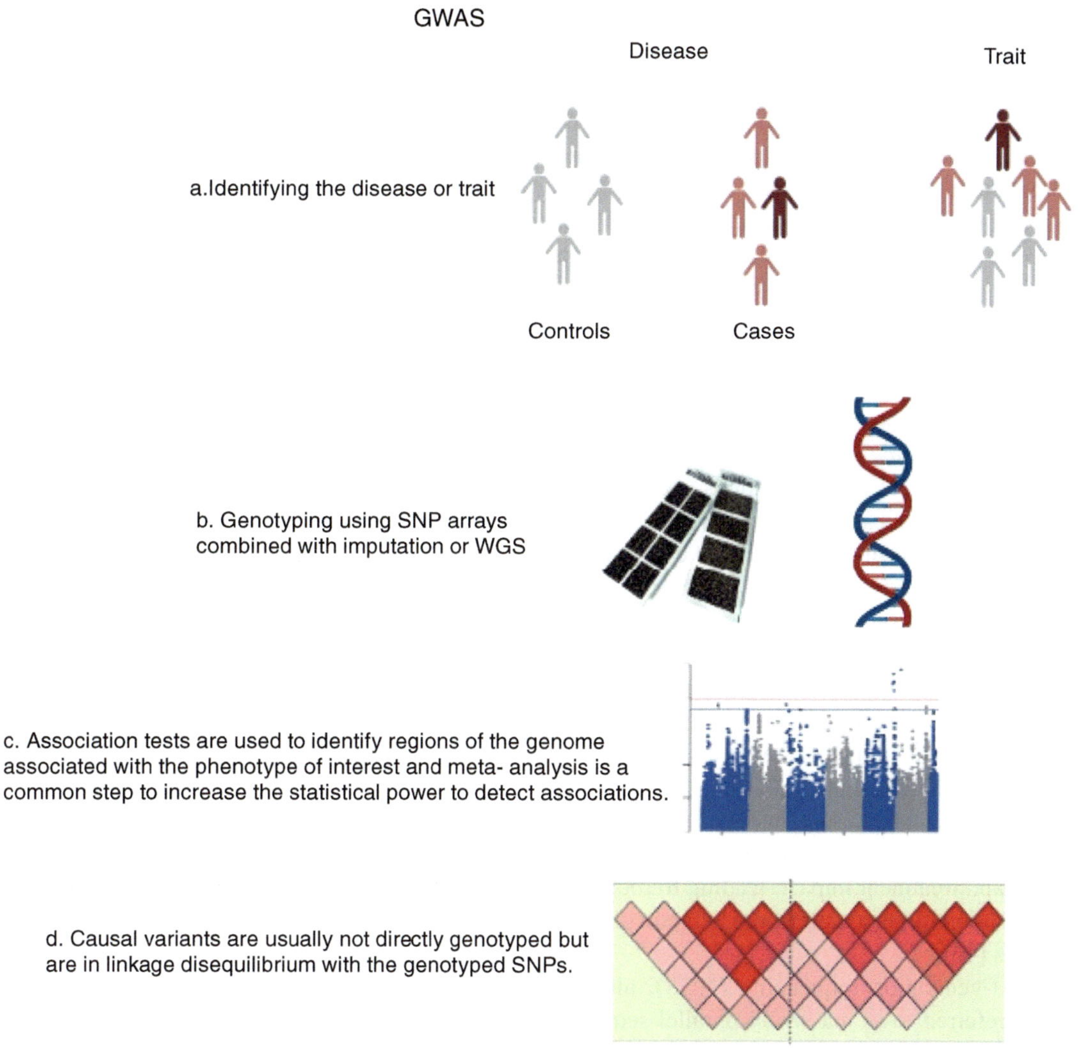

Fig. 2.1 Analysis of GWAS results

advancements. Although it often takes significant time to translate scientific discoveries into practical applications [12], there is an increasing number of examples demonstrating the clinical potential of GWAS findings. Recent advancements, such as the development of large-scale biological databases, sophisticated patient characterization methods such as proteomics approaches [13], and computational tools for data analysis, have greatly enhanced the possibility of broadly applying a broad precision medicine approach. Consequently, GWAS and next-generation sequencing (NGS) techniques enable the identification of preclinical disease stages, the formulation of appropriate differential diagnoses, and the provision of early and optimized therapeutic strategies, which can replace traditional treatments [13]. The noteworthy aspect is that the identification of genetic variants through GWAS provides a means to identify individuals at an elevated risk for specific diseases. This discovery has the potential to significantly enhance patient outcomes by enabling early detection, prevention, or targeted treatment approaches. By leveraging this genetic information, healthcare professionals can proactively intervene and implement strategies to mitigate disease progression, leading to improved patient management and ultimately better health outcomes [14]. This chapter aims to comprehensively gather and synthesize relevant research on the utilization and implications of GWAS data in clinics. It seeks to explore various aspects, including the GWAS concept, clinical applications in both Mendelian and complex disorders, and some application of GWAS in pharmacogenomics for drug response and toxicity.

There is a significant challenge in identifying and elucidating the specific genes and mutations that underlie various human disorders. The early detection of these causal mutations can facilitate the timely identification of a range of disorders. In today's era, healthcare costs have risen substantially. However, if genetic disorders, carrier status, and predispositions to conditions such as cancer, cardiovascular diseases, and neurodegenerative disorders can be identified early, it has the potential to reduce healthcare expenses [15]. A deeper understanding of DNA sequencing can provide valuable insights into the genetic factors underlying diseases. The insights gained from this knowledge have significantly valuable diagnostic and treatment impact, leading to the gradual adoption of personalized medicine in clinical settings [16].

Next-generation sequencing (NGS), alternatively referred to as massively parallel sequencing or high-throughput sequencing, represents a groundbreaking technology enabling the simultaneous sequencing of millions of DNA/RNA sequences [17]. By leveraging NGS, it becomes possible to resolve billions of sequencing reactions within a matter of days, spanning from library preparations to the generation of final results. Nevertheless, the interpretation of such immense volumes of data continues to pose a challenge in extracting clinically relevant insights [16]. NGS presents numerous advantages compared to previous sequencing methods. These include enhanced throughput achieved through sample multiplexing, improved sensitivity in identifying low-frequency variants, accelerated turnaround time for processing large sample volumes, and reduced costs. Undoubtedly, NGS stands as a genuine revolution in sequencing technology following the traditional sequencing era [18]. The advent of NGS has not only enabled groundbreaking genetic and genomic discoveries but has also revolutionized our comprehension of the molecular mechanisms underlying diseases and potential treatment modalities. The continuous advancements in NGS methods have played a pivotal role in this transformative process, such as the use of fluorescent dyes instead of radiolabeling and capillary array electrophoresis instead of gel electrophoresis, which have introduced automation and enabling the sequencing of several thousand base pairs in a single run, significantly amplifying the sequencing volume [19]. NGS approaches have the remarkable capability to generate billions of nucleotides within a short timeframe while maintaining a cost-effective approach [20–22]. NGS has found extensive applications in medical laboratories and has seamlessly integrated itself into personalized care and treatment outcomes. It is widely utilized in various facets of medical practice, including prognosis, diagnosis, and therapeutic approach [23, 24]. This review presents a comprehensive overview of the current status and impact of NGS technology in clinical settings and its potential as a future diagnostic approach in personalized medicine.

2.2 GWAS Design and the Scope of Genetic Variation

Genetic variants play a pivotal role in shaping the distinctiveness of each individual and can also influence our personalized predisposition to various diseases. Through comprehensive analysis of human SNPs, researchers have identified intriguing SNP markers associated with specific disorders [25]. SNPs are the most prevalent form of genetic variation in the human genome, resulting from the substitution or insertion/deletion of a single base. These polymorphisms account for over ninety percent of all known genetic variations [26]. SNPs can be categorized into two types based on their impact on traits. The first type is non-synonymous SNPs, where a change in the base sequence alters the encoded protein sequence, potentially affecting protein function. Non-synonymous SNPs often directly contribute to biological trait changes and disease development. The second type is synonymous SNPs, where a change in the coding sequence does not affect the amino acid sequence of the translated protein. However, due to linkage disequilibrium, synonymous SNPs have become widely utilized and important markers in gene association research, playing a significant role in candidate gene studies and whole-genome association research [27].

Before undertaking a GWAS, it is crucial to carefully consider the scope of existing consent documents. When retrospectively extending clinical trials or observational studies to test genetic hypotheses, the applicability of the original informed consent to the proposed genetic study raises noteworthy concerns. Although family-based designs can be utilized for GWAS projects, wherein the obtained data is analyzed using family-based tests of association, several initial advancements in GWAS have been accomplished through retrospective case-control studies [27]. The utilization of case-control designs brings benefits in the form of efficient recruitment of independent sampling units, ease of analysis, and enhanced statistical power for detecting associations with common genetic variants that have modest-to-low relative risks. However, it is imperative to emphasize the significance of proper sample matching in case-control designs to mitigate biases that could potentially result in the identification of numerous SNPs that appear to be significant [27]. It is of utmost importance to closely match cases and controls in terms of their ethnicity to prevent confounding and spurious associations. In this regard, GWAS studies adhere to the same design principles employed in any observational or clinical trial [27]. Additionally, the power to detect associations in GWAS is influenced by factors such as effect size, sample size (number of cases, controls, or families), and the disease model assumed for the tested association. These factors are further affected by the disease's prevalence, the frequency of disease alleles, and the genotypic relative risk (GRR) [27]. One potential source of error in any GWAS is the mishandling or mislabeling of samples within the laboratory. Implementing appropriate sample tracking and evaluation measures can help reduce the risk of such errors [27] (Fig. 2.1).

2.3 GWAS in Complex, Multigenic Diseases

Most human diseases with substantial genetic contributions do not adhere to a Mendelian pattern of inheritance. Instead, they are complex disorders influenced by multiple genes. The evidence supporting genetic factors' involvement in disease often stems from the observation of disease clustering within families. Twin concordance studies play a crucial role in estimating disease heritability, which quantifies the extent to which genetic factors contribute to overall phenotypic variation [28, 29]. For complex, non-Mendelian diseases, the significance of genetic factors in disease development has been established through epidemiological and twin studies. Biomedical research has experienced a surge in efforts to identify genetic variations for personalized management of complex diseases. However, despite initial optimism, the clinical implications of these risk variants are still subject to debate [30]. Personal genomics holds the promise of discovering prognostic and predictive tools for personalized prevention and treatment of diseases. Yet, despite the numerous risk alleles discovered through GWAS, many scientists now question the clinical utility of these DNA variants [31, 32]. The clinical utility of single, common, low-penetrance genes is restricted because their individual risk contributions are modest. Consequently, their potential as markers for risk prediction is limited. Nevertheless, there are hints that the combination of these risk alleles has the potential to categorize the general population into high-risk and low-risk groups. Nevertheless, the polygenic model of diseases that incorporates multiple genetic factors still requires validation through large-scale clinical studies [33].

The initial value of GWAS findings lies in their ability to provide insights into disease causes, therapeutic targets, and gene function. These discoveries have uncovered new etiologic pathways and therapeutic possibilities not previously associated with complex diseases. For example, the autophagy pathway was implicated in inflammatory bowel disease, and the HLA-C locus was found to influence viral load control in HIV infection [34].

One of the most intriguing aspects of GWA studies is the potential genetic connections between diseases that were previously considered unrelated. For instance, variants near CDKN2A/B were found to be associated with both type 2 diabetes and coronary disease, while variants near PTPN2 were linked to both Crohn's disease and type 1 diabetes. These findings open up new avenues of research to explore similarities in the causes, progression, or treatment of these conditions [35]. Before GWA studies, type 2 diabetes had already been associated with a few common variants. However, these studies have identified an additional 32 loci, which together account for approximately 10% of the heritable risk of the disease. Although the contribution of these variants to disease risk is still relatively low, they provide valuable information, especially in earlier stages of life, and complement known risk factors [36].

Moreover, these variants are shedding light on both familiar and novel aspects of disease development. Some of the identified loci involve genes that are already known to be associated with monogenic forms of diabetes, genes involved in insulin signaling (IRS1), or genes targeted by antidiabetic medications (KCNJ11, PPARG). Analysis of the effects of these variants in individuals without diabetes has revealed that many of them increase the risk of developing diabetes by impairing beta cell function, while others clearly affect insulin action [37].

In summary, GWA studies offer valuable insights into disease causes, therapeutic targets, and gene function. They reveal new pathways, uncover genetic connections between seemingly unrelated diseases, and enhance our understanding of disease development [38].

2.3.1 GWAS in Cancer

The potential of GWAS in relation to cancer risk was initially promising, particularly in improving our ability to identify individuals in the general population who could benefit from modified screening strategies [39, 40].

The road to personalized cancer management is lengthy and riddled with numerous challenges and obstacles [10, 41]. It is feasible to envision the accomplishment of cancer genetic mapping through the utilization of advanced technologies, alongside the existence of genotyping platforms containing more than 1 million SNPs and CNVs. Nevertheless, depending solely on a cancer genetic atlas has its constraints. It is crucial to undertake functional investigations that delve into the interplay among genetic risk variants. Additionally, comprehending the interactions between genetic and environmental risk factors is imperative before implementing a prevention strategy built upon risk stratification [33]. At present, personalized prevention of breast cancer remains more of an aspiration than an actuality.

Initial optimism surrounding GWAS was based on the belief that the identification of genetic risk variants would provide substantial discriminatory accuracy [31]. Nonetheless, the impact of these identified SNPs on risk elevation is minimal, and recent research has demonstrated that this genetic data does not outperform simpler and more economical clinical information derived from family history and conventional risk models [42]. Enhancing clinical outcomes for patients diagnosed with advanced stages of the disease holds promise through the potential of targeted screening and primary prevention for both the general population and individuals. Nevertheless, within the realm of personalized prevention, there exist two pivotal inquiries that require attention. The initial query revolves around the feasibility of precisely classifying individuals into high, intermediate, and low-risk categories. The second question pertains to the implementation of personalized prevention strategies and how to determine the most suitable preventive interventions for each individual.

Progress has already been made towards these goals. For example, individuals who possess heritable high-penetrance mutations in genes like BRCA1, BRCA2, mismatch-repair genes, and the CDH1 gene confront a notably heightened risk (approximately 75%) of developing various types of cancer. These include breast, ovarian, colorectal, endometrial, and diffuse-type stomach cancer [10]. Prophylactic surgery or enhanced surveillance is presently advised for these high-risk individuals, who can be identified using readily accessible genetic testing. Nonetheless, it is important to note that these mutations, which have significant implications for cancer risk, are uncommon in the broader population. They account for less than 0.3% and less than 3–10% of newly diagnosed cancers annually at these specific sites, respectively [10].

2.3.2 GWAS in Psychological and Neurological Disorders

Specialized therapies are required to address the disabilities that arise from neurodegenerative processes, which typically begin several years before the appearance of clinical symptoms. As a result, it is of utmost importance to intervene against these pathological mechanisms at the earliest possible stage, prior to the manifestation of clinical signs [43]. However, despite notable progress in biomedical research, our understanding of the underlying mechanisms behind these pathologies remains incomplete, and a universal therapeutic approach does not seem to be the most effective resolution. Neurodegenerative diseases, including Alzheimer's disease (AD) and Parkinson's disease (PD), exhibit clinical heterogeneity and have a strong genetic component. Consequently, patients with the same pathology may present variable clinical courses, making it challenging to apply a uniform treatment to all individuals [44, 45]. Notably, these subgroups exhibit distinct molecular characteristics, prompting researchers and clinicians to focus on discovering and categorizing these variants. Therefore, it becomes essential to identify biomarkers that function as biological labels, enabling the identification and classification of each specific subtype. Once this milestone is achieved, it will become possible to accurately determine the precise sub-variant affecting an individual patient and tailor the treatment accordingly, based on this valuable information [46].

Neurological diseases, in fact, present significant opportunities to realize the potential of precision medicine, where individual healthcare is tailored based on factors such as genetics and lifestyle [47]. Effective collaboration between different domains of basic and translational research is essential to transform precision medicine into an attainable goal in neurological diseases. Additionally, the inclusion of a substantial number of samples from healthy individuals, who can serve as controls for comparison with patient data, is paramount. By integrating genetic information, clinical data, and patient imaging, we can gradually unravel the puzzle and gain a clearer understanding of the diseases [48].

In this context, the "Italian IRCSS Network of Neuroscience and Neurorehabilitation" serves as an excellent example of a web-based platform. Its primary objective is to standardize and optimize clinical care for patients with neurodegenerative diseases, along with the therapeutic strategies employed. This platform plays a crucial role in advancing precision medicine in neurology [49]. Although GWAS arrays demonstrate high efficiency in genotyping SNPs, they possess limitations in accurately detecting various structural variations prevalent in the human genome, including insertions, deletions, inversions, and copy number variants [50]. These structural variants have been strongly associated with various conditions, particularly neurodevelopmental diseases like autism and schizophrenia [51]. They also play a role in interindividual variations in drug metabolism [52]. However, detecting these variants using GWAS methods can be challenging [3]. Gaining insights into the molecular and cellular mechanisms that drive the pathogenesis of these structural variants opens up fresh possibilities for personalized treatments grounded in mechanisms. Such treatments hold the potential to decrease morbidity and mortality within the field. However, the realization of personalized approaches for predicting, diagnosing, and treating central nervous system diseases and disorders is presently distant. Substantial advancements are required to grasp the fundamental mechanisms governing cell death, regeneration, and the interactions between neurons and glial cells in both healthy and diseased brains and spinal cords. Only with this progress can we bridge the gap and achieve effective personalized approaches in the field.

Psychiatric disorders are prevalent worldwide and contribute significantly to the global burden of disease. While current pharmacological treatments for these disorders show some effectiveness, a substantial number of patients do not achieve remission [53]. In the study of major depressive disorder, the genes related to the serotonergic neural system have received extensive attention for several reasons. This system is a primary target for many antidepressant medications, and individuals with major depressive disorder often exhibit abnormalities in various components of the serotonin system [54, 55]. Specifically, polymorphisms in the 5-hydroxytryptamine (5HT; serotonin) transporter (5-HTT) gene have been associated with the development of major depressive disorder [56]. GWAS offers a unique approach as they do not require prior hypotheses and have the potential to identify novel molecular factors involved in the etiology of these disorders. Moreover, these studies have not only identified associated SNPs but have also revealed biological pathways that provide initial insights into the underlying mechanisms [57, 58].

Examples include the association of genes related to immune function and glial cell function with schizophrenia and the connection between genes involved in the structural elements of the postsynaptic density and both schizophrenia and bipolar disorder [59, 60]. Furthermore, genes associated with the glutamatergic pathway have been implicated in various psychiatric disorders such as schizophrenia, bipolar disorder, autism spectrum disorders, and major depressive disorder [60, 61]. The involvement of Fragile X mental retardation protein (FMRP) targets has been observed in both schizophrenia and autism spectrum disorders. Furthermore, a diverse array of psychiatric disorders, including schizophrenia, bipolar disorder, autism spectrum disorders, major depressive disorder, and attention-deficit/hyperactivity disorder (ADHD), have been associated with genes related to calcium channel signaling [62, 63].

These findings provide potential avenues for developing therapeutics that target the underlying biology of these disorders, aligning with the concept of "precision medicine." There have been successful examples of translating genetic discoveries into targeted treatments, such as imatinib (Gleevec) for chronic myelogenous leukemia. While the impact of low penetrance, inherited alleles detected through common and rare variant GWAS is not as straightforward, there is compelling evidence that GWAS findings can identify viable treatment targets. For instance, variants in the gene encoding HMG-CoA reductase, the target of statins, have been associated with LDL levels. Large-scale GWAS of rheumatoid arthritis have also revealed genes targeted by approved drugs for the disorder [64]. Similar progress is being made in the field of psychiatric disorders, particularly in the case of schizophrenia. Notably, one of the strongest associations observed in recent GWAS of schizophrenia was with variants in the dopamine 2 receptor gene (DRD2), which is the target of all approved antipsychotic drugs. However, it is important to note that some risk loci are more amenable to drug targeting than others, with gain-of-function mutations representing promising targets where inhibiting the gene products could potentially be therapeutic [64].

2.3.3 GWAS in Cardiovascular Disease

GWAS has proven valuable in identifying numerous genetic variants linked to cardiovascular diseases. However, they have been limited in their ability to elucidate the underlying mechanisms behind these associations. Consequently, GWAS findings have not been effective in predicting preclinical diagnoses providing guidance for potential prevention strategies, assessing risk stratification, or offering specific therapeutic interventions. While GWAS has successfully identified genetic variants associated with cardiovascular diseases, further research is needed to understand the mechanisms linking these variants to the diseases and to develop actionable insights for clinical applications [65]. GWAS-defined risk variants have the potential to be utilized in the reclassification of individuals initially classified as low-risk based on established risk factors for conditions like cardiovascular disease, such as age, sex, and hypertension. Reclassification becomes particularly relevant in clinical settings where risk thresholds have been defined and intervention guidelines have been established, acknowledging that these thresholds, while clinically useful, are inherently subjective and open to debate [66]. In one study, a genotype score was developed using a combination of genetic variants in a cohort previously categorized by their risk of cardiovascular events. The aim was to assess reclassification based on this genotype score [67]. According to the findings, within the cohort, 26% were initially categorized as having an intermediate risk. Among them, 35% were subsequently reclassified into either low- or high-risk categories. This reclassification revealed a noteworthy enhancement in the statistics of reclassification, surpassing the reliance solely on established risk factors. Reclassification holds significant importance in clinical decision-making, especially for individuals initially classified as intermediate risk based on standard guidelines. It should be noted that the actual reclassification typically involves minor adjustments in the risk score, leading to the shifting of borderline subjects between adjacent risk categories. Nonetheless, this reclassification can have meaningful implications for treatment and management decisions [66].

Cardiovascular pharmacogenetics is a dynamic field that provides valuable clinical insights into personalized and individualized drug prescriptions. DNA sequence variations have been shown to influence the clinical responses of several commonly used cardiovascular drugs, including statins and clopidogrel [68, 69]. Research focused on statins has primarily investigated the genetic factors contributing to dose-related myopathy, a side effect that limits patient tolerance or compliance with cholesterol-lowering treatment [70, 71]. While myopathy is rarely life-threatening, it is crucial to address its impact. One particularly common variant, allele rs4149056C, in the SLCO1B1 gene, has been identified as increasing the risk of myopathy. Homozygous carriers of the C allele have a higher risk compared to heterozygous carriers [71]. These data have been replicated in studies involving individuals taking simvastatin [70]. However, the risk of myopathy appears to be lower with other statins such as atorvastatin and pravastatin. Clopidogrel has also been extensively studied in the field of pharmacogenetics to uncover the molecular basis of its effectiveness or lack thereof. Clopidogrel is administered as a prodrug that requires activation by the enzyme CYP2C19, responsible for metabolizing several drugs. Pharmacogenetic studies have focused on understanding the genetic variations that influence the activation and efficacy of clopidogrel [72, 73]. Overall, cardiovascular pharmacogenetics provides valuable information for tailoring drug prescriptions on an individual level, taking into account genetic variations that can impact the effectiveness and safety of commonly used cardiovascular medications.

2.4 GWAS in Pharmacogenetic, Drug Response and Toxicity

In the years ahead, the pharmaceutical industry is anticipated to show substantial interest in conducting pharmacogenetic GWA studies. These studies aim to identify markers that can aid in patient stratification during clinical trials. The extensive pharmacogenetic data obtained from such studies will serve as a catalyst for the progress of personalized medicine. Particularly, the application of pharmacogenetic GWA studies, coupled with the early adoption of personalized therapy, holds significant relevance in the selection of expensive or long-term medications for life-threatening conditions. Additionally, it becomes crucial when the therapeutic window is narrow or concerns arise regarding adverse events, such as in cancer chemotherapy [1]. Recent efforts combining GWA studies with drug development have yielded promising results. An instance of a potential therapeutic approach for inflammatory bowel disease involves targeting the inhibition of IL23R. IL23R has various variants, some associated with elevated risk and others with reduced risk. The identification of specific variants in individual patients may eventually aid in determining the appropriateness of IL23-specific therapy [74].

GWA studies have also been employed in the field of drug toxicity. Variants predisposing individuals to paclitaxel-induced peripheral neuropathy and aspirin-exacerbated respiratory disease have been identified using the GWA approach [75, 76]. As the thousands of GWA study results from previous years are further explored and additional GWA studies are conducted, more clinical applications are expected to emerge. With the increasing integration of genomic approaches in clinical care beyond major medical centers, the pace of detecting clinically relevant GWA results and their translation into the clinic is likely to accelerate [77, 78].

Although GWA studies may not directly unveil causal genes or disease mechanisms, subsequent functional experiments conducted after GWA studies can offer insights into novel targets and pathways for therapeutic intervention. By choosing drug targets that are supported by genetic evidence, the success rate of drug development can be enhanced while reducing both time and costs involved. Furthermore, this approach can contribute to the advancement of drug repurposing efforts [79]. GWA studies have proven beneficial for several diseases, such as type 2 diabetes, rheumatoid arthritis, ankylosing spondylitis, psoriasis, osteoporosis, schizophrenia, and dyslipidemia. These studies have facilitated the discovery of new drugs or candidate drugs currently employed in clinical settings or undergoing evaluation in clinical trials [6]. Furthermore, the genetic variants identified through GWA studies have played a significant role in guiding drug selection, determining optimal dosage, and preventing adverse drug reactions [7]. An exemplary illustration of this is the utilization of single nucleotide polymorphisms (SNPs) near the IL28B gene, also known as IFNL3. These SNPs have proven instrumental in predicting the probability of a positive treatment response to pegylated interferon-α and ribavirin therapy in individuals with hepatitis C virus infection [80, 81].

2.5 Limitations of the GWAS

GWA studies offer a powerful and innovative approach to identifying genetic variants associated with various diseases. However, it's important to acknowledge their limitations, such as the potential for false positive results and their limited sensitivity in detecting rare variants. These factors need to be taken into consideration when interpreting GWA study findings. For the foreseeable future, the primary utility of GWA studies is expected to lie in the exploration of biological pathways underlying disease causation, as well as normal health and development. Understanding these pathways can provide valuable insights into disease mechanisms and help uncover potential therapeutic targets. The clinical application of GWA findings requires robust evidence that incorporating these genetic markers into testing provides additional information beyond known risk factors. Moreover, it is crucial to determine whether effective interventions are available based on these findings. The associated costs of implementing such testing should be justified by improved patient outcomes. Additionally, it is essential to ensure that obtaining this genetic information does not have significant adverse consequences for patients and their families. While many GWA findings are currently several steps away from mainstream clinical use, the functional investigation and experimental application of these discoveries hold great promise for advancing disease prevention and treatment. As our understanding of the biological significance of these genetic variants deepens, it is expected that new breakthroughs will emerge, leading to improved strategies for addressing various diseases.

2.6 Overview of NGS Technologies and Extension of Sequencing

2.6.1 Platforms

The previous method of DNA sequencing, referred to as Sanger sequencing, employed the sequencing by synthesis (SBS) approach. This technique involved the use of a radioactively labeled DNA strand that was complementary to the template strand. Additionally, it used the dideoxy chain termination technique to halt DNA synthesis at specific nucleotides. Subsequently, the resulting DNA fragments were subjected to analysis using polyacrylamide gel electrophoresis. This technique allowed for the separation of the fragments based on their size, providing valuable insights into the DNA sequence. This sequencing approach, known as 1st NGS, underwent further advancements and automation and was subsequently made commercially accessible [82, 83]. The subsequent generation of sequencing methods, referred to as high throughput and massively parallel sequencing or 2nd GS, eliminated the need for separating sequencing reactions into lanes or capillaries. Instead, it facilitated the simultaneous occurrence of billions of sequencing processes on a slide surface, such as glass or beads. This advancement significantly improved throughput and reduced costs compared to 1st GS. Major companies like Roche Life Sciences, Thermo Fisher Scientific, Illumina, and Applied Biosystems developed and commercialized the most popular 2nd NGS platforms, collectively known as NGS platforms, revolutionizing DNA sequencing [84]. NGS technologies can be categorized into two groups: SBS or sequencing by synthesis and SBH or sequencing by hybridization. Based on sequencing by synthesis, Illumina sequencing technology is the most widely used approach, responsible for generating approximately 90% of the world's sequence data [85]. The Roche GS-FLX 454 Genome Sequencer, launched in 2004 as the 454 Sequencer machine, was the pioneering commercial system of its kind [86, 87]. It enabled the successful sequencing of the complete genome of an individual, James D. Watson, marking a significant achievement. In 2008, Roche introduced an upgraded version called the 454 GS FLX Titanium system, which improved the average read length to 700 bp and accuracy to 99.997%. Within a 24-h timeframe, this platform had the capacity to generate 0.7 gigabases (Gb) of data per run. The GS Junior bench-top sequencer system, on the other hand, exhibited an average read length of 700 base pairs (bp), a throughput of 70 megabases (Mb), and a runtime ranging from 10 to 18 h. However, Roche made a strategic decision to redirect its efforts away from gene sequencing, leading to the discontinuation of the 454 Life Sciences sequencing services by the conclusion of 2013. Consequently, this study will not delve further into Roche NGS instruments [88, 89].

In the Illumina sequencing method, the reaction system consists of a blend of DNA polymerase, primers, and four dNTPs, each labeled with a specific fluorescent marker corresponding to their bases. The addition of each base occurs sequentially, facilitated by the 3′-OH of the dNTPs. Once the sequencing reaction is complete, the DNA polymerase and unused dNTPs are removed. A buffer solution is then introduced for fluorescence excitation, and the resulting fluorescence signal is captured using optical tools. These signals are utilized for base calling. To prepare for the subsequent round of sequencing, a chemical reagent is employed to quench the fluorescence signal and eliminate the protective group from the 3′-OH of the dNTPs. It is important to note that all sequencing records produced within a given experiment possess an identical length. Contemporary sequencing platforms have the capability to generate DNA sequences in a paired-end fashion, such as 22 × 300 bp, allowing for the reading of both ends of a fragment [90]. Errors in sequencing can occur due to dephasing and signal decay resulting from incorrect cleavage of fluorescent labels or terminating moieties. It is worth noting that the average error rates for these sequencing platforms range from 1% to 1.5% [91]. In addition to Illumina, another sequencing technology called Ion Torrent has been introduced, which offers faster and more cost-effective sequencing and is predominantly used in clinical settings [16, 92]. Illumina and Ion's Torrent platforms employ different underlying chemistry to determine the base sequences. Ion Torrent sequencing involves the incorporation of a DNA template, along with a single bead, into a droplet called a bead emulsion. Within each reaction unit, the template undergoes emulsion PCR to amplify it clonally until it covers the entire surface of the corresponding bead. This step is similar to Illumina's flow cell and ensures sufficient signal during sequencing. The covered beads are then loaded onto a chip containing millions of micro-wells, with each well housing a single bead. The sequencing process relies on pH changes, where the release of a hydrogen ion (H^+) upon the addition of each nucleotide is detected by observing a pH shift. In this process, the chip is sequentially exposed to individual nucleotides, one at a time, and this iterative procedure is repeated multiple times [16, 93, 94].

QIAGEN offers the GeneReader platform, which enables the generation of next-generation sequencing (NGS) data. This technology utilizes fluorescent signals to detect nucleotides, which are then matched with templates that have been clonally amplified using the GeneRead QIAcube. The GeneReader system requires trained individuals with expertise in Molecular Biology (MB) techniques and familiarity with the GeneReader platform. It is designed to provide a comprehensive workflow that addresses the challenges typically encountered during sample preparation. The utilization of the GeneReader system allows for the comprehensive management of all stages involved in sample processing and sequencing. This includes DNA extraction, library preparation, sequencing, bioinformatics data analysis, and the assessment of clinical implications and evidence. This includes the utilization of the "QCI Analyze" and "QCI Interpret" tools, which facilitate the analysis of biological data, variant calling, annotation, read mapping, and alignment visualization. To ensure the quality of data, the GeneReader system guarantees a run-level quality assurance of over 85% and a Q30 score. This validation process helps minimize the occurrence of indications of false-positive and false-negative associated with each variant [95, 96].

Compared to 2nd GS sequencing systems that rely on PCR amplification of template and suffer from sequencing errors, single-molecule sequencing (SMS) offers a potential solution by sequencing individual molecules without amplification. Additionally, the time required to produce results using traditional methods is prolonged due to multiple scanning and washing cycles, leading to the loss of synchronicity and noisy sequencing data, resulting in shorter reads [97]. An exemplary SMS approach is known as single-molecule real-time sequencing (SMRT), which was developed by Pacific Biosciences (PacBio). SMRT employs sequencing by synthesis chemistry, which is akin to certain 2nd GS techniques, but with reduced reliance on starting material and PCR amplification. This results in a lower error rate and the generation of long reads, all achieved in less time [98, 99]. SMRT has the capability to generate kb-long reads, exemplified by the latest PacBio sequencing machine, the Sequel IIe System released on October 5, 2020. In a mere 30-h timeframe, this system has the ability to generate over 4 million reads with an accuracy exceeding 99%. Comparative studies have demonstrated that PacBio HiFi Sequencing, performed by SMRT technology compared to Nanopore and Illumina systems, offers improved contiguity (N50), accuracy, and completeness (genome size). Furthermore, PacBio HiFi Sequencing has been reported to be cost-effective when compared to its competitors [99, 100]. The advantages of 3rdGS sequencing over 2nd GS sequencing are numerous. These advantages include greater throughput, direct haplotype detection, elongated read lengths, improved consensus accuracy for identifying rare variants, whole chromosome phasing, and the ability to work with small sample sizes. These unique features make third-generation sequencing highly valuable in clinical diagnostics [101].

The integration of nanopore technology into SMS has led to the emergence of fourth-generation sequencing (4G sequencing). This cutting-edge approach enables real-time sequencing without the need for amplification or repeated

Table 2.1 Comparison of sequencing instruments

Platform	Amplification	Data output (Gb)	Sequencing technology	Read length	Limitations and error rate	Run time (h)
Sanger	PCR with dNTPs and ddNTPs	≤0.002	Irreversible chain termination	≤900	0.001%	~4
Illumina (NovaSeq 6000)	Bridge amplification	6000	Sequencing by synthesis	300 PE	– Short read length – Difficult sample preparation – Long run time – GC-rich regions – 0.2%	44
Thermo Fisher Scientific Ion Torrent	Emulsion PCR	50	Sequencing by synthesis	600 SE	– Short read length – Difficult sample preparation – Long run time – Homopolymers – 1% indel	12
QIAGEN (GeneReader)	Fluorescent signals	Not available	Sequencing by synthesis	100 SE	– Short read length – Difficult sample preparation – Long run time	Not available
Roche 454	Emulsion PCR	0.7	Sequencing by synthesis	600	Not precisely determining homopolymer length	24
PacBio (HiFi Reads)	Not available	66.5	Real-time sequencing	25 kb	10–15%	Real time
Nanopore (PromethION)	Not available	14000	Real-time sequencing	4 mb	5–20%	Real time

cycles, earning it the designation of 4thGS or in situ sequencing technology. This method allows the order of nucleotides in cells to be accurately determined, opening up new possibilities in DNA sequencing [102]. 4G sequencing sets itself apart from previous generations in two significant ways. Firstly, it allows for the observation of the spatial distribution of DNA reads within a sample, providing valuable insights into tissue heterogeneity based on known markers. Secondly, it enables the simultaneous analysis of a large number of cells. For instance, cost-effective and robust approaches for single-cell RNA sequencing have been developed, capable of sequencing multiple cells using minimal starting material [103]. The technique does have a limitation when it comes to sequencing single cells due to the complex nature of tissues composed of numerous cells. This presents challenges in both the technical and computational aspects of single-cell sequencing. However, there is anticipation that in situ sequencing will be utilized to extract clinical data generated by conventional NGS techniques. Targeted in situ sequencing methods can be utilized to directly identify validated biomarkers in samples, while non-targeted techniques offer the potential to develop molecular profiles for disease classification at the molecular level or to cater to individual patient needs. The integration of in situ sequencing with conventional NGS methods holds promise for accelerating the advancement of these techniques, eventually becoming indispensable tools for personalized medicine [97] (Table 2.1).

2.6.2 Extend of Sequencing and NGS Bioinformatics

Driver mutations are defined as mutations that give cells a proliferative or survival advantage [104], while passenger mutations are randomly distributed mutations in the genome that do not provide immediate benefits or phenotypes [105].

Bioinformatic tools should be considered more predictive rather than definitive in nature. NGS allows for the rapid generation of WGS data in a relatively short turnaround time. Additionally, sequencing techniques are available for sequencing all protein-coding genes (Whole Exome Sequencing) or selected protein-coding genes (Targeted Exome Sequencing). In a clinical context, targeted gene panel sequencing is widely used due to its ability to quickly assess disease-related gene panels. While WGS and WES have broad applications, their implementation in the clinical setting presents challenges. Therefore, gene panel assessment can facilitate crucial disease diagnosis, guide treatment strategies, and provide valuable prognostic information within a shorter timeframe. NGS has significantly contributed to the identification of potential disease-associated candidates in both complex and Mendelian diseases [106]. However, variant detection through bioinformatic methods focuses primarily on identifying novel mutations and genes, which are then prioritized for subsequent functional testing. With advancements in sequencing platforms, it is now possible to sequence the entire human genome within one week at a relatively affordable cost [107, 108]. Consequently, the large volume of NGS data generated requires computational and bioinformatics skills to effectively manage, analyze, and interpret the data. NGS bioinformatics is divided into primary, secondary, and tertiary analyses, each aiming to process and analyze NGS data. While the overall goal of these analyses remains the same across NGS platforms, each platform has its unique characteristics and specificities (Fig. 2.2) [107].

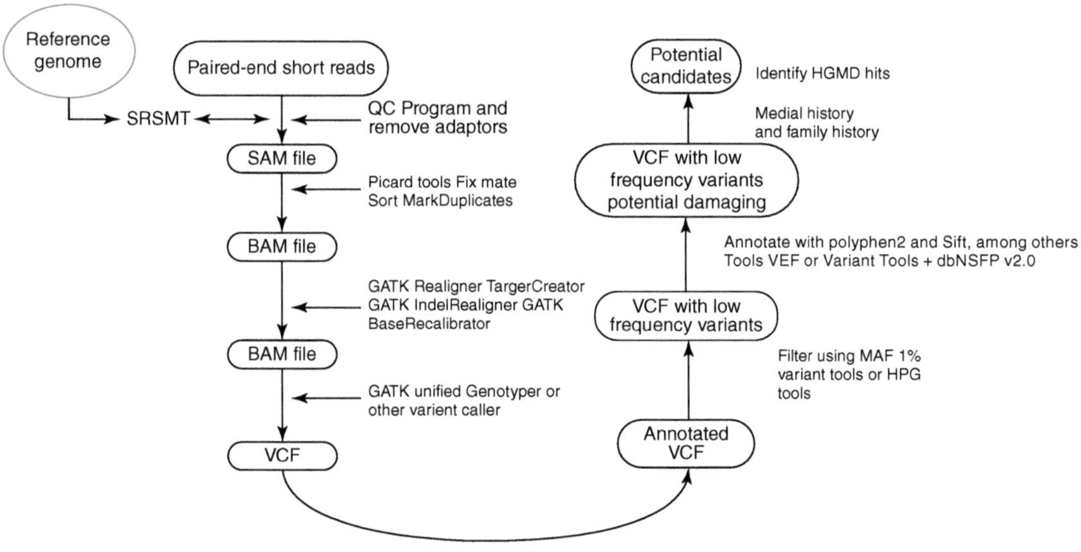

Fig. 2.2 Overview of next-generation sequencing (NGS) bioinformatics workflow. The bioinformatics workflow for NGS can be conceptualized in three main steps. Initially, raw data is detected and subjected to analysis. The subsequent step involves aligning the reads to the reference human genome or conducting de novo assembly, followed by variant calling. Lastly, multiple tasks are executed in the tertiary step, encompassing variant annotation, variant filtering, prioritization, data visualization, and reporting. The sequencing machine generates paired-end short reads, which undergo a rigorous quality control procedure. Following this, adaptor removal is performed on the reads, and specialized tools for short-read sequencing are utilized to map them to the human reference genome. The resulting alignments, in the form of sequence alignment/map (SAM), are then subjected to cleaning using tools like Pickard. Subsequently, the cleaned alignments are converted into the binary format of SAM known as BAM and undergo further processing using tools such as the Genome Analysis Toolkit, enhancing the alignment quality. Quality control reports are generated at this stage, and genetic variants are extracted using variant callers. The variants are documented in the variant call format (VCF) and undergo annotation and filtering. Additionally, low-frequency variants that are known or predicted to be detrimental are validated and incorporated into a comprehensive report for healthcare professionals or genetic counselors

2.7 NGS in Oncology

Disruptions to healthy cellular biology caused by acquired genomic and epigenetic changes collectively drive the distinctive capabilities of cancer cells [109, 110]. In oncology, precision medicine is founded on the understanding that each cancer patient has a unique pathophysiology resulting from the interplay between their constitutional characteristics, the specific genomic alterations fueling the cancer, and the environmental influences affecting both. Effective treatment approaches require a comprehensive understanding of these elements to design customized treatments for patients, although achieving absolute individualization currently remains impractical. Nevertheless, a precision medicine approach provides oncologists with another valuable tool. Precision medicine primarily focuses on identifying genomic alterations that can be targeted with drugs and treating patients accordingly. The Memorial Sloan Kettering-Integrated Mutation Profiling of Actionable Cancer Targets (MSK-IMPACT) panel was FDA-approved in November 2017 for in vitro diagnosis of variants within the genome [111–113]. This test analyzes samples obtained from patients with solid malignant neoplasms, including both tumor and matched normal samples, aiming to identify genetic alterations in 468 unique genes that may guide treatment options. Another test approved by the FDA is the NantHealth Omics Core, which utilizes a WES platform. It provides information on tumor mutational burden and somatic alterations, including single nucleotide variants (SNVs) and insertions/deletions (indels), in 468 genes relevant to cancer. The Illumina TruSight Oncology 500 (TSO 500) test, launched in October 2018, targets 523 genes for the detection of SNVs and indels, as well as 55 genes for identifying fusion events and splice variants. Moreover, the TSO 500 test can identify biomarkers associated with immunotherapy, such as tumor mutational burden and microsatellite instability. Illumina is currently seeking FDA approval to use this test as a diagnostic approach.

NGS in oncology has diverse applications, including the identification and enrollment of patients into suitable clinical trials. Two types of clinical trials are commonly used: umbrella trials and basket trials [114, 115]. Umbrella trials involve enrolling patients with cancer and assigning them to different treatment arms based on the mutations found in their cancer. On the other hand, basket trials enroll patients with different types of cancer but with the same genetic abnormality to evaluate the efficacy of a new drug targeting the shared genetic modification. Thousands of cancer biopsies undergo NGS to identify genetic abnormalities that may respond to targeted therapies. In the context of cancer treatment, minimal residual disease (MRD) refers to the small population of cancer cells that remain in the body even after undergoing therapy. Assessing MRD can provide valuable insights into treatment effectiveness, relapse risk prediction, confirmation or monitoring of remission, and early detection of relapses [116]. Adaptive Biotechnologies has developed the ClonoSEQ test, which utilizes NGS to evaluate and monitor MRD, specifically in patients with B-cell acute lymphoblastic leukemia and multiple myeloma [117, 118]. Another promising approach known as liquid biopsy involves analyzing circulating cancer cells, circulating cell-free cancer DNA/RNA, or exosomes present in bodily fluids like blood, urine, or cerebrospinal fluid. Liquid biopsy holds promise for detecting MRD, real-time monitoring of disease progression, selecting appropriate therapies, and early-stage cancer diagnosis [119]. However, the clinical utility of liquid biopsy requires more evidence and addressing certain challenges. These challenges include low cancer-derived DNA fraction in plasma samples, necessitating modified sample preparation methods, deeper sequence coverage for improved sensitivity, and potential confounding results from clonal hematopoiesis [120–123]. Liquid biopsies also face issues such as signal loss due to dilution of cell-free DNA and the potential misinterpretation of mutational events originating from clonal hematopoiesis in the bone marrow as cancer-related mutations [124].

Currently, a significant focus in cancer genomics revolves around somatic mutations, copy number alterations, and gene rearrangements within cancer cells. However, it is evident that the tumor microenvironment plays a crucial role in tumor growth and treatment response. While studying the transcriptional profile of this ecosystem has shown the potential of gene signatures for better prognosis and classification, the effective integration of these signatures into clinical practice remains limited [125, 126]. By employing transcriptional profiling, researchers can gain valuable insights into how regulatory mechanisms within a tumor, such as changes in promoter methylation or chromatin states, impact gene expression. RNA sequencing offers the potential to uncover allele-specific expression, validate altered gene expression levels, and identify alternative splicing events. Various commercial entities have made efforts to incorporate information about complete or targeted transcriptomes in their clinical reports [24, 127]. However, only a few studies have successfully demonstrated the clinical utility of transcriptome sequencing, primarily due to challenges such as sample degradation, which poses a more significant concern for RNA analyses compared to DNA analyses [128, 129].

The utilization of circulating tumor DNA (ctDNA), which is known as tumor-derived cell-free DNA, holds promise for cancer diagnosis and monitoring of relapse or progression [130]. It can potentially guide therapeutic decisions and monitor patients with existing cancer. Cancer patients often exhibit elevated levels of ctDNA in their plasma. However, the proportion of ctDNA within the pool of circulating cell-free DNA remains relatively small, posing a challenge to its current application as a biomarker. Researchers are actively investigating somatic mutation detection in ctDNA for cancer diagnosis, optimal treatment guidance, and surveillance of progression or recurrence. ctDNA enters the bloodstream as a result of apoptosis or necrosis, with DNA fragmentation being a characteristic feature of apoptosis mediated by the activation of caspase proteins [131]. A study conducted by Mouliere et al. revealed that ctDNA fragments were predominantly shorter than 167 bp across 18 different cancer types, with a notable enrichment observed in fragments ranging from 250 bp to 320 bp in size [132]. Due to the relatively low proportion of tumor DNA present within normal circulating free DNA background levels, the analysis of ctDNA necessitates the utilization of highly sensitive methods for detection and enrichment. Liquid biopsy-based examination of ctDNA presents a potential alternative for managing metastatic cancers when obtaining tumor tissue is not feasible [133].

2.8 NGS in Neurology

The significance of genetic factors in neurological diseases has been widely acknowledged for a considerable period, and initial molecular diagnostic tools have achieved considerable success in identifying genes linked to these disorders [134]. Out of the approximately 20,000 genes present in the human genome, more than 80% are actively expressed in the brain [135]. According to the Online Mendelian Inheritance in Man catalog, out of the more than 5000 genetic disorders that have identified molecular causes, approximately 40% of them are associated with the brain or the nervous system. In clinical neurology, obtaining a molecular diagnosis plays a crucial role in guiding clinical management. For instance, recognizing ataxia resulting from an ATM gene mutation or intellectual disability caused by a PTEN gene mutation might indicate the need for cancer screening, as these mutations are associated with specific forms of cancer. Likewise, in cases where a genetic diagnosis of adrenomyeloneuropathy or cerebrotendinous xanthomatosis is being established, it may be necessary to conduct testing to identify specific endocrine dysfunctions. This information is crucial for determining appropriate treatment approaches. NGS has ample evidence supporting its use in reducing the time taken for a patient to receive a diagnosis, often described as the "diagnostic odyssey."

Precision health involves utilizing patient-specific information to customize individualized care. It tackles a notable challenge in previous approaches to diagnosing neurogenetic disorders: the unaccounted genetic and phenotypic heterogeneity [136]. This heterogeneity often hinders clinicians' ability to predict the effectiveness of a particular therapy for a specific individual, even if the treatment has shown benefits in group studies. Neglecting to address this heterogeneity is thought to contribute to the lack of success in drug trials, the occurrence of avoidable adverse effects, inefficient use of available therapies, substantial financial waste, and minimal or no benefit to patients and their families [137–139]. The field of neurology, specifically in the realm of neurometabolic disorders, is undergoing a notable shift towards patient-tailored therapies, which has the potential to revolutionize the field of medicine [136]. Prominent examples of patient-tailored therapies in neurology include the implementation of a ketogenic diet for glucose transporter type-1 deficiency [140] and the utilization of bone marrow transplantation or gene therapy for X-linked adrenoleukodystrophy [141, 142], riboflavin treatment for Brown-Vialetto-Van Laere syndrome [143, 144], and levodopa treatment for dopa-responsive dystonia [145]. According to a study conducted at the University of British Columbia [146], genetic diagnosis resulted in treatment modifications for 44% of pediatric patients diagnosed with neurometabolic diseases. Precision approaches have also significantly advanced the diagnosis and treatment of epilepsy syndromes. Challenges in managing refractory epilepsy, such as the trial-and-error selection of drugs, can lead to the use of multiple medications, seizure-related injuries, side effects, and increased costs [147]. However, in genetic epilepsy syndromes caused by single-gene Mendelian mutations, the effectiveness of specific anti-epileptic drugs can be directly linked to the underlying mutation. For instance, in the case of Dravet syndrome, the use of sodium channel blockers for treatment is contraindicated [148, 149]. Clinical pharmacogenetics, which utilizes genomic data to guide patient treatment, is an area of great interest. For example, certain genetic variations in CYP2C9 can affect the metabolism of phenytoin and increase the risk of adverse events [150], while the HLA-B*15:02 variant is correlated with potential complications of phenytoin and carbamazepine in specific populations [150, 151]. Pharmacogenetics has applications beyond epilepsy and is an area that continues to evolve in the clinical field. Currently, there are various pharmacogenetic profiling tests and services available to assess a patient's likelihood of response to specific drugs or the risk of developing side effects. These tests cover a wide range of medication categories, including psychiatric, cardiac, and pain medications. As research progresses in these areas, the use of patient genetic data to predict drug efficacy and minimize side effects is expected to expand over time [152].

2.9 NGS in Cardiology

In recent years, the understanding that genetics plays a substantial role in developing various cardiovascular diseases (CVDs) has significantly expanded [153, 154]. These conditions can be classified into monogenic and polygenic disorders based on their genetic causes. Monogenic disorders are caused by the inheritance of one or two specific genetic mutations and often run in families. On the other hand, polygenic disorders involve multiple genetic mutations and have a lower tendency to cluster within families [155]. To address the genetic complexity of CVDs, several NGS strategies have been adopted. One commonly employed approach involves the use of targeted gene panels that focus on specific disorders. This targeted approach simplifies the analysis process and helps reduce costs for laboratories. However, advancements in NGS technology and increased sequencing capacity have enabled the simultaneous and accurate analysis of a larger number of genes. Various NGS methodologies currently concentrate on sequencing the coding genes and nearby intronic regions of 5000–7000 clinically significant genes, known as Clinically Exome Sequencing. Alternatively, some approaches encompass sequencing approximately 20,000 protein-producing genes, referred to as WES, as discussed previously. These NGS techniques enable comprehensive analysis of the genetic information associated with cardiovascular diseases [156].

The selection of genes for inclusion in NGS panels for cardiovascular diseases is guided by expert panels and cardiology societies. For instance, Ashley et al. conducted whole-genome sequencing (WGS) of a patient with a family history of vascular disease and sudden early death. Despite the absence of clinically significant medical records, the analysis revealed rare variants in three genes (TMEM43, DSP, and MYBPC3) known to be associated with sudden cardiac death. Additionally, the patient carried a heterozygous null mutation in the CYP2C19 gene, indicating a potential resistance to clopidogrel. The authors suggested that WGS could provide valuable and clinically relevant information for individual patients, particularly in terms of pharmacogenetic variants that could inform personalized medical care. Overall, personal genomic analysis represents a field of genomics that offers practical medical therapies based on genome analysis. The integration of predictive and preventive care will advance healthcare practices [157].

Existing treatments for cardiovascular diseases (CVDs) primarily involve conventional pharmacotherapy, interventional procedures, and surgeries [158–160]. While these approaches alleviate symptoms and improve prognosis, they have limitations. Traditional treatments can have adverse effects on organs such as the liver and kidneys, and cardiac surgeries are constrained by their complexity and the risk of postoperative complications [161, 162]. In light of the current scenario, there is a growing demand for innovative therapeutic approaches that offer enhanced effectiveness and reduced side effects for CVDs. In the era of targeted therapy, the effective management of CVDs necessitates the utilization of biomarkers that can provide valuable insights into diagnosis, treatment management, and prognosis. Specifically, for treatment selection, the identification and implementation of appropriate predictive biomarkers play a critical role in guiding personalized treatment decisions [163]. In cardiology, it is important to note that the term "predictive biomarker" may not have a clear and consistent definition. Sometimes, it is erroneously used to refer to biomarkers that solely predict the onset or progression of a disease without considering their role in treatment selection or response. The accurate characterization and understanding of biomarkers in cardiology are crucial for their appropriate application in guiding patient management and optimizing treatment outcomes [164]. As per its definition, a predictive biomarker serves the purpose of forecasting how a disease will progress in response to a particular treatment. It aids in predicting the patient's response to a specific therapeutic approach. By utilizing predictive biomarkers in cardiology, healthcare professionals can make well-informed decisions about treatment selection and customization, optimizing the chances of successful outcomes for individual patients. This concept is successfully applied in oncology and hematology, where specialized treatment regimens target hereditary or somatic genetic abnormalities and predictive biomarkers are used to identify patients who would benefit from such targeted treatments [165–168]. With the increasing application of genomic analysis in patient care, the identification of gene alterations as potential biomarkers for targeted treatments is increasingly evident across various medical fields, including cardiology. Patients with different genetic causes of heart failure, such as cardiac amyloidosis, hypertrophic cardiomyopathy (HCM), and Fabry disease, already have access to disease-specific medications [169, 170]. Moreover, monoclonal antibodies like evolocumab and alirocumab received FDA approval in 2015 as adjunctive therapies for homozygous familial hypercholesterolemia (HoFH) and heterozygous familial hypercholesterolemia (HeFH) [171]. Clinical studies that are currently underway have the objective of broadening the application of gene-informed therapy selection and are expected to lead to the development of targeted therapies accompanied by predictive biomarkers in the near future. The ongoing research in the field of CVDs is primarily directed towards exploring the potential of protein drugs, gene editing technologies, nucleic acid medicines, and cell therapy as targeted therapeutic approaches. These advancements hold promise for optimizing treatment outcomes and improving the management of CVDs through more personalized and effective interventions.

2.10 NGS in Pharmacogenetics and Drug Therapy

Undiscovered variations in genes related to drug metabolism and response play a significant role in individual variations in drug therapy. NGS has been utilized to extensively profile pharmacogenes involved in drug pharmacokinetics and pharmacodynamics [172–174]. Notably, a substantial majority of coding region variants in phase I and II metabolic enzymes, drug transporters, and nuclear receptors are rare (approximately 93% with a minor allele frequency [MAF] < 1%) or very rare (approximately 83% with a MAF < 0.01%), and predominantly nonsynonymous (56% to 65%). These rare variants are estimated to contribute to approximately 30–40% of functional variability in these genes [175, 176]. Similarly, a study examining 202 drug target genes in 14,002 individuals revealed that over 95% of the identified variants were rare, with MAF below 0.5%, and approximately 90% of them had not been previously reported [177].

Despite the discovery of numerous pharmacogenetic biomarkers, only a limited number have been implemented in clinical practice, primarily due to the stringent requirements for genetic testing. Before a genetic test can be used clinically, it needs to undergo evaluation for its analytical validity, clinical validity, and clinical utility, as defined by the ACCE framework [32, 178]. Analytical validity refers to the test's ability to accurately and reliably measure the target genotype, which can be challenging for genes with complex genetic structures. Clinical validity assesses the test's ability to accurately detect or predict a patient's phenotype based on their genotype. While current genotyping and sequencing technologies are generally robust for pharmacogenetic testing, defining the drug-effect phenotype can be complex, as many drug effects are influenced by multiple variants within the same gene, including rare variants, as well as variants in multiple genes within an individual [179]. Additionally, the variable prevalence of SNVs among different populations can complicate the translation of pharmacogenetic traits [180]. For example, genetic polymorphisms in CYP2C9 and VKORC1 that determine race-specific anticoagulant response to warfarin have been reported [181]. Furthermore, unlike most Mendelian traits, drug-effect phenotypes are influenced by a combination of genetic, clinical (comorbidity), and environmental factors (drug interactions, diet), which can result in a lower positive predictive value for pharmacogenetic tests [182]. Lastly, clinical utility assesses whether the use of a genetic test improves health outcomes by enabling individualized drug dosing or treatment and identifying patients at risk of drug toxicity [183].

2.11 Limitations of NGS

Performing routine analysis of NGS data necessitates the collaboration of multidisciplinary teams. These teams typically consist of clinical and biomedical bioinformaticians, computational biologists, molecular pathologists, programmers, statisticians, biologists, and clinicians. However, many institutions may not have the financial means to hire an experienced bioinformatician. Therefore, it is crucial to develop an in-house team of individuals with diverse backgrounds and varying levels of expertise, including graduate students and postdoctoral researchers. It is particularly important to have personnel with specialized skills in computing systems, programming, biology, and statistics. The collaborative atmosphere nurtures the exchange of ideas and expedites the analysis of data. With the continuous application of NGS technology across diverse biological and clinical domains, the establishment of standardized quality metrics for the produced NGS data becomes imperative. These metrics play a vital role in ensuring the data's precision, dependability, and reproducibility, enabling researchers and clinicians to make well-informed decisions based on reliable genomic information. By implementing uniform quality metrics, the field can uphold stringent standards and promote consistency in the analysis of NGS data across various laboratories and research settings [184]. These metrics encompass validation and comparison across different platforms, ensuring data reliability, robustness, and reproducibility and assessing the quality of assemblers used in data analysis [185].

2.12 Conclusion

GWA studies provide a valuable tool for identifying genetic variants associated with diseases. However, their implementation in clinical practice necessitates robust evidence, effective interventions, cost-effectiveness, and consideration of potential adverse consequences. Although GWA findings are not currently widely used clinically, further research and experimental applications are anticipated to yield significant advancements in disease prevention and treatment. In order to apply the GWA study's findings to clinical practice, it will be necessary to develop predictive models that accurately assess the genetic architecture of complex traits. These models should provide a higher level of precision in evaluating the impacts of factors like epistasis, genocopies, phenocopies, and penetrance. These predictive models aim to better understand how these various genetic factors interact and contribute to the development of complex traits. By incorporating these factors into the models, we can improve the accuracy and precision of predicting the likelihood or risk of developing specific traits or diseases based on an individual's genetic makeup.

NGS data analysis has emerged as a powerful tool in clinical practice, with significant implications for personalized medicine in various fields. The comprehensive analysis of DNA sequences using NGS has revolutionized our understanding of diseases, enabling the identification of genetic variants associated with cancer, neurological disorders, cardiovascular diseases, and pharmacogenetics. These insights have facilitated the development of targeted therapies and personalized treatment strategies. However, several challenges must be overcome to fully realize the potential of NGS in clinical settings. Standardization of quality metrics, data interpretation, and cost-effectiveness are critical areas that require attention. Collaborative efforts involving multidisciplinary teams are necessary to optimize the use of NGS technology and ensure its seamless integration into routine clinical practice. Despite the challenges, the future of medicine holds immense promise, with NGS as a cornerstone of precision diagnostics and individualized therapeutic interventions. Continued research, technological advancements, and improved accessibility will further propel the field of NGS toward realizing its full potential in transforming patient care and improving health outcomes.

Conflict of Interest Authors have no conflict of interest to declare.

References

1. Kingsmore SF, Lindquist IE, Mudge J, Gessler DD, Beavis WD. Genome-wide association studies: progress and potential for drug discovery and development. Nat Rev Drug Discov. 2008;7(3):221–30.
2. Shaw G. Polymorphism and single nucleotide polymorphisms (SNPs). BJU Int. 2013;112(5):664–5.
3. McCarroll SA. Extending genome-wide association studies to copy-number variation. Hum Mol Genet. 2008;17(R2):R135–42.
4. Gibbs RA, Belmont JW, Hardenbol P, Willis TD, Yu F, Yang H, et al. The international HapMap project. 2003.
5. Visscher PM. Five years of GWAS discovery. Am J Hum Genet. 2012;90:7–24.
6. Visscher PM, Wray NR, Zhang Q, Sklar P, McCarthy MI, Brown MA, et al. 10 years of GWAS discovery: biology, function, and translation. Am J Hum Genet. 2017;101(1):5–22.
7. Giacomini KM, Yee SW, Mushiroda T, Weinshilboum RM, Ratain MJ, Kubo M. Genome-wide association studies of drug response and toxicity: an opportunity for genome medicine. Nat Rev Drug Discov. 2017;16(1):70.
8. Roukos DH. Genome-wide association studies: how predictable is a person's cancer risk? Expert Rev Anticancer Ther. 2009;9(4):389–92.
9. Rockman MV. Reverse engineering the genotype–phenotype map with natural genetic variation. Nature. 2008;456(7223):738–44.
10. Roukos DH, Murray S, Briasoulis E. Molecular genetic tools shape a roadmap towards a more accurate prognostic prediction and personalized management of cancer. Cancer Biol Ther. 2007;6(3):308–12.
11. Goldhirsch A, Wood W, Gelber R, Coates A, Thürlimann B, Senn H-J, et al. Progress and promise: highlights of the international expert consensus on the primary therapy of early breast cancer 2007. Ann Oncol. 2007;18(7):1133–44.

12. Morris ZS, Wooding S, Grant J. The answer is 17 years, what is the question: understanding time lags in translational research. J R Soc Med. 2011;104(12):510–20.
13. Alaaeddine R, Fayad M, Nehme E, Bahmad HF, Kobeissy F. The emerging role of proteomics in precision medicine: applications in neurodegenerative diseases and neurotrauma. In: Personalised medicine: lessons from neurodegeneration to cancer. Cham: Springer; 2017. p. 59–70.
14. Tam V, Patel N, Turcotte M, Bossé Y, Paré G, Meyre D. Benefits and limitations of genome-wide association studies. Nat Rev Genet. 2019;20(8):467–84.
15. Gonzalez-Garay ML. The road from next-generation sequencing to personalized medicine. Pers Med. 2014;11(5):523–44.
16. Vestergaard LK, Oliveira DN, Høgdall CK, Høgdall EV. Next generation sequencing technology in the clinic and its challenges. Cancer. 2021;13(8):1751.
17. Shaw KRM, Maitra A. The status and impact of clinical tumor genome sequencing. Annu Rev Genomics Hum Genet. 2019;20:413–32.
18. Consortium APG, Consortium APG, André F, Arnedos M, Baras AS, Baselga J, et al. AACR Project GENIE: powering precision medicine through an international consortium. Cancer Discov. 2017;7(8):818–31.
19. Lohmann K, Klein C. Next generation sequencing and the future of genetic diagnosis. Neurotherapeutics. 2014;11:699–707.
20. Rabbani B, Nakaoka H, Akhondzadeh S, Tekin M, Mahdieh N. Next generation sequencing: implications in personalized medicine and pharmacogenomics. Mol BioSyst. 2016;12(6):1818–30.
21. Roh SW, Abell GC, Kim K-H, Nam Y-D, Bae J-W. Comparing microarrays and next-generation sequencing technologies for microbial ecology research. Trends Biotechnol. 2010;28(6):291–9.
22. Voelkerding KV, Dames S, Durtschi JD. Next generation sequencing for clinical diagnostics-principles and application to targeted resequencing for hypertrophic cardiomyopathy: a paper from the 2009 William Beaumont Hospital Symposium on Molecular Pathology. J Mol Diagn. 2010;12(5):539–51.
23. Baum JE, Zhang P, Hoda RS, Geraghty B, Rennert H, Narula N, et al. Accuracy of next-generation sequencing for the identification of clinically relevant variants in cytology smears in lung adenocarcinoma. Cancer Cytopathol. 2017;125(6):398–406.
24. Beaubier N, Tell R, Huether R, Bontrager M, Bush S, Parsons J, et al. Clinical validation of the Tempus xO assay. Oncotarget. 2018;9(40):25826.
25. Lee C, Morton CC. Structural genomic variation and personalized medicine. N Engl J Med. 2008;358(7):740–1.
26. Fareed M, Afzal M. Single nucleotide polymorphism in genome-wide association of human population: a tool for broad spectrum service. Egypt J Med Hum Genet. 2013;14(2):123–34.
27. Peng Z. A brief overview of GWAS: discover genetic variations of diseases and phenotypes. In: E3S web of conferences. Les Ulis: EDP Sciences; 2020.
28. Manolio TA, Collins FS, Cox NJ, Goldstein DB, Hindorff LA, Hunter DJ, et al. Finding the missing heritability of complex diseases. Nature. 2009;461(7265):747–53.
29. Claussnitzer M, Cho JH, Collins R, Cox NJ, Dermitzakis ET, Hurles ME, et al. A brief history of human disease genetics. Nature. 2020;577(7789):179–89.
30. Hunter DJ, Khoury MJ, Drazen JM. Letting the genome out of the bottle-will we get our wish? N Engl J Med. 2008;358(2):105.
31. Pennisi E. Breakthrough of the year. Human genetic variation. Science. 2007;318(5858):1842–3.
32. Burke W. Genetic tests: clinical validity and clinical utility. Curr Protoc Hum Genet. 2014;81(1):9.
33. Roukos DH. Personal genomics and genome-wide association studies: novel discoveries but limitations for practical personalized medicine. Ann Surg Oncol. 2009;16:772–3.
34. Manolio TA, Brooks LD, Collins FS. A HapMap harvest of insights into the genetics of common disease. J Clin Invest. 2008;118(5):1590–605.
35. Manolio TA, Collins FS. The HapMap and genome-wide association studies in diagnosis and therapy. Annu Rev Med. 2009;60:443–56.
36. Florez JC, Hirschhorn J, Altshuler D. The inherited basis of diabetes mellitus: implications for the genetic analysis of complex traits. Annu Rev Genomics Hum Genet. 2003;4(1):257–91.
37. Lyssenko V, Jonsson A, Almgren P, Pulizzi N, Isomaa B, Tuomi T, et al. Clinical risk factors, DNA variants, and the development of type 2 diabetes. N Engl J Med. 2008;359(21):2220–32.
38. Hirschhorn JN, Gajdos ZK. Genome-wide association studies: results from the first few years and potential implications for clinical medicine. Annu Rev Med. 2011;62:11–24.
39. Stadler ZK, Thom P, Robson ME, Weitzel JN, Kauff ND, Hurley KE, et al. Genome-wide association studies of cancer. J Clin Oncol. 2010;28(27):4255.
40. Sud A, Kinnersley B, Houlston RS. Genome-wide association studies of cancer: current insights and future perspectives. Nat Rev Cancer. 2017;17(11):692–704.
41. Diamandis M, White NM, Yousef GM. Personalized medicine: marking a new epoch in cancer patient management. Mol Cancer Res. 2010;8(9):1175–87.
42. Gail MH. Discriminatory accuracy from single-nucleotide polymorphisms in models to predict breast cancer risk. J Natl Cancer Inst. 2008;100(14):1037–41.
43. Simon DK, Tanner CM, Brundin P. Parkinson disease epidemiology, pathology, genetics, and pathophysiology. Clin Geriatr Med. 2020;36(1):1–12.
44. Bekris LM, Yu C-E, Bird TD, Tsuang DW. Genetics of Alzheimer disease. J Geriatr Psychiatry Neurol. 2010;23(4):213–27.

45. Klein C, Westenberger A. Genetics of Parkinson's disease. Cold Spring Harb Perspect Med. 2012;2(1):a008888.
46. Tan L, Jiang T, Tan L, Yu J-T. Toward precision medicine in neurological diseases. Ann Transl Med. 2016;4(6):104.
47. Strafella C, Caputo V, Galota MR, Zampatti S, Marella G, Mauriello S, et al. Application of precision medicine in neurodegenerative diseases. Front Neurol. 2018;9:701.
48. Strianese O, Rizzo F, Ciccarelli M, Galasso G, D'Agostino Y, Salvati A, et al. Precision and personalized medicine: how genomic approach improves the management of cardiovascular and neurodegenerative disease. Gene. 2020;11(7):747.
49. Giardina E, Caltagirone C, editors. The IRCCS Network of Neuroscience and Neurorehabilitation: the Italian platform for care and research about neurodegenerative disorders. Eur J Neurol. 2018;6(1):58–9.
50. Tuzun E, Sharp AJ, Bailey JA, Kaul R, Morrison VA, Pertz LM, et al. Fine-scale structural variation of the human genome. Nat Genet. 2005;37(7):727–32.
51. Girirajan S, Rosenfeld JA, Coe BP, Parikh S, Friedman N, Goldstein A, et al. Phenotypic heterogeneity of genomic disorders and rare copy-number variants. N Engl J Med. 2012;367(14):1321–31.
52. He Y, Hoskins JM, McLeod HL. Copy number variants in pharmacogenetic genes. Trends Mol Med. 2011;17(5):244–51.
53. Wium-Andersen IK, Vinberg M, Kessing LV, McIntyre RS. Personalized medicine in psychiatry. Nord J Psychiatry. 2017;71(1):12–9.
54. Risch SC, Nemeroff CB. Neurochemical alterations of serotonergic neuronal systems in depression. J Clin Psychiatry. 1992;53:3–7.
55. Krishnan V, Nestler EJ. The molecular neurobiology of depression. Nature. 2008;455(7215):894–902.
56. Ogilvie AD, Battersby S, Fink G, Harmar A, Goodwin G, Bubb V, et al. Polymorphism in serotonin transporter gene associated with susceptibility to major depression. Lancet. 1996;347(9003):731–3.
57. Ripke S, Neale B, Corvin A, Walters J, Farh K, Holmans P, et al. Schizophrenia Working Group of the Psychiatric Genomics Consortium Biological insights from 108 schizophrenia-associated genetic loci. Nature. 2014;511(7510):421–7.
58. Duncan LE, Holmans PA, Lee PH, O'Dushlaine CT, Kirby AW, Smoller JW, et al. Pathway analyses implicate glial cells in schizophrenia. PLoS One. 2014;9(2):e89441.
59. Pedroso I, Lourdusamy A, Rietschel M, Nöthen MM, Cichon S, McGuffin P, et al. Common genetic variants and gene-expression changes associated with bipolar disorder are over-represented in brain signaling pathway genes. Biol Psychiatry. 2012;72(4):311–7.
60. Purcell SM, Moran JL, Fromer M, Ruderfer D, Solovieff N, Roussos P, et al. A polygenic burden of rare disruptive mutations in schizophrenia. Nature. 2014;506(7487):185–90.
61. Nurnberger JI, Koller DL, Jung J, Edenberg HJ, Foroud T, Guella I, et al. Identification of pathways for bipolar disorder: a meta-analysis. JAMA Psychiatry. 2014;71(6):657–64.
62. Uddin M, Tammimies K, Pellecchia G, Alipanahi B, Hu P, Wang Z, et al. Brain-expressed exons under purifying selection are enriched for de novo mutations in autism spectrum disorder. Nat Genet. 2014;46(7):742–7.
63. Consortium. Identification of risk loci with shared effects on five major psychiatric disorders: a genome-wide analysis. Lancet. 2013;381(9875):1371–9.
64. Okada Y, Wu D, Trynka G, Raj T, Terao C, Ikari K, et al. Genetics of rheumatoid arthritis contributes to biology and drug discovery. Nature. 2014;506(7488):376–81.
65. Lenfant C. Prospects of personalized medicine in cardiovascular diseases. Metabolism. 2013;62:S6–S10.
66. Manolio TA. Genomewide association studies and assessment of the risk of disease. N Engl J Med. 2010;363(2):166–76.
67. Kathiresan S, Melander O, Anevski D, Guiducci C, Burtt NP, Roos C, et al. Polymorphisms associated with cholesterol and risk of cardiovascular events. N Engl J Med. 2008;358(12):1240–9.
68. Roden DM, Johnson JA, Kimmel SE, Krauss RM, Medina MW, Shuldiner A, et al. Cardiovascular pharmacogenomics. Circ Res. 2011;109(7):807–20.
69. Degoma EM, Rivera G, Lilly SM, Usman MHU, Mohler ER. Personalized vascular medicine: individualizing drug therapy. Vasc Med. 2011;16(5):391–404.
70. Voora D, Shah SH, Spasojevic I, Ali S, Reed CR, Salisbury BA, et al. The SLCO1B1* 5 genetic variant is associated with statin-induced side effects. J Am Coll Cardiol. 2009;54(17):1609–16.
71. Group SC. SLCO1B1 variants and statin-induced myopathy—a genomewide study. N Engl J Med. 2008;359(8):789–99.
72. Mega JL, Close SL, Wiviott SD, Shen L, Hockett RD, Brandt JT, et al. Cytochrome p-450 polymorphisms and response to clopidogrel. N Engl J Med. 2009;360(4):354–62.
73. Peters BJ, Harmsze AM, ten Berg JM, Maitland-Van Der Zee A-H, Tjoeng MM, De Boer A, et al. CYP2C19 and ABCB1 genes and individualized treatment with clopidogrel. Pharmacogenomics. 2011;12(2):141–4.
74. Manolio TA. Bringing genome-wide association findings into clinical use. Nat Rev Genet. 2013;14(8):549–58.
75. Baldwin RM, Owzar K, Zembutsu H, Chhibber A, Kubo M, Jiang C, et al. A genome-wide association study identifies novel loci for paclitaxel-induced sensory peripheral neuropathy in CALGB 40101. Clin Cancer Res. 2012;18(18):5099–109.
76. Park BL, Kim T-H, Kim J-H, Bae JS, Pasaje CFA, Cheong HS, et al. Genome-wide association study of

aspirin-exacerbated respiratory disease in a Korean population. Hum Genet. 2013;132:313–21.
77. Manolio TA, Chisholm RL, Ozenberger B, Roden DM, Williams MS, Wilson R, et al. Implementing genomic medicine in the clinic: the future is here. Genet Med. 2013;15(4):258–67.
78. Crews KR, Hicks JK, Pui CH, Relling MV, Evans WE. Pharmacogenomics and individualized medicine: translating science into practice. Clin Pharmacol Therap. 2012;92(4):467–75.
79. Nelson MR, Tipney H, Painter JL, Shen J, Nicoletti P, Shen Y, et al. The support of human genetic evidence for approved drug indications. Nat Genet. 2015;47(8):856–60.
80. Ge D, Fellay J, Thompson AJ, Simon JS, Shianna KV, Urban TJ, et al. Genetic variation in IL28B predicts hepatitis C treatment-induced viral clearance. Nature. 2009;461(7262):399–401.
81. Tanaka Y, Nishida N, Sugiyama M, Kurosaki M, Matsuura K, Sakamoto N, et al. Genome-wide association of IL28B with response to pegylated interferon-α and ribavirin therapy for chronic hepatitis C. Nat Genet. 2009;41(10):1105–9.
82. Hunkapiller T, Kaiser R, Koop B, Hood L. Large-scale and automated DNA sequence determination. Science. 1991;254(5028):59–67.
83. Smith LM, Sanders JZ, Kaiser RJ, Hughes P, Dodd C, Connell CR, et al. Fluorescence detection in automated DNA sequence analysis. Nature. 1986;321(6071):674–9.
84. Pervez MT, Abbas SH, Moustafa MF, Aslam N, Shah SSM. A comprehensive review of performance of next-generation sequencing platforms. Biomed Res Int. 2022;2022:3457806.
85. Quail MA, Smith M, Coupland P, Otto TD, Harris SR, Connor TR, et al. A tale of three next generation sequencing platforms: comparison of Ion Torrent, Pacific Biosciences and Illumina MiSeq sequencers. BMC Genomics. 2012;13(1):1–13.
86. Meder B, Laufer C, Hassel D, Just S, Marquart S, Vogel B, et al. A single serine in the carboxyl terminus of cardiac essential myosin light chain-1 controls cardiomyocyte contractility in vivo. Circ Res. 2009;104(5):650–9.
87. Wang G-S, Cooper TA. Splicing in disease: disruption of the splicing code and the decoding machinery. Nat Rev Genet. 2007;8(10):749–61.
88. Chacko E, Ranganathan S. Comprehensive splicing graph analysis of alternative splicing patterns in chicken, compared to human and mouse. BMC Genomics. 2009;10(1):1–11.
89. Matlin AJ, Clark F, Smith CW. Understanding alternative splicing: towards a cellular code. Nat Rev Mol Cell Biol. 2005;6(5):386–98.
90. Lahens NF, Ricciotti E, Smirnova O, Toorens E, Kim EJ, Baruzzo G, et al. A comparison of illumina and ion torrent sequencing platforms in the context of differential gene expression. BMC Genomics. 2017;18:1–13.
91. Cao Y, Fanning S, Proos S, Jordan K, Srikumar S. A review on the applications of next generation sequencing technologies as applied to food-related microbiome studies. Front Microbiol. 2017;8:1829.
92. Meder B, Haas J, Keller A, Heid C, Just S, Borries A, et al. Targeted next-generation sequencing for the molecular genetic diagnostics of cardiomyopathies. Circulation. 2011;4(2):110–22.
93. Lara-Pezzi E, Dopazo A, Manzanares M. Understanding cardiovascular disease: a journey through the genome (and what we found there). Dis Model Mech. 2012;5(4):434–43.
94. Herman DS, Lam L, Taylor MR, Wang L, Teekakirikul P, Christodoulou D, et al. Truncations of titin causing dilated cardiomyopathy. N Engl J Med. 2012;366(7):619–28.
95. Norton N, Li D, Rieder MJ, Siegfried JD, Rampersaud E, Züchner S, et al. Genome-wide studies of copy number variation and exome sequencing identify rare variants in BAG3 as a cause of dilated cardiomyopathy. Am J Hum Genet. 2011;88(3):273–82.
96. Clark MJ, Chen R, Lam HY, Karczewski KJ, Chen R, Euskirchen G, et al. Performance comparison of exome DNA sequencing technologies. Nat Biotechnol. 2011;29(10):908–14.
97. Frese KS, Katus HA, Meder B. Next-generation sequencing: from understanding biology to personalized medicine. Biology. 2013;2(1):378–98.
98. Bentley D. Coupling RNA polymerase II transcription with pre-mRNA processing. Curr Opin Cell Biol. 1999;11(3):347–51.
99. Gupta PK. Single-molecule DNA sequencing technologies for future genomics research. Trends Biotechnol. 2008;26(11):602–11.
100. Klumpp J, Fouts DE, Sozhamannan S. Next generation sequencing technologies and the changing landscape of phage genomics. Bacteriophage. 2012;2(3):190–9.
101. Maniatis T, Reed R. An extensive network of coupling among gene expression machines. Nature. 2002;416(6880):499–506.
102. Cordero F, Beccuti M, Donatelli S, Calogero A, R. Large disclosing the nature of computational tools for the analysis of next generation sequencing data. Curr Top Med Chem. 2012;12(12):1320–30.
103. Philips A, Cooper T. RNA processing and human disease. Cell Mol Life Sci. 2000;57:235–49.
104. Pon JR, Marra MA. Driver and passenger mutations in cancer. Annu Rev Pathol. 2015;10:25–50.
105. McFarland CD, Mirny LA, Korolev KS. Tug-of-war between driver and passenger mutations in cancer and other adaptive processes. Proc Natl Acad Sci. 2014;111(42):15138–43.
106. Bailey MH, Tokheim C, Porta-Pardo E, Sengupta S, Bertrand D, Weerasinghe A, et al. Comprehensive characterization of cancer driver genes and mutations. Cell. 2018;173(2):371–85.
107. Pereira R, Oliveira J, Sousa M. Bioinformatics and computational tools for next-generation sequenc-

ing analysis in clinical genetics. J Clin Med. 2020;9(1):132.

108. ten Bosch JR, Grody WW. Keeping up with the next generation: massively parallel sequencing in clinical diagnostics. J Mol Diagn. 2008;10(6):484–92.
109. Hanahan D, Weinberg RA. The hallmarks of cancer. Cell. 2000;100(1):57–70.
110. Hanahan D, Weinberg RA. Hallmarks of cancer: the next generation. Cell. 2011;144(5):646–74.
111. Zehir A, Benayed R, Shah RH, Syed A, Middha S, Kim HR, et al. Mutational landscape of metastatic cancer revealed from prospective clinical sequencing of 10,000 patients. Nat Med. 2017;23(6):703–13.
112. Frampton GM, Fichtenholtz A, Otto GA, Wang K, Downing SR, He J, et al. Development and validation of a clinical cancer genomic profiling test based on massively parallel DNA sequencing. Nat Biotechnol. 2013;31(11):1023–31.
113. Cheng DT, Mitchell TN, Zehir A, Shah RH, Benayed R, Syed A, et al. Memorial sloan kettering-integrated mutation profiling of actionable cancer targets (MSK-IMPACT): a hybridization capture-based next-generation sequencing clinical assay for solid tumor molecular oncology. J Mol Diagn. 2015;17(3):251–64.
114. Cunanan KM, Iasonos A, Shen R, Begg CB, Gönen M. An efficient basket trial design. Stat Med. 2017;36(10):1568–79.
115. Strzebonska K, Waligora M. Umbrella and basket trials in oncology: ethical challenges. BMC Med Ethics. 2019;20:1–10.
116. Luskin MR, Murakami MA, Manalis SR, Weinstock DM. Targeting minimal residual disease: a path to cure? Nat Rev Cancer. 2018;18(4):255–63.
117. Wu D, Emerson RO, Sherwood A, Loh ML, Angiolillo A, Howie B, et al. Detection of minimal residual disease in B lymphoblastic leukemia by high-throughput sequencing of IGH. Clin Cancer Res. 2014;20(17):4540–8.
118. Wood B, Wu D, Crossley B, Dai Y, Williamson D, Gawad C, et al. Measurable residual disease detection by high-throughput sequencing improves risk stratification for pediatric B-ALL. Blood. 2018;131(12):1350–9.
119. Berger MF, Mardis ER. The emerging clinical relevance of genomics in cancer medicine. Nat Rev Clin Oncol. 2018;15(6):353–65.
120. Chin R-I, Chen K, Usmani A, Chua C, Harris PK, Binkley MS, et al. Detection of solid tumor molecular residual disease (MRD) using circulating tumor DNA (ctDNA). Mol Diagn Ther. 2019;23(3):311–31.
121. Rossi G, Ignatiadis M. Promises and pitfalls of using liquid biopsy for precision medicine. Cancer Res. 2019;79(11):2798–804.
122. Heitzer E, Haque IS, Roberts CE, Speicher MR. Current and future perspectives of liquid biopsies in genomics-driven oncology. Nat Rev Genet. 2019;20(2):71–88.
123. Merker JD, Oxnard GR, Compton C, Diehn M, Hurley P, Lazar AJ, et al. Circulating tumor DNA analysis in patients with cancer: American Society of Clinical Oncology and College of American Pathologists joint review. Arch Pathol Lab Med. 2018;142(10):1242–53.
124. Razavi P, Li BT, Brown DN, Jung B, Hubbell E, Shen R, et al. High-intensity sequencing reveals the sources of plasma circulating cell-free DNA variants. Nat Med. 2019;25(12):1928–37.
125. Guinney J, Dienstmann R, Wang X, De Reynies A, Schlicker A, Soneson C, et al. The consensus molecular subtypes of colorectal cancer. Nat Med. 2015;21(11):1350–6.
126. Raphael BJ, Hruban RH, Aguirre AJ, Moffitt RA, Yeh JJ, Stewart C, et al. Integrated genomic characterization of pancreatic ductal adenocarcinoma. Cancer Cell. 2017;32(2):185–203.
127. Cuppen E, Elemento O, Rosenquist R, Nikic S, Ijzerman M, Zaleski ID, et al. Implementation of whole-genome and transcriptome sequencing into clinical cancer care. JCO Precis Oncol. 2022;6:e2200245.
128. Rusch M, Nakitandwe J, Shurtleff S, Newman S, Zhang Z, Edmonson MN, et al. Clinical cancer genomic profiling by three-platform sequencing of whole genome, whole exome and transcriptome. Nat Commun. 2018;9(1):3962.
129. Weiss GJ, Byron SA, Aldrich J, Sangal A, Barilla H, Kiefer JA, et al. A prospective pilot study of genome-wide exome and transcriptome profiling in patients with small cell lung cancer progressing after first-line therapy. PLoS One. 2017;12(6):e0179170.
130. Dagogo-Jack I, Shaw AT. Tumour heterogeneity and resistance to cancer therapies. Nat Rev Clin Oncol. 2018;15(2):81–94.
131. Keller L, Belloum Y, Wikman H, Pantel K. Clinical relevance of blood-based ctDNA analysis: mutation detection and beyond. Br J Cancer. 2021;124(2):345–58.
132. Mouliere F, Chandrananda D, Piskorz AM, Moore EK, Morris J, Ahlborn LB, et al. Enhanced detection of circulating tumor DNA by fragment size analysis. Sci Transl Med. 2018;10(466):4921.
133. Ignatiadis M, Dawson S-J. Circulating tumor cells and circulating tumor DNA for precision medicine: dream or reality? Ann Oncol. 2014;25(12):2304–13.
134. Pihlstrøm L, Wiethoff S, Houlden H. Genetics of neurodegenerative diseases: an overview. Handb Clin Neurol. 2018;145:309–23.
135. Negi SK, Guda C. Global gene expression profiling of healthy human brain and its application in studying neurological disorders. Sci Rep. 2017;7(1):897.
136. Wolf SM, Thyagarajan B, Fogel BL. The need to develop a patient-centered precision medicine model for adults with chronic disability. Expert Rev Mol Diagn. 2017;17(5):415–8.
137. Berm EJ, Looff M, Wilffert B, Boersma C, Annemans L, Vegter S, et al. Economic evaluations of pharmacogenetic and pharmacogenomic screening tests: a systematic review. Second update of the literature. PLoS One. 2016;11(1):e0146262.

138. Verbelen M, Weale ME, Lewis CM. Cost-effectiveness of pharmacogenetic-guided treatment: are we there yet? Pharmacogenomics J. 2017;17(5):395–402.
139. Ginsburg GS, Phillips KA. Precision medicine: from science to value. Health Aff. 2018;37(5):694–701.
140. Wang D, Pascual JM, De Vivo D. Glucose transporter type 1 deficiency syndrome. In: Adam MP, Feldman J, Mirzaa GM, Pagon RA, Wallace SE, Bean LJH, Gripp KW, Amemiya A, editors. GeneReviews®. Seattle: University of Washington; 2018.
141. Kühl J-S, Suarez F, Gillett GT, Hemmati PG, Snowden JA, Stadler M, et al. Long-term outcomes of allogeneic haematopoietic stem cell transplantation for adult cerebral X-linked adrenoleukodystrophy. Brain. 2017;140(4):953–66.
142. Eichler F, Duncan C, Musolino PL, Orchard PJ, De Oliveira S, Thrasher AJ, et al. Hematopoietic stem-cell gene therapy for cerebral adrenoleukodystrophy. N Engl J Med. 2017;377(17):1630–8.
143. Fan J, Fogel BL. Successful treatment of a genetic childhood ataxia due to riboflavin transporter deficiency. Cerebellum Ataxias. 2018;5(1):1–3.
144. Manole A, Jaunmuktane Z, Hargreaves I, Ludtmann MH, Salpietro V, Bello OD, et al. Clinical, pathological and functional characterization of riboflavin-responsive neuropathy. Brain. 2017;140(11):2820–37.
145. Wijemanne S, Jankovic J. Dopa-responsive dystonia—clinical and genetic heterogeneity. Nat Rev Neurol. 2015;11(7):414–24.
146. Tarailo-Graovac M, Shyr C, Ross CJ, Horvath GA, Salvarinova R, Ye XC, et al. Exome sequencing and the management of neurometabolic disorders. N Engl J Med. 2016;374(23):2246–55.
147. Strzelczyk A, Griebel C, Lux W, Rosenow F, Reese J-P. The burden of severely drug-refractory epilepsy: a comparative longitudinal evaluation of mortality, morbidity, resource use, and cost using German health insurance data. Front Neurol. 2017;8:712.
148. Miller I. SCN1A-related seizure disorders. In: GeneReviews®. Seattle: University of Washington; 1993.
149. Zhang D, Liu X, Deng X. Genetic basis of pediatric epilepsy syndromes. Exp Ther Med. 2017;13(5):2129–33.
150. Silvado CE, Terra VC, Twardowschy CA. CYP2C9 polymorphisms in epilepsy: influence on phenytoin treatment. Pharmacog Personal Med. 2018;11:51–8.
151. Shi Y-W, Min F-L, Zhou D, Qin B, Wang J, Hu F-Y, et al. HLA-A* 24: 02 as a common risk factor for antiepileptic drug–induced cutaneous adverse reactions. Neurology. 2017;88(23):2183–91.
152. Mukerjee G, Huston A, Kabakchiev B, Piquette-Miller M, van Schaik R, Dorfman R. User considerations in assessing pharmacogenomic tests and their clinical support tools. NPJ Genom Med. 2018;3(1):26.
153. Kathiresan S, Srivastava D. Genetics of human cardiovascular disease. Cell. 2012;148(6):1242–57.
154. Khera AV, Emdin CA, Drake I, Natarajan P, Bick AG, Cook NR, et al. Genetic risk, adherence to a healthy lifestyle, and coronary disease. N Engl J Med. 2016;375(24):2349–58.
155. Wilde AA, Semsarian C, Márquez MF, Shamloo AS, Ackerman MJ, Ashley EA, et al. European heart rhythm association (EHRA)/heart rhythm society (HRS)/Asia pacific heart rhythm society (APHRS)/Latin American Heart Rhythm Society (LAHRS) expert consensus statement on the state of genetic testing for cardiac diseases. Europace. 2022;24(8):1307–67.
156. Scherr CL, Kalke K, Ramesh S, Fakhari H, Dellefave-Castillo LM, Smith ME, et al. Integrating clinical genetics in cardiology: current practices and recommendations for education. Genet Med. 2022;24(5):1054–61.
157. Ashley EA, Butte AJ, Wheeler MT, Chen R, Klein TE, Dewey FE, et al. Clinical assessment incorporating a personal genome. Lancet. 2010;375(9725):1525–35.
158. Xu M, Song J. Targeted therapy in cardiovascular disease: a precision therapy era. Front Pharmacol. 2021;12:623674.
159. Stehlik J, Kobashigawa J, Hunt SA, Reichenspurner H, Kirklin JK. Honoring 50 years of clinical heart transplantation in circulation: in-depth state-of-the-art review. Circulation. 2018;137(1):71–87.
160. Stein CA, Castanotto D. FDA-approved oligonucleotide therapies in 2017. Mol Ther. 2017;25(5):1069–75.
161. Roth S, Torregroza C, Huhn R, Hollmann MW, Preckel B. Perioperative cardioprotection: clinical implications. Anesth Analg. 2020;131(6):1751–64.
162. Lassiter G, Melancon C, Rooney T, Murat A-M, Kaye JS, Kaye AM, et al. Ozanimod to treat relapsing forms of multiple sclerosis: a comprehensive review of disease, drug efficacy and side effects. Neurol Int. 2020;12(3):89–108.
163. Mishina E. Monitoring biomarker. Silver Spring: Food and Drug Administration; 2016.
164. Pahwa P, Chu L, Karunanayake C, Aich P, Hecker M, Saxena A, et al. Predictive biomarkers of cardiovascular disease in adult Canadian population. J Diabetes Metab Disord. 2021;20:1199–209.
165. Jürgensmeier JM, Eder JP, Herbst RS. New strategies in personalized medicine for solid tumors: molecular markers and clinical trial designs. Clin Cancer Res. 2014;20(17):4425–35.
166. Malone ER, Oliva M, Sabatini PJ, Stockley TL, Siu LL. Molecular profiling for precision cancer therapies. Genome Med. 2020;12(1):1–19.
167. Özdoğan M, Papadopoulou E, Tsoulos N, Tsantikidi A, Mariatou V-M, Tsaousis G, et al. Comprehensive tumor molecular profile analysis in clinical practice. BMC Med Genet. 2021;14(1):1–21.
168. Chakravarty D, Johnson A, Sklar J, Lindeman NI, Moore K, Ganesan S, et al. Somatic genomic testing in patients with metastatic or advanced cancer: ASCO provisional clinical opinion. J Clin Oncol. 2022;40(11):1231–58.

169. Maack C, Tardiff JC. Targeted therapies for cardiac diseases. Nat Rev Cardiol. 2022;19(6):343–4.
170. Lehman SJ, Crocini C, Leinwand LA. Targeting the sarcomere in inherited cardiomyopathies. Nat Rev Cardiol. 2022;19(6):353–63.
171. Ilut S, Pirlog BO, Pirlog R, Nutu A, Vacaras V, Armean SM. Recent advances on the roles of PCSK-9 inhibitors in the management of acute ischemic stroke patients. Int J Mol Sci. 2022;23(18):10221.
172. Gordon AS, Fulton RS, Qin X, Mardis ER, Nickerson DA, Scherer S. PGRNseq: a targeted capture sequencing panel for pharmacogenetic research and implementation. Pharmacogenet Genomics. 2016;26(4):161–8.
173. Han S, Park J, Lee J, Lee S, Kim H, Han H, et al. Targeted next-generation sequencing for comprehensive genetic profiling of pharmacogenes. Clin Pharmacol Therap. 2017;101(3):396–405.
174. Twist GP, Gaedigk A, Miller NA, Farrow EG, Willig LK, Dinwiddie DL, et al. Constellation: a tool for rapid, automated phenotype assignment of a highly polymorphic pharmacogene, CYP2D6, from whole-genome sequences. NPJ Genom Med. 2016;1(1):1–10.
175. Fujikura K, Ingelman-Sundberg M, Lauschke VM. Genetic variation in the human cytochrome P450 supergene family. Pharmacogenet Genomics. 2015;25(12):584–94.
176. Kozyra M, Ingelman-Sundberg M, Lauschke VM. Rare genetic variants in cellular transporters, metabolic enzymes, and nuclear receptors can be important determinants of interindividual differences in drug response. Genet Med. 2017;19(1):20–9.
177. Nelson MR, Wegmann D, Ehm MG, Kessner D, St Jean P, Verzilli C, et al. An abundance of rare functional variants in 202 drug target genes sequenced in 14,002 people. Science. 2012;337(6090):100–4.
178. Sanderson S, Zimmern R, Kroese M, Higgins J, Patch C, Emery J. How can the evaluation of genetic tests be enhanced? Lessons learned from the ACCE framework and evaluating genetic tests in the United Kingdom. Genet Med. 2005;7(7):495–500.
179. Gaedigk A. Complexities of CYP2D6 gene analysis and interpretation. Int Rev Psychiatry. 2013;25(5):534–53.
180. Zhou Y, Ingelman-Sundberg M, Lauschke VM. Worldwide distribution of cytochrome P450 alleles: a meta-analysis of population-scale sequencing projects. Clin Pharmacol Therap. 2017;102(4):688–700.
181. Limdi NA, Beasley TM, Crowley MR, Goldstein JA, Rieder MJ, Flockhart DA, et al. VKORC1 polymorphisms, haplotypes and haplotype groups on warfarin dose among African–Americans and European–Americans. Pharmacogenomics. 2008;9(10):1445–58.
182. DeGorter M, Xia C, Yang J, Kim R. Drug transporters in drug efficacy and toxicity. Annu Rev Pharmacol Toxicol. 2012;52:249–73.
183. Lesko L, Zineh I, Huang SM. What is clinical utility and why should we care? Clin Pharmacol Ther. 2010;88:729–33.
184. Shendure J, Ji H. Next-generation DNA sequencing. Nat Biotechnol. 2008;26(10):1135–45.
185. Gullapalli RR, Desai KV, Santana-Santos L, Kant JA, Becich MJ. Next generation sequencing in clinical medicine: challenges and lessons for pathology and biomedical informatics. J Pathol Inf. 2012;3(1):40.

Personalized Cancer Therapies 3

Fatemeh Khatami, George P. Patrinos, Shekoufeh Nikfar, Hamid Reza Aghaei Meybodi, Bagher Larijani, Marziyeh Nosrati, Mandana Hasanzad, and Negar Sarhangi

What Will You Learn in This Chapter?
The chapter explores how tailored approaches based on molecular biology have a tremendous impact on cancer treatment, delving into this rapidly changing field. Precision oncology aims to maximize therapeutic success while minimizing side effects by customizing medicines to the unique molecular and genetic properties of individual tumors. A comprehensive overview of the molecular foundations sheds light on the principles underlying this innovative methodology.

F. Khatami
Urology Research Center, Tehran University of Medical Sciences, Tehran, Iran

G. P. Patrinos
Department of Pharmacy, School of Health Sciences, University of Patras, Patras, Greece

Department of Genetics and Genomics, College of Medicine and Health Sciences, United Arab Emirates University, Al-Ain, Abu Dhabi, UAE

Zayed Center for Health Sciences, United Arab Emirates University, Al-Ain, Abu Dhabi, UAE

Faculty of Medicine and Health Sciences, Department of Pathology, Clinical Bioinformatics Unit, Erasmus University Medical Center, Rotterdam, The Netherlands

S. Nikfar
Personalized Medicine Research Center, Endocrinology and Metabolism Clinical Sciences Institute, Tehran University of Medical Sciences, Tehran, Iran

Faculty of Pharmacy, Department of Pharmacoeconomics and Pharmaceutical Administration, Tehran University of Medical Sciences, Tehran, Iran

H. R. Aghaei Meybodi
Evidence Based Medicine Research Center, Endocrinology and Metabolism Clinical Sciences Institute, Tehran University of Medical Sciences, Tehran, Iran

B. Larijani
Endocrinology and Metabolism Research Center, Endocrinology and Metabolism Clinical Sciences Institute, Tehran University of Medical Sciences, Tehran, Iran

M. Nosrati
Personalized Medicine Research Center, Endocrinology and Metabolism Clinical Sciences Institute, Tehran University of Medical Sciences, Tehran, Iran

M. Hasanzad
Personalized Medicine Research Center, Endocrinology and Metabolism Clinical Sciences Institute, Tehran University of Medical Sciences, Tehran, Iran

N. Sarhangi (✉)
Personalized Medicine Research Center, Endocrinology and Metabolism Clinical Sciences Institute, Tehran University of Medical Sciences, Tehran, Iran

Medical Genomics Research Center, Tehran Medical Sciences, Islamic Azad University, Tehran, Iran

M. Hasanzad (ed.), *A Glimpse at Medicine in the Future*,
https://doi.org/10.1007/978-981-97-4974-4_3

The U.S. Food and Drug Administration's (FDA) approval of tailored treatments that target particular genetic variations and biomarkers in cancer is reviewed, with a focus on recent accomplishments. The importance of these approvals in moving precision oncology from abstract ideas to real-world clinical applications is discussed in this chapter.

The challenges involved in precision oncology and tackling issues such as genetic evolution, tumor heterogeneity, and the lack of reliable prognostic biomarkers are all highlighted. The financial implications, interpretational challenges, and restricted availability of molecular profiling technologies, which provide a more complex picture of the obstacles to general use, are discussed.

Precision medicine will be effortlessly incorporated into standard cancer care, with a focus on global collaboration, ongoing research, and technological developments. However, it is emphasized that novel approaches to the design of clinical trials are necessary to verify the efficacy of customized care.

Rationale and Importance

The rationale and importance of personalized treatment in cancer is the ability of customized cancer treatment to maximize therapeutic outcomes, reduce side effects, and offer a patient-centered strategy that takes into account the complex genetic landscape of cancer. This paradigm shift toward precision oncology promises to transform the way that cancer is treated with its increased effectiveness, targetedness, and individualization of treatment. Personalized cancer treatment is based on the knowledge that tumors and cancer patients have inherent molecular variation.

Precision oncology (personalized treatment) aims to tailor therapeutic strategies according to the distinct molecular and genetic characteristics of specific tumors and patients, as opposed to traditional one-size-fits-all approaches.

Indeed, personalized treatment makes it possible to determine the precise molecular changes that drive tumor development, which in turn enables the choice of treatments that are more likely to be efficient for a particular patient. To overcome resistance and extend the efficacy of treatment, personalized treatment techniques can adjust to the evolving molecular landscape of malignancies.

3.1 Introduction

Cancer is the main reason of death worldwide, resulting in nearly 10 million deaths in 2020 (about one in six deaths) [1]. It is frequently observed in clinical practice that the same medication doses result in significant variations in both efficacy and toxicity from person to person [2, 3]. The use of genomics in oncology has significantly improved in recent years, providing insights into treatment decisions for many patients [4].

Genomes of cancer cells exhibit subtle yet meaningful variations compared to the surrounding healthy tissue, and numerous drug targets for treating cancer are also found in normal cells [5].

In cancer genetics, it is crucial to consider both acquired (somatic) and inherited (germline) variations. These two types of genetic variations play a role in determining the efficacy and safety of a drug [6].

A germline mutation, or germinal mutation, is any detectable variation within germ cells (cells that, when fully developed, become sperm and ova) and can be considered to identify individuals w ith the greatest likelihood of severe adverse events. At the same time, somatic mutation describes any alteration at the cellular level in somatic tissues occurring after fertilization is linked to the efficacy of the treatment [7]. Germline mutations have a considerable impact on the response to chemotherapy and targeted anticancer agents in the field of oncology. These variations are mostly linked to the drug's pharmacokinetics, which in turn contribute to the occurrence of treatment-related adverse events in patients [6, 8]. Hence, germline pharmacogenomic markers can be used to identify patients at the greatest risk of experiencing severe adverse events that may result in discontinuation and failure of treatment.

The use of prospective tumor sequencing is increasing, transforming the cancer treatment approach from site-specific cytotoxic therapy to targeted treatments, focusing on the tumor's molecular characteristics [9, 10]. The concept of somatic mutations is the underpinning of precision oncology as the foundation of cancer development. Precision oncology (Personalized treatment) is an emerging era in tumor prevention and treatment, and molecular profiling of tumors is used to identify inter- and intra-tumor alterations in every patient diagnosed with cancer [11].

The term "precision oncology" encompasses a wide range of strategies in cancer medicine, from the use of targeted treatment generally to the utilization of next-generation sequencing data to select treatment appropriate for individuals based on their specific characteristics, regardless of the type of cancer [12]. The field of precision oncology is transitioning from singular genomic assessments to a multi-omic strategy, aiming to enhance comprehension of tumor biology and expand treatment possibilities [13].

Precision medicine in cancer treatment involves tailoring therapies to the specific molecular or cellular characteristics of each patient, focusing on individualized factors such as gene or protein expression patterns. It represents a departure from a one-size-fits-all approach and has rapidly become a vital component of clinical practice in oncology [14]. The goal of precision medicine is to personalize therapy by targeting the specific oncogenic factors driving the tumor and shaping its immune environment, enhancing treatment efficacy, and minimizing adverse effects by specifically addressing molecules involved in cancer growth and progression (Fig. 3.1) [15]. It results in improved patient care and enhanced quality of life through the maintenance of organ function.

The variability observed inter- and intraindividual during cancer therapy, along with the need for effective tumor-specific treatments while minimizing toxicity in normal cells, are key considerations in cancer treatment. There is an extreme need to identify biomarkers in personalized treatments, ensuring the administration of the correct dose to the right patient at the appropriate time.

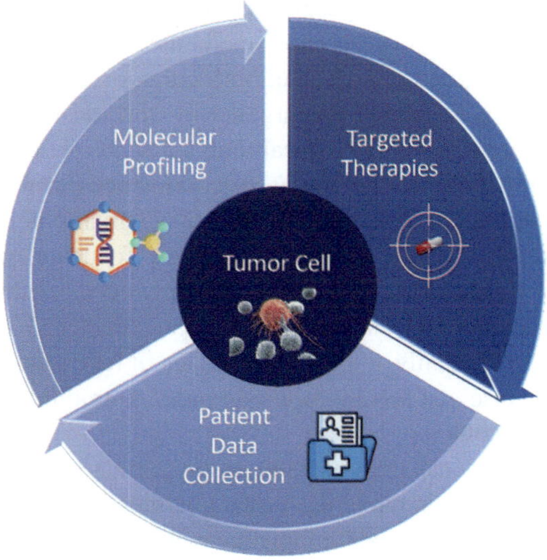

Fig. 3.1 The personalized journey from patient data to targeted therapies. This figure shows the complex journey including tumor cells, targeted therapies, molecular profiling, and the crucial foundation of patient data collection, paving the way for personalized cancer therapy

3.2 Molecular Basis of Personalized Treatment in Cancer

Cancer is a dynamic condition that typically becomes more heterogeneous as the disease progresses. This means that the tumor can consist of a different collection of cells containing unique molecular characteristics and varying levels of sensitivity to treatment.

Tumor heterogeneity may lead to a non-uniform distribution of genetically different subgroups of tumor cells within and between areas affected by the disease (spatial heterogeneity) or changes over time in the molecular composition of cancer cells (temporal heterogeneity). Heterogeneity is the enhancing force behind resistance, making it crucial to accurately evaluate tumor heterogeneity to develop therapeutic efficacy [15]. Precision medicine has already transformed cancer care: both common and rare malignancies can be targeted by specific therapies to improve clinical outcomes in patients [13]. Targeted therapy, as an exciting and promising area of cancer treatment, shows great potential in the fight against cancer by focusing on specific molecules or proteins involved in the growth and spread of cancer cells [16]. Consequently, it has the advantage of being more precise and less harmful compared to traditional chemotherapy.

Personalized treatment in cancer, also known as precision oncology, bases treatment decisions on knowledge of the molecular causes of particular cancers, allowing clinicians to customize medications to the specific needs of each patient. The identification of certain biomarker changes within cancer cells that promote their growth and survival forms the biological basis of individualized treatment. Biomarkers refer to factors (such as a mutation or protein level) linked to a disease's outcome, such as progression or death. Prognostic biomarkers are connected with a disease outcome regardless of treatment, while predictive biomarkers involve tumor-specific molecular alterations like protein overexpression or mutations, influencing a drug's clinical impact, where the effect is contingent on the presence of the biomarker [17].

The completion of the human genome sequencing project in 2001 is considered a significant catalyst for the advancement of targeted therapy and precision medicine [18]. This achievement provided a comprehensive comprehension of the genetic composition of the human body, enabling the identification of specific genetic mutations and variations that play a role in the onset and advancement of diverse diseases. With this knowledge of the human genome, researchers have been able to develop therapies that specifically focus on the genetic alterations associated with a particular disease. The combination of this knowledge and technological advancements has resulted in a surge of precision medicine trials in recent times. These trials encompass the utilization of protein markers to pinpoint specific genetic alterations, molecular profiling to identify potential therapeutic targets, and tailored combinations of treatments to address multiple genetic changes.

The progress in genomics and the comprehension of the immune system's involvement in cancer have paved the way for the development of targeted treatments that specifically address particular molecular alterations or biological traits. Nevertheless, genomics has also revealed that cancer is a complex condition that needs a shift in treatment approach, moving beyond a narrow focus on tumor types and towards personalized, gene-directed treatment based on biomarker testing for per-individual [19, 20].

There has been a notable increase in the number of tumor-specific molecular abnormalities that can be targeted with drugs in the last 10 years, resulting in significant improvements in survival through therapies tailored to specific biomarkers in various cancer types. As a result, molecular pathology has become essential not only for providing insights into tumor diagnosis and prognosis but also for guiding treatment decisions in clinical practice. The emergence of next-generation sequencing technologies and the proliferation of comprehensive tumor molecular profiling initiatives in institutions worldwide have brought about a revolutionary transformation in precision oncology [13].

Numerous recognized genetic abnormalities are linked to heightened cancer risks, and commercially accessible genetic tests cover various genes, such as *BRCA1* and *BRCA2*, linked to breast and ovarian cancer. While these tests currently guide preventative surgeries, ongoing developments in testing are poised to influence the surgical care of both cancer patients and their relatives [21].

Molecular pathology tools like immunohistochemistry (IHC) and fluorescence in situ hybridization (FISH) are fundamental precision medicine tools in characterizing tumor biology and guiding therapeutic decisions in daily practice [22]. As druggable gene aberrations grow, next-generation sequencing (NGS) has become integral, replacing conventional techniques. NGS allows simultaneous analysis of a broad spectrum of genomic alterations, including mutations, copy number variations (CNV), translocations, and fusions in multiple genes, providing efficient, cost-effective tumor analysis compared to single-biomarker analyses. Despite the high sensitivity and specificity of conventional techniques, NGS offers advantages in comprehensiveness and efficiency, especially given the genomic complexity of tumors [22, 23].

The clinical utility and scope of precision medicine should ideally increase with the expansion of other molecular characterization methods beyond genomics, such as transcriptomics, epigenetics, immunophenotyping, other molecular profiling technologies, and the evaluation of medication combinations beyond monotherapy techniques [13].

3.3 Recent FDA-Approved Personalized Drug

The progression toward integrating precision medicine into healthcare delivery is an ongoing process, marked by continual advancements and enhanced insights in gene profiling, biomarker development, and the application of artificial intelligence (AI) and data analytics. These innovations steadily narrow the gap between these tailored therapies and individual patients.

U.S. Food and Drug Administration (FDA) approvals for personalized medicines have seen a notable increase, with recent research indicating a rise from less than 10% to over 25% of all approvals in the past decade. These medications rely on biomarkers or mutations to discern alternative treatment strategies for specific patients, taking into account not only their medical conditions but also their social, economic, and environmental circumstances [24].

3.3.1 Lung Cancer

Non-small cell lung cancer (NSCLC) stands out as a notable example of precision medicine within the realm of solid tumor malignancies. The clinical approach to NSCLC now relies on a combination of surgical, chemotherapeutic, and radiation treatments determined by pathologic findings and clinical staging. Additionally, targeted therapies guided by molecular profiling play a crucial role [25]. As the importance of molecular testing continues to grow, the preservation of tissue for this purpose, alongside ensuring an accurate histologic diagnosis, becomes a significant consideration. This is particularly vital in advanced-stage NSCLC, where small biopsy samples or aspirates often represent the sole available specimens. Utilizing NGS panels emerges as a potent method for supplying information pertinent to both standard-of-care and investigational treatment options [25, 26].

The current recommendations from the College of American Pathologists, International Association for the Study of Lung Cancer, and Association for Molecular Pathology advise the assessment of *epidermal growth factor receptor* (*EGFR*), *ALK*, and *ROS1* for all patients with metastatic NCLC, regardless of clinical characteristics [27]. These guidelines do not suggest the routine use of stand-alone assays for other genes, such as *BRAF*, *KRAS*, *RET*, *ERBB2* (*HER2*), and *MET*, except in the context of a clinical trial. However, in the United States, combination therapy is approved for NSCLC patients with the *BRAF* p.V600E mutation, potentially prompting consideration of a stand-alone assay for this target. It is worth noting that multiplexed genetic sequencing panels are preferred over multiple single-gene tests to explore additional treatment options beyond *EGFR*, *ALK*, and *ROS1*. Laboratories employing NGS are advised to include *BRAF*, *KRAS*, *RET*, *ERBB2*, and *MET* in their assessments.

The National Comprehensive Cancer Network (NCCN) guidelines offer specific treatment recommendations for all genes except *KRAS* [28]. These treatment approaches are recommended for individuals with metastatic NSCLC exhibiting sensitizing molecular changes.

The FDA has granted approval for the first targeted treatment designed for patients with non-small cell lung cancer carrying an exon 20 insertion variant in the *EGFR* gene. The FDA noted that approximately 2–3% of NSCLCs possess this *EGFR* exon 20 insertion, which promotes metastasis by stimulating rapid cancer cell growth. Before the approval of amivantamab-vmjw, there were no specific drugs available for patients with this particular genetic subtype of lung cancer. Individuals with this subtype often do not respond well to chemotherapy or EGFR-inhibiting drugs, which have demonstrated effectiveness in treating other *EGFR* gene variants. Consequently, many patients experienced unfavorable outcomes [29].

Approximately 10–15% of patients diagnosed with NSCLC exhibit tumors driven by somatic mutations in the *EGFR* gene [30, 31]. While these mutations are linked to a roughly 70% response rate to certain EGFR tyrosine kinase inhibitors such as afatinib (Gilotrif), erlotinib (Tarceva), and gefitinib (Iressa), a significant number of NSCLCs develop resistance, termed "acquired resistance," typically after 9–12 months. The primary cause of resistance is the emergence of an EGFR p.T790M mutation, identified as a pathogenic variation in the "gatekeeper" residue [32].

Osimertinib, formerly known as AZD9291 (AstraZeneca; [33]), and rociletinib are the approved tyrosine kinase inhibitors targeting the frequently acquired resistance mutation *EGFR* p.T790M. AZD9291 and rociletinib, the *EGFR* p.T790M-targeted drugs with the most extensive data, demonstrate similarly high response rates exceeding 50%. However, they exhibit distinct patterns of adverse events that may set certain members of this drug class apart.

Tyrosine kinase inhibitors (TKIs) erlotinib, gefitinib, afatinib, and osimertinib are all FDA–approved for the treatment of both *EGFR* exon 19 deletions and *EGFR* p.L858R mutations, whereas afatinib is the only TKI approved for uncommon *EGFR* driver pathogenic variants (e.g., p.G719X, p.L861Q) [34]. Crizotinib holds FDA approval for treating NSCLC with *ALK* or *ROS1* rearrangements. Alectinib, ceritinib, and brigatinib are also approved for ALK-rearranged NSCLC.

The combination of the *BRAF* inhibitor dabrafenib and MEK inhibitor trametinib has recently gained approval for NSCLC featuring *BRAF* p.V600E mutations. The anti-programmed cell death ligand 1 (PD-L1) antibody pembrolizumab is FDA-approved for NSCLC, serving as a first-line therapy for cases exhibiting high PD-L1 expression without an *EGFR* mutation or *ALK* rearrangement. Additionally, pembrolizumab is authorized for patients who have progressed while on platinum-based chemotherapy or those with *EGFR* or *ALK* alterations who have experienced progression during an FDA-approved targeted therapy, provided the tumor proportion score is 1% or higher [35].

According to the NCCN guidelines, it is recommended to take into account novel targeted therapeutic options, which include cabozantinib or vandetanib for *RET* rearrangements, ado-trastuzumab emtansine for *ERBB2* mutations, and crizotinib for *MET* exon14 skipping mutations or high-level MET stimulation [25]. Additionally, the guidelines emphasize the association between KRAS and a worse prognosis as well as decreased responsiveness to EGFR tyrosine kinase inhibitor therapy. Given the unlikely occurrence of overlapping targetable variations, the discovery of a *KRAS* mutation may suggest that additional molecular testing is not warranted.

Several other potentially targetable gene variations are undergoing assessment in different clinical trials, encompassing *FGFR1* amplification, *FGFR3* fusions, *NTRK1* fusions, and mutations in *PIK3CA* and *AKT1*. The routine evaluation of these genes is becoming more feasible through the increasing use of NGS testing, aligning with national guidelines [25]. This approach extends to the evaluation of many genes currently under investigation in ongoing trials.

3.3.2 Breast Cancer

Breast cancer stands as the most commonly diagnosed and second-leading cause of mortality among women globally [36]. It is caused by the clonal proliferation of genetically aberrant ductal epithelial cells that are exposed to hormones or have inherited susceptibility genes [36].

Over time, treatment advancements have significantly improved survival rates for breast cancer patients. The specific treatment plan may vary depending on factors such as cancer type, stage, the patient's overall health, and individual preferences. Surgical procedures, radiation therapy, chemotherapy, and hormone therapy represent diverse approaches to breast cancer treatment. Targeted endocrine therapies, also known as hormonal therapy, including aromatase inhibitors and selective estrogen receptor (ER) modulators, are established as standard treatment options [37].

Precision medicine in the context of breast cancer involves an approach to diagnosis, treatment, and prevention that considers both inherent genetic makeup and the genetic or other markers found within the cancer cells. This genetic information can aid in predicting or diagnosing diseases and influencing treatment decisions.

The emphasis on precision medicine, especially genome medicine within breast cancer, plays a significant role in therapeutic strategies. The identification of pathogenic variations and the use of molecular targeted agents are vital components of developing precision medicine for breast cancer therapy. Specific druggable mutations in breast cancer have been pinpointed [38].

Advancements in molecular medicine have played a pivotal role in the evolution of precision medicine. Current progress in omics technologies, such as NGS, has specifically targeted precision medicine approaches for breast cancer and triple-negative breast cancer (TNBC) [12, 38].

Innovative therapeutic strategies for Breast Cancer and TNBC include targeted approaches like immune checkpoint inhibitors (ICIs), epidermal growth factor receptor inhibitors (EGFRis), poly (ADP-ribose) polymerase inhibitors (PARPis), antibody–drug conjugates (ADCs), oncolytic viruses (OVs), glucose transporter-1 inhibitor (GLUT1i), and signaling pathway targeting [37]. Combination therapy strategies hold the potential to enhance the precision-medicine treatment of metastatic and TNBC patients [39].

Breast cancers often exhibit overexpression of HER2 and EGFR. Trastuzumab monoclonal antibodies (mAbs) show effectiveness against HER2+ breast cancer, especially when combined with taxanes, platinum compounds, and vinorelbine, demonstrating both safety and promising efficiency. The combination of taxanes with trastuzumab is considered the optimal first-line therapy for metastatic breast cancer overexpressing HER2. The FDA has approved trastuzumab in combination with chemotherapy as an adjuvant therapy for HER2+ breast cancer.

Lapatinib, a tyrosine-kinase inhibitor (TKI), has proven efficacy against both HER2- and EGFR-overexpressing breast tumors. It inhibits proliferation and hampers the activation of *EGFR*, *HER2*, *AKT*, and *ERK1/2* both in vitro and in vivo [38, 40].

The mTOR inhibitor everolimus has received FDA approval and is frequently employed in combination with the aromatase inhibitor exemestane for patients with hormone receptor-positive (HR+) advanced breast cancer following nonsteroidal aromatase inhibitor therapy [41].

Approximately 5–10% of breast cancers exhibit germline *BRCA* mutations, referred to as *gBRCA* [40]. PARP inhibitors, such as olaparib and talazoparib, trigger synthetic lethality in cells with impaired *BRCA1/2*. Synthetic lethality occurs when mutations in both genes collectively result in cell death, whereas mutations in either gene alone do not. The interplay between *PARP* and *BRCA* is characterized as synthetic lethal [42].

The FDA approved *BRCA* analysis CDx as a companion diagnostic to test for *BRCA* mutations [43].

Several multiplex gene panels are available for estimating the risk of relapse after definitive surgery, primarily used in ER-positive breast cancers [38]. MammaPrint (Agendia) and Oncotype DX (Genomic Health, Inc., Redwood City, CA) are the most prevalent used multigene methods in the world.

MammaPrint (Agendia) and Oncotype DX (Genomic Health, Inc., Redwood City, CA) are globally utilized multigene assays. MammaPrint examines 1391 genes in fresh/frozen or formalin-fixed paraffin-embedded (FFPE) samples, classifying patients into high- to low-risk for relapse based on the results of 70 genes [44, 45], proving particularly useful in distinguishing high-risk from low-risk cases [46].

Oncotype DX, using FFPE samples, is another widely adopted multigene assay examining 21 genes (including five reference genes) to estimate the 10-year relapse risk and the benefit of adjuvant endocrine therapy and chemotherapy. The results are expressed as a recurrence score (RS) ranging from 0 to 100.

PAM50-based Prosigna (NanoString Technologies) evaluates a 50-gene signature to estimate the risk of recurrence (ROR) [47].

A clinically validated multianalyte gene expression test called EndoPredict (Myriad Genetics) can predict distant metastases in ER-positive, HER2-negative breast cancer patients receiving endocrine treatment alone [48]. It is the combined analysis of 12 genes using reverse transcription-quantitative real-time PCR (RT-qPCR) (RT-qPCR).

Curebest 95GC breast (Sysmex), a 95-gene classifier, predicts the prognosis for ER-positive and lymph node metastasis-negative resected breast cancer who have only been treated with adjuvant endocrine therapy [49]. These assays are currently incorporated into practice guidelines and extensively utilized in clinics.

3.4 Colorectal Cancer

Despite substantial advancements in creating targeted therapies based on biomarkers for various cancer types, precision oncology for colorectal cancer (CRC) has been slower to progress. The challenge arises from the genetic diversity of CRC tumors and the scarcity of viable therapeutic targets [50]. While the application of molecular biomarkers to steer the development of targeted therapies and tailor treatment to each patient is complex in CRC, the landscape for metastatic CRC (mCRC) treatment is swiftly evolving towards a more personalized or precision-oriented approach. This involves considering the inherent biology of CRC cells and the molecular characteristics of individual tumors to effectively customize treatment. In the context of treating metastatic colorectal cancer, molecular profiling has become a crucial factor in informing clinical decisions.

FOLFOX and FOLFIRI (fluorouracil, leucovorin plus oxaliplatin, or irinotecan, respectively) are the standard base regimens used for the treatment of mCRC. Biologic agents, such as the EGFR targeted therapies, cetuximab and panitumumab, and the vascular endothelial growth factor monoclonal antibody, bevacizumab, are safe and effective in the first-line setting [51].

EGFR serves as a transmembrane tyrosine kinase receptor that governs the signaling pathways of serine/threonine-specific protein kinase (AKT), JNK, and mitogen-activated protein kinase (MAPK)/ERK, which play crucial roles in DNA synthesis, cell proliferation, apoptosis, and motility. Elevated EGFR levels are linked to the advancement of tumors in various cancer types, including colorectal cancer (CRC). Inhibiting EGFR through the utilization of monoclonal antibodies like cetuximab or panitumumab, combined with chemotherapy formulations such as 5-FU, leucovorin plus oxaliplatin (FOLFOX), or 5-FU, leucovorin plus irinotecan (FOLFIRI), leads to improved treatment responses in patients with metastatic CRC (mCRC). These therapeutic approaches can be personalized using an FDA-approved pharmacogenetic tool that assesses a patient's EGFR expression level or detects ProtoOncogene mutations.

The examination of mutations in genes that code for proteins downstream of EGFR has enabled the identification of biomarkers that can forecast resistance to therapies targeting *EGFR* [52].

Targeting RAS with precision cancer medicines is an active area of research. The European Society for Medical Oncology (ESMO) [53], the European Society of Pathology (ESP), and the Association of Clinical Pathologists Molecular Pathology and Diagnostics Group in the United Kingdom have endorsed the *KRAS/NRAS* mutation test for individuals with metastatic colorectal cancer (mCRC) [54].

The most prevalent *KRAS* mutations are identified in exon 2, specifically in codon 12 or 13. Numerous studies have confirmed that the presence of *KRAS* mutations in exon 2 serves as a predictor of resistance to anti-EGFR therapies. As part of the current standard of care, testing tumor samples for *KRAS* exon 2 mutations is considered a cost-effective method to anticipate resistance to these therapies [55, 56].

The most prevalent *KRAS* mutations are identified in exon 2, specifically in codon 12 or 13. Numerous studies have confirmed that the presence of *KRAS* mutations in exon 2 serves as a predictor of resistance to anti-EGFR therapies. As part of the current standard of care, testing tumor samples for *KRAS* exon 2 mutations is considered a cost-effective method to anticipate resistance to these therapies [52].

The European Society of Pathology recommended *KRAS* exon 2 testing for patients with metastatic colorectal cancer (mCRC) in 2008, and the American Society for Clinical Oncology (ASCO) endorsed it in 2009. Data on *KRAS* mutations in exon 2 prompted the FDA to revise its approval of anti-EGFR therapies in 2009, advising their use exclusively in mCRC patients without mutations at codons 12 or 13 [57, 58].

BRAF, an oncogene within the RAF gene family responsible for encoding a serine-threonine protein kinase in the *RAS-RAF-MAPK* cascade, is found in around 10% of colorectal cancers. However, this percentage can vary based on the study population [59]. The most notable and widespread pathogenic variation involves a single substitution, p.V600E, in the kinase domain. Numerous clinical studies have indicated that the presence of this mutation serves as an indicator of resistance to anti-EGFR therapies and is a significant marker of poor prognosis.

Braftovi™ (Encorafenib) is an oral small-molecule BRAF kinase inhibitor, and Mektovi® (binimetinib) is an oral small-molecule MEK inhibitor, both targeting key enzymes in the MAPK signaling pathway. In the United States, the combination of Braftovi™ + Mektovi® is approved for treating advanced melanoma and colon cancers with a *BRAF* p.V600E or *BRAF* p.V600K mutation, as identified by an FDA-approved test [60].

Encorafenib has received approval, in combination with the monoclonal anti-EGFR antibody cetuximab, for treating adult metastatic CRC patients who carry a *BRAF* p.V600E mutation [61].

MAPK pathway reactivation through EGFR-mediated feedback is suggested as a potential resistance mechanism to BRAF inhibitors in CRC patients with *BRAF* mutation [62]. Therefore, the rationale for using a combination is to block the EGFR pathway and enhance the effectiveness of the BRAF-targeting encorafenib.

Additional potential biomarkers have been identified that could assist in identifying colorectal cancer (CRC) patients suitable for anti-EGFR treatment. These biomarkers encompass *PIK3CA*, *PTEN*, Human Epidermal Growth Factor Receptor 2 (HER2), HER3, and the EGFR ligands EREG and AREG. While these biomarkers are not currently ready for clinical application, utilizing a combination of multiple biomarkers might enhance predictive accuracy compared to using a single one [52, 63]. Further prospective studies are required to validate combinations of predictive biomarkers for EGFR-targeted treatment.

3.5 Challenge of Precision Oncology

Several obstacles affect precision oncology's application and efficacy in cancer treatment. Tumor heterogenicity within and between patients makes it difficult to determine a single, effective target for treatment, and the outcomes of targeted therapies might differ greatly. New pathogenic variants and changing genetic profiles as a result of tumor evolution may cause resistance to targeted medicines, requiring ongoing monitoring and treatment plan changes. Lack of Predictive Biomarkers: Identifying reliable predictive biomarkers is crucial for guiding treatment decisions in precision oncology. However, such biomarkers are still not well-established for many cancers, limiting the ability to tailor treatments based on individual patient characteristics. Targetable mutation identification requires molecular profiling, such as genomic sequencing. However, all patients, especially those in settings with limited resources, may not have easy access to thorough molecular profiling due to the uneven availability of these technologies.

It is difficult to analyze the massive amount of genetic and molecular data generated in precision oncology. Proficiency in data interpretation and sophisticated computational and bioinformatics tools are necessary to accurately integrate and interpret this information [64]. Many targeted therapies are expensive, posing challenges for widespread adoption. The cost of developing and producing these drugs, combined with the need for individualized treatment plans, can contribute to financial burdens on healthcare systems and patients [65]. Ethical concerns around patient privacy, consent, and the possibility of genetic discrimination are brought up by precision oncology. Furthermore, in many places, there may not be a clear legal framework for processing genetic and genomic information.

The resolution of these obstacles necessitates continuous investigation, innovations in technology, global cooperation, and a comprehensive strategy that takes into account not only the genetic components of cancer but also the wider medical and societal environments in which patients receive care.

3.6 The Future of Precision Oncology

Precision oncology has a bright future thanks to developments in several fields. This future is largely shaped by the development of genomic technologies, the integration of multi-omics data to get a deeper knowledge of cancer, and the use of artificial intelligence (AI) and machine learning (ML) to analyze large genomic and clinical datasets. Liquid biopsies provide a less invasive way to track the development of cancer by concentrating on circulating tumor DNA. Immunotherapy and precision oncology will interact to provide tailored immune system improvement, with key biomarkers directing therapeutic choices. Innovative clinical trial designs, like umbrella and basket studies, will speed up the development of new treatments. Further research aiming at the discovery of new biomarkers will ensure that patient classification is refined to ensure customized therapy based on unique tumor characteristics. The incorporation of real-world evidence will impact healthcare policies and population-level health outcomes by providing important insights into the long-term safety and efficacy of precision oncology. Initiatives for international cooperation and data exchange are essential for increasing the generalizability of research results and hastening the detection of uncommon genetic alterations. With the progress of precision oncology, patients are anticipated to take on a more proactive role, enabled by enhanced education and information accessibility. However, issues like data privacy, moral dilemmas, and healthcare inequities must be addressed to guarantee fair access and advantages for every patient. All things considered, precision oncology has enormous potential to change cancer treatment into a more customized, efficient, and low-toxicity strategy. To guarantee fair access and advantages for all patients, it will be necessary to solve issues like data privacy, moral dilemmas, and healthcare inequities.

3.7 Conclusion

In summary, "Precision Oncology" is at the vanguard of cancer treatment, transforming patient care by traversing the complex field of molecular biology. This approach opens up a new age of personalized medicine by customizing medicines based on the unique genetic and molecular characteristics of individual tumors. The incorporation of precision medicine using molecular profiling technologies is increasingly combined with standard clinic pathological assessments to improve the diagnosis, prognosis, and prediction of clinical outcomes. Advancing in precision medicine will necessitate not only technical expansion beyond genomics but also the education and involvement of end-users, including clinicians and patients. This involves increasing access to genotype–drug pairs through adaptive and innovative clinical trial designs and promoting data sharing to maximize knowledge acquisition.

The recent FDA-approved personalized cancer medications are evidence of the real progress being made in this area. These medications, which are intended to focus on particular genetic changes or biomarkers, represent an effective example of how precision oncology concepts have been successfully applied to real-world clinical settings. The approvals represent a major advancement in the search for better patient outcomes by pointing to a move towards more focused and efficient therapies.

Although precision oncology is not without its challenges, precision oncology holds immense promise in the future. Clinical trial designs are crucial for evaluating new precision treatments and expanding personalized approaches across different cancer types, representing a shift from generic to tailored strategies for each patient.

The integration of precision oncology into routine clinical practice provides a glimmer of hope for more efficient, customized, and ethical cancer treatment as we approach the next stage of the disease's treatment. The pursuit of fully realizing the potential of precision oncology carries significant promise for a time when cancer treatment will be tailored to the individual patient.

References

1. de Bono JS, Ashworth A. Translating cancer research into targeted therapeutics. Nature. 2010;467(7315):543–9.
2. Evans WE, Relling MV. Pharmacogenomics: translating functional genomics into rational therapeutics. Science. 1999;286(5439):487–91.
3. Ruwali M. Pharmacogenetics and cancer treatment: progress and prospects. In: Molecular medicine. London: IntechOpen; 2019.
4. Cayún JP, Quiñones LA. Chapter 22: Current status of cancer pharmacogenomics. In: Faintuch J, Faintuch S, editors. Precision medicine for investigators, practitioners and providers. Academic Press: Cambridge; 2020. p. 233–43.
5. Relling MV, Dervieux T. Pharmacogenetics and cancer therapy. Nat Rev Cancer. 2001;1(2):99–108.
6. Gillis NK, Patel JN, Innocenti F. Clinical implementation of germ line cancer pharmacogenetic variants during the next-generation sequencing era. Clin Pharmacol Therap. 2014;95(3):269–80.
7. O'Donnell PH, Ratain MJ. Germline pharmacogenomics in oncology: decoding the patient for targeting therapy. Mol Oncol. 2012;6(2):251–9.
8. Hertz DL, McLeod HL. Use of pharmacogenetics for predicting cancer prognosis and treatment exposure, response and toxicity. J Hum Genet. 2013;58(6):346–52.
9. MacConaill LE, Van Hummelen P, Meyerson M, Hahn WC. Clinical implementation of comprehensive strategies to characterize cancer genomes: opportunities and challenges. Cancer Discov. 2011;1(4):297–311.
10. Ong FS, Das K, Wang J, Vakil H, Kuo JZ, Blackwell W-LB, et al. Personalized medicine and pharmacogenetic biomarkers: progress in molecular oncology testing. Expert Rev Mol Diagn. 2012;12(6):593–602.
11. Hoeben A, Joosten EA, van den Beuken-van Everdingen MH. Personalized medicine: recent progress in cancer therapy. Cancer. 2021;13(2):242.
12. Collins FS, Varmus H. A new initiative on precision medicine. N Engl J Med. 2015;372(9):793–5.
13. Malone ER, Oliva M, Sabatini PJ, Stockley TL, Siu LL. Molecular profiling for precision cancer therapies. Genome Med. 2020;12(1):1–19.
14. Schwartzberg L, Kim ES, Liu D, Schrag D. Precision oncology: who, how, what, when, and when not? Am Soc Clin Oncol Educ Book. 2017;37:160–9.
15. Dagogo-Jack I, Shaw AT. Tumour heterogeneity and resistance to cancer therapies. Nat Rev Clin Oncol. 2018;15(2):81–94.
16. Naithani N, Atal AT, Tilak T, Vasudevan B, Misra P, Sinha S. Precision medicine: uses and challenges. Med J Armed Forces India. 2021;77(3):258–65.
17. Ballman KV. Biomarker: predictive or prognostic? J Clin Oncol. 2015;33(33):3968–71.
18. Venter JC, Adams MD, Myers EW, Li PW, Mural RJ, Sutton GG, et al. The sequence of the human genome. Science. 2001;291(5507):1304–51.
19. Hainsworth JD, Meric-Bernstam F, Swanton C, Hurwitz H, Spigel DR, Sweeney C, et al. Targeted therapy for advanced solid tumors on the basis of molecular profiles: results from mypathway, an open-label, phase IIa multiple basket study. J Clin Oncol. 2018;36(6):536–42.
20. Von Hoff DD, Stephenson JJ Jr, Rosen P, Loesch DM, Borad MJ, Anthony S, et al. Pilot study using molecular profiling of patients' tumors to find potential targets and select treatments for their refractory cancers. J Clin Oncol. 2010;28(33):4877–83.
21. Lee YT, Tan YJ, Oon CE. Molecular targeted therapy: treating cancer with specificity. Eur J Pharmacol. 2018;834:188–96.
22. Dietel M, Jöhrens K, Laffert M, Hummel M, Bläker H, Pfitzner B, et al. A 2015 update on predictive molecular pathology and its role in targeted cancer therapy: a review focussing on clinical relevance. Cancer Gene Ther. 2015;22(9):417–30.

23. Moorcraft SY, Gonzalez D, Walker BA. Understanding next generation sequencing in oncology: a guide for oncologists. Crit Rev Oncol. 2015;96(3):463–74.
24. Personalized Medicine Coalition (PMC). Personalized medicine at FDA: the scope and significance of progress in 2021. https://www.personalizedmedicinecoalition.org/Userfiles/PMC-Corporate/file/Personalized_Medicine_at_FDA_The_Scope_Significance_of_Progress_in_2021.pdf.
25. Brown NA, Aisner DL, Oxnard GR. Precision medicine in non–small cell lung cancer: current standards in pathology and biomarker interpretation. Am Soc Clin Oncol Educ Book. 2018;38:708–15.
26. Liam CK, Mallawathantri S, Fong KM. Is tissue still the issue in detecting molecular alterations in lung cancer? Respirology. 2020;25(9):933–43.
27. Lindeman NI, Cagle PT, Aisner DL, Arcila ME, Beasley MB, Bernicker EH, et al. Updated molecular testing guideline for the selection of lung cancer patients for treatment with targeted tyrosine kinase inhibitors: guideline from the College of American Pathologists, the International Association for the Study of Lung Cancer, and the Association for Molecular Pathology. Arch Pathol Lab Med. 2018;142(3):321–46.
28. Network NCC. Clinical practice guidelines in oncology: non-small cell lung cancer. 2007. http://www.nccn.org/professionals/physician_gls/PDF/nscl.pdf.
29. Kuehn BM. New targeted non-small cell lung cancer therapy approved. JAMA. 2021;326(1):20.
30. Lynch TJ, Bell DW, Sordella R, Gurubhagavatula S, Okimoto RA, Brannigan BW, et al. Activating mutations in the epidermal growth factor receptor underlying responsiveness of non-small-cell lung cancer to gefitinib. N Engl J Med. 2004;350(21):2129–39.
31. Gandara DR, Hammerman PS, Sos ML, Lara PN Jr, Hirsch FR. Squamous cell lung cancer: from tumor genomics to cancer therapeutics. Clin Cancer Res. 2015;21(10):2236–43.
32. Riely GJ, Yu HA. EGFR: the paradigm of an oncogene-driven lung cancer. Clin Cancer Res. 2015;21(10):2221–6.
33. Cross DA, Ashton SE, Ghiorghiu S, Eberlein C, Nebhan CA, Spitzler PJ, et al. AZD9291, an irreversible EGFR TKI, overcomes T790M-mediated resistance to EGFR inhibitors in lung cancer. Cancer Discov. 2014;4(9):1046–61.
34. Solassol I, Pinguet F, Quantin X. FDA- and EMA-approved tyrosine kinase inhibitors in advanced EGFR-mutated non-small cell lung cancer: safety, tolerability, plasma concentration monitoring, and management. Biomolecules. 2019;9(11):668.
35. Guaitoli G, Zullo L, Tiseo M, Dankner M, Rose AA, Facchinetti F. Non-small-cell lung cancer: how to manage BRAF-mutated disease. Drugs Context. 2023;12:2022.
36. Giaquinto AN, Sung H, Miller KD, Kramer JL, Newman LA, Minihan A, et al. Breast cancer statistics, 2022. CA Cancer J Clin. 2022;72(6):524–41.
37. Lumachi F, Luisetto G, Mm Basso S, Basso U, Brunello A, Camozzi V. Endocrine therapy of breast cancer. Curr Med Chem. 2011;18(4):513–22.
38. Naito Y, Urasaki T. Precision medicine in breast cancer. Chin Clin Oncol. 2018;7(3):29.
39. Subhan MA, Parveen F, Shah H, Yalamarty SSK, Ataide JA, Torchilin VP. Recent advances with precision medicine treatment for breast cancer including triple-negative sub-type. Cancer. 2023;15(8):2204.
40. O'Donovan N, Crown J. EGFR and HER-2 antagonists in breast cancer. Anticancer Res. 2007;27(3A):1285–94.
41. Baselga J, Campone M, Piccart M, Burris HA, Rugo HS, Sahmoud T, et al. Everolimus in postmenopausal hormone-receptor–positive advanced breast cancer. N Engl J Med. 2012;366(6):520–9.
42. Helleday T. The underlying mechanism for the PARP and BRCA synthetic lethality: clearing up the misunderstandings. Mol Oncol. 2011;5(4):387–93.
43. BRACAnalysis CDx. https://myriad.com/products-services/companion-diagnostics/bracanalysis-cdx/. Accessed 28 April 2018.
44. Van't Veer LJ, Dai H, Van De Vijver MJ, He YD, Hart AA, Mao M, et al. Gene expression profiling predicts clinical outcome of breast cancer. Nature. 2002;415(6871):530–6.
45. Van De Vijver MJ, He YD, Van't Veer LJ, Dai H, Hart AA, Voskuil DW, et al. A gene-expression signature as a predictor of survival in breast cancer. N Engl J Med. 2002;347(25):1999–2009.
46. Drukker CA, Bueno-de-Mesquita J, Retèl VP, van Harten WH, van Tinteren H, Wesseling J, et al. A prospective evaluation of a breast cancer prognosis signature in the observational RASTER study. Int J Cancer. 2013;133(4):929–36.
47. Parker JS, Mullins M, Cheang MC, Leung S, Voduc D, Vickery T, et al. Supervised risk predictor of breast cancer based on intrinsic subtypes. J Clin Oncol. 2009;27(8):1160.
48. Kronenwett R, Bohmann K, Prinzler J, Sinn BV, Haufe F, Roth C, et al. Decentral gene expression analysis: analytical validation of the Endopredict genomic multianalyte breast cancer prognosis test. BMC Cancer. 2012;12:1–10.
49. Naoi Y, Kishi K, Tanei T, Tsunashima R, Tominaga N, Baba Y, et al. Development of 95-gene classifier as a powerful predictor of recurrences in node-negative and ER-positive breast cancer patients. Breast Cancer Res Treat. 2011;128:633–41.
50. Di Nicolantonio F, Vitiello PP, Marsoni S, Siena S, Tabernero J, Trusolino L, et al. Precision oncology in metastatic colorectal cancer—from biology to medicine. Nat Rev Clin Oncol. 2021;18(8):506–25.
51. Sunakawa Y, Schirripa M, Lenz H-J. Understanding the FOLFOXIRI-regimen to optimize treatment for metastatic colorectal cancer. Crit Rev Oncol. 2016;100:117–26.
52. Tran NH, Cavalcante LL, Lubner SJ, Mulkerin DL, LoConte NK, Clipson L, et al. Precision medicine in colorectal cancer: the molecular profile

alters treatment strategies. Therap Adv Med Oncol. 2015;7(5):252–62.

53. Van Cutsem E, Cervantes A, Nordlinger B, Arnold D. Metastatic colorectal cancer: ESMO clinical practice guidelines for diagnosis, treatment and follow-up. Ann Oncol. 2014;25:1–9.
54. Wong NA, Gonzalez D, Salto-Tellez M, Butler R, Diaz-Cano SJ, Ilyas M, et al. RAS testing of colorectal carcinoma—a guidance document from the Association of Clinical Pathologists Molecular Pathology and Diagnostics Group. J Clin Pathol. 2014;67(9):751–7.
55. Karapetis CS, Khambata-Ford S, Jonker DJ, O'Callaghan CJ, Tu D, Tebbutt NC, et al. K-ras mutations and benefit from cetuximab in advanced colorectal cancer. N Engl J Med. 2008;359(17):1757–65.
56. Amado RG, Wolf M, Peeters M, Van Cutsem E, Siena S, Freeman DJ, et al. Wild-type KRAS is required for panitumumab efficacy in patients with metastatic colorectal cancer. J Clin Oncol. 2023;41(18):3278–86.
57. Allegra CJ, Jessup JM, Somerfield MR, Hamilton SR, Hammond EH, Hayes DF, et al. American Society of Clinical Oncology provisional clinical opinion: testing for KRAS gene mutations in patients with metastatic colorectal carcinoma to predict response to anti–epidermal growth factor receptor monoclonal antibody therapy. J Clin Oncol. 2009;27(12):2091–6.
58. Van Krieken J, Jung A, Kirchner T, Carneiro F, Seruca R, Bosman F, et al. KRAS mutation testing for predicting response to anti-EGFR therapy for colorectal carcinoma: proposal for an European quality assurance program. Virchows Arch. 2008;453:417–31.
59. Davies H, Bignell GR, Cox C, Stephens P, Edkins S, Clegg S, et al. Mutations of the BRAF gene in human cancer. Nature. 2002;417(6892):949–54.
60. Safaee Ardekani G, Jafarnejad SM, Tan L, Saeedi A, Li G. The prognostic value of BRAF mutation in colorectal cancer and melanoma: a systematic review and meta-analysis. PLoS One. 2012;7(10):e47054.
61. BRAFTOVI® (encorafenib) capsules, for oral use: FDA packaging insert. U.S. Food and Drug Administration. https://www.accessdata.fda.gov/drugsatfda_docs/label/2018/210496lbl.pdf.
62. Corcoran RB, Ebi H, Turke AB, Coffee EM, Nishino M, Cogdill AP, et al. EGFR-mediated reactivation of MAPK signaling contributes to insensitivity of BRAF-mutant colorectal cancers to RAF inhibition with vemurafenib. Cancer Discov. 2012;2(3):227–35.
63. Yang ZY, Wu XY, Huang YF, Di MY, Zheng DY, Chen JZ, et al. Promising biomarkers for predicting the outcomes of patients with KRAS wild-type metastatic colorectal cancer treated with anti-epidermal growth factor receptor monoclonal antibodies: a systematic review with meta-analysis. Int J Cancer. 2013;133(8):1914–25.
64. Roychowdhury S, Iyer MK, Robinson DR, Lonigro RJ, Wu Y-M, Cao X, et al. Personalized oncology through integrative high-throughput sequencing: a pilot study. Sci Transl Med. 2011;3(111):111–21.
65. Prasad V, De Jesús K, Mailankody S. The high price of anticancer drugs: origins, implications, barriers, solutions. Nat Rev Clin Oncol. 2017;14(6):381–90.

Machine Learning and Artificial Intelligence in Modern Medicine

4

Parham Rahmani, Parham Mortazavi, Hamid Reza Aghaei Meybodi, George P. Patrinos, Negar Sarhangi, Shekoufeh Nikfar, Bagher Larijani, Nazli Khodayari, and Mandana Hasanzad

What We Learn in This Chapter?

This chapter describes the transformative landscape of Machine learning (ML) and artificial intelligence (AI) in medicine, exploring its diverse applications and the potential to revolutionize healthcare. We are on a journey through the complex methods and techniques applied by AI systems, examining how they are being harnessed to enhance diagnosis, patient care, and biomedical research. From assisting radiologists in identifying

P. Rahmani
Faculty of Pharmacy, Department of Pharmacoeconomics and Pharmaceutical Administration, Tehran University of Medical Sciences, Tehran, Iran

Endocrinology and Metabolism Research Center, Endocrinology and Metabolism Clinical Sciences Institute, Tehran University of Medical Sciences, Tehran, Iran

P. Mortazavi
Isfahan Cardiovascular Research Center, Cardiovascular Research Institute, Isfahan University of Medical Sciences, Isfahan, Iran

H. R. Aghaei Meybodi · B. Larijani
Endocrinology and Metabolism Research Center, Endocrinology and Metabolism Clinical Sciences Institute, Tehran University of Medical Sciences, Tehran, Iran

G. P. Patrinos
Department of Pharmacy, School of Health Sciences, University of Patras, Patras, Greece

Department of Genetics and Genomics, College of Medicine and Health Sciences, United Arab Emirates University, Al-Ain, Abu Dhabi, UAE

Zayed Center for Health Sciences, United Arab Emirates University, Al-Ain, Abu Dhabi, UAE

Faculty of Medicine and Health Sciences, Department of Pathology, Clinical Bioinformatics Unit, Erasmus University Medical Center, Rotterdam, The Netherlands

N. Sarhangi · M. Hasanzad (✉)
Personalized Medicine Research Center, Endocrinology and Metabolism Clinical Sciences Institute, Tehran University of Medical Sciences, Tehran, Iran

S. Nikfar
Personalized Medicine Research Center, Endocrinology and Metabolism Clinical Sciences Institute, Tehran University of Medical Sciences,, Tehran, Iran

Faculty of Pharmacy, Department of Pharmacoeconomics and Pharmaceutical Administration, Tehran University of Medical Sciences, Tehran, Iran

N. Khodayari
Division of Pulmonary, Critical Care and Sleep Medicine, Department of Medicine, University of Florida, Gainesville, FL, USA

M. Hasanzad (ed.), *A Glimpse at Medicine in the Future*,
https://doi.org/10.1007/978-981-97-4974-4_4

subtle abnormalities in scans to guiding surgeons through complex procedures, AI proves its worth in various medical domains. We also provide a summary of the ethical and social implications of AI in healthcare, addressing concerns regarding data privacy, algorithmic bias, and the potential displacement of human workers. Ultimately, we'll explore the future of AI in medicine, examining how it is poised to shape the delivery of healthcare and the way we understand and treat diseases.

Rationale and Importance?
The rapid advancements in artificial intelligence (AI) have opened up a new era in medicine, promising to transform the way we diagnose, treat, and manage diseases. AI can be employed to automate repetitive tasks, freeing up clinicians to focus on more complex conditions. As AI continues to evolve, its role in healthcare will undoubtedly grow, leading to more effective medical care.

4.1 Introduction

Artificial Intelligence (AI) refers to developing computer systems that perform tasks, such as problem-solving or decision-making, that typically require human intelligence. Over the last few decades, new technologies, such as AI and Machine learning (ML), have become more widespread in many aspects of various fields, such as science and technology, industry, and even our daily routine life [1]. AI encompasses the field of creating intelligent machines using algorithms or predefined rules. These machines aim to replicate human cognitive functions, including learning and problem-solving. AI systems can proactively anticipate and address problems, operating in an intelligent and adaptable manner [2, 3]. AI replicates the features of human intelligence and executes processes within the computer system. Rapid learning, prediction outcomes, drawing inferences, and the capacity for self-correction are among the brilliant features of this remarkable technology, which can be highly beneficial to the daily problem of medical care [4]. ML is a subset of AI characterized by a machine's capacity to learn from a designated training dataset and acquired knowledge, enabling it to predict outcomes for data points beyond the initial training set. ML leverages statistical principles to construct mathematical models, primarily making inferences based on provided samples [5, 6].

These advancements can revolutionize numerous aspects of healthcare, including medical care and operational processes within the scope of clinical providers, health facilities, financial contributors, and pharmaceutical industries [7]. One crucial aspect of AI is creating a digital database for healthcare providers to provide diagnostic, treatment, and routine medical care usage. It leads to a reduction in health documentation time. Based on ultimate specifications, medical experts, collaborating with software and hardware professionals, construct a platform facilitating data collection and managing healthcare tasks. In conclusion, this innovative technology has helped healthcare providers better evaluate diagnoses, treatment methodologies, and post-treatment monitoring. The integration of AI in healthcare administrative processes promises to streamline workflows, reduce operational costs, and enhance the efficiency of healthcare delivery [8, 9].

However, the digital transformation of medical data and the rapid adoption of technology are pushing for reform and improvement in the utilization of AI in medicine. Nevertheless, challenges such as integrating diverse data sources, addressing ethical considerations, ensuring data security, optimizing model performance, establishing robust regulatory infrastructure, and mitigating bias could delay the growth of AI in healthcare systems [10]. As we explore the impacts of AI on the healthcare industry, it becomes essential to solve these challenges alongside the unprecedented opportunities, shaping a path toward a healthcare landscape where the symbiosis of artificial and human intelligence maximizes the potential for improved patient care and innovative medical practices.

This chapter will elucidate how these advancements impact and improve the experiences of each stakeholder, fostering a deeper connection between technology and the human touch in healthcare. Additionally, by unfolding in a logical sequence and guiding readers through the intricacies of AI in medicine—from an introduction to its historical roots, challenges, and the burgeoning importance of AI—each section contributes to a comprehensive narrative, aiming to explain this transformative collaboration discussed in this chapter.

4.2 The Methods of AI

AI encompasses various methods such as reasoning, knowledge representation, solution search, and machine learning (ML). ML involves using algorithms to identify patterns in categorized data sets. A subset of ML is deep learning (DL), which utilizes artificial neural networks (ANNs). ANNs consist of interconnected computing elements called neurons, resembling the biological neurons in the human brain and replicating the transmission of electrical impulses [11]. These networks consist of nodes that receive inputs and convert them to outputs, either individually or interconnected, using algorithms to solve problems [12]. ANNs can take different forms, including multilayer perceptron (MLP) networks, recurrent neural networks (RNNs), and convolutional neural networks (CNNs). These networks can be trained using supervised or unsupervised training methods [13]. A thorough examination of the different sub-domains of AI and their applicability across numerous industries needs a thorough investigation of the subject. Natural Language Processing (NLP) refers to a machine's ability to read, comprehend, and interpret language. Once the computer recognizes the user's intended message, it can react appropriately, enabling applications such as chatbots, language translation, and voice assistants [14]. Computer Vision algorithms aim to comprehend images by dissecting them and examining various aspects. This enables machines to classify and learn from collections of photos, leading to improved results based on prior observations. Applications of computer vision include object recognition, image segmentation, and autonomous vehicles [15]. Cognitive computing algorithms simulate the functioning of the human brain by analyzing text, audio, images, and other inputs to provide desired results. These algorithms enable tasks such as sentiment analysis, recommendation systems, and personalized healthcare [16].

In conclusion, the application of ML, DL, NLP, computer vision, and cognitive computing play crucial roles in improving diagnostic accuracy and treatment decisions.

4.3 Applications of Artificial Intelligence in Medicine

Medical science-generated information currently has a very broad scope; it is rapidly expanding and will continue to do so in both volume and variety. Simultaneously, the future promise for AI in medicine and health is enormous and growing as AI technologies are developed by industry, academia, government, and individuals (Fig. 4.1). The incorporation of AI-based technology into medical practice is predicted to result in significant changes in many fields of medicine and healthcare [17]. In telemedicine, patients in non-emergency situations can interact with hospital AI systems to analyze symptoms, input vital signs, and determine the need for medical assistance. This approach focuses on addressing the most urgent cases, thereby reducing the workload on medical professionals. AI has revolutionized assisted diagnosis, enabling the rapid analysis of MRI scans to detect tumors and other malignancies. Using computer vision and convolutional neural networks allows AI to outpace radiologists significantly, with a notably narrower margin of error. Robotic surgery, characterized by a minimal margin of error and the ability to perform surgeries reliably 24 h without fatigue, represents a breakthrough. The precision of robotic surgery reduces invasiveness compared to traditional techniques, potentially reducing patient recovery time in the hospital. In vital stats monitoring, continual assessment of an individual's health is crucial. Although wearable technology is increasingly common, the data it generates requires analysis to yield meaningful insights.

Fig. 4.1 Artificial intelligence (AI) is transforming the healthcare landscape, empowering healthcare providers to deliver more precise care. This picture highlights the diverse applications of AI in medicine, encompassing many usages from disease diagnosis to drug discovery and pharmaceutical manufacturing

4.4 Primary Areas of Use for AI in Healthcare

Today, artificial intelligence (AI) has been gradually developed and implemented into practically every field of medicine, from primary care to rare conditions, emergency medicine, biomedical research, and public health. Several aspects of health administration management and policy are expected to gain from new AI-mediated solutions [18]. Healthcare AI solutions are frequently categorized based on stakeholder user groups, including (1) patients and citizens, (2) clinicians and caregivers, (3) healthcare administrators, and (4) public health experts and policymakers. AI tools in healthcare are comprehensively classified and applied in the following categories: clinical practice, biomedical research, public health, and health administration.

4.5 AI in Clinical Practice

The potential for AI applications in the clinical setting is extensive, ranging from diagnostic process automation to treatment decision-making and clinical research. Data for diagnosis and therapy can be obtained from a variety of sources, including clinical notes, laboratory testing, pharmacy data, medical imaging, and genomics data. Image analysis (e.g., radiology, ophthalmology, dermatology, and pathology) and signal processing (e.g., ECG, audiology, and electroencephalography) will be heavily influenced by AI. Aside from test and picture interpretation, AI could be used to combine and array results with other clinical data to improve clinical procedures [19].

4.6 Diagnostic Application

Despite tremendous progress in medicine, the accurate diagnosis of diseases continues to pose a formidable global challenge. The intricate interaction of disease mechanisms and their symptoms presents an ongoing barrier to the development of early diagnostic tools. The integration of AI into medical diagnosis promises a host of benefits, including enhanced accuracy, cost reduction, time efficiency, and a marked reduction in human errors. AI has the capacity to provide real-time assistance and invaluable insights to clinicians, thus supporting their decision-making process. ML harnesses the power of data to enhance decision-making, streamline workflow management, and automate tasks in a manner that is both cost-effective and timely. Deep learning, employing CNN and sophisticated data mining techniques, holds tremendous potential in uncovering intricate patterns within vast datasets. These AI-driven tools have demonstrated tremendous promise in the realm of disease detection, prediction, and classification [20]. For example, state-of-the-art AI systems have been used to analyze mammograms, resulting in a significant decrease in both false positives and false negatives [21]. In contrast to their human counterparts, AI-driven diagnoses showed increased sensitivity in detecting breast cancer masses and outperformed radiologists in accurately identifying early-stage breast cancer [21]. In dermatology, AI, supported by CNN methods, has exhibited remarkable precision in diagnosing melanoma cases and suggesting appropriate treatment options [22, 23]. Furthermore, ML algorithms have been successfully applied to detect diabetic retinopathy, estimate cardiovascular risk factors, and examine chest radiography for early pneumonia detection [24–26]. The scope of AI's potential goes beyond image analysis, covering a range of medical conditions. A noteworthy study investigated the use of various ML techniques to achieve early detection of acute appendicitis, producing impressive accuracy in predicting the need for surgical intervention. Interestingly, similar techniques can be used to detect infections such as COVID-19, using blood samples or image analysis. Remarkably, analogous techniques may be employed to detect infections such as COVID-19, utilizing blood specimens or image analysis [27]. By using ML techniques, medical imaging modalities like X-rays, CT scans, and MRIs can be carefully examined to identify anomalies and accelerate accurate diagnoses. AI transformative impact also reaches the domain of clinical laboratory testing. AI has enormous potential to transform clinical microbiology by improving the accuracy, speed, and efficiency of laboratory procedures. ML systems have been created to detect and identify microorganisms, diagnose diseases, and forecast clinical outcomes, relying on diverse data sources such as genomic data, gene sequencing, metagenomic sequencing, and microscopic imaging [28, 29]. Furthermore, AI has demonstrated remarkable efficacy in automating blood cultures, susceptibility testing, and molecular platforms, leading to rapid and optimal selection of antibiotic treatments [29, 30]. In the context of emergency medical services, AI can perform related tasks with improved efficiency and heightened accuracy. AI algorithms can facilitate patient triage by prioritizing urgent cases, thereby minimizing waiting times and optimizing patient flow [31, 32]. AI-driven decision support systems can give healthcare professionals recommendations in real-time, assisting in diagnosis and treatment choices and reducing the possibility of diagnostic errors [33].

In breast imaging, digital mammography, as the primary screening modality, exhibits limited sensitivity in dense breasts and raises false positive rates. To address these limitations, computer-aided tools have been developed to assist radiologists in early cancer detection [34]. AI has gained significant traction in breast imaging, with approximately 14.8% of all AI applications in this domain [35]. Studies have demonstrated the efficacy of AI in improving radiologist performance in mammography interpretation. Chest imaging is a popular area for AI algorithms, with 54 FDA-approved algorithms [36]. Most algorithms focus on chest CT, particularly lung nodule detection [35]. Chest radiographs are also targeted, with 8 algorithms for pneumothorax detection and 3 for pleural effusion detection. Other uses include line and tube detection, pneumoperitoneum detection, lung nodule detection, and image quality improvement. AI has been shown to improve sensitivity for pneumonia, nodules, pneumothorax, and pleural effusion detection.

The field of neuroradiology faces misdiagnosis challenges, ranging from 5 to 51% for subarachnoid hemorrhages and 30–42% for ischemic stroke [37], making AI application in this field crucial. AI algorithms can significantly impact this issue, with over 70 FDA-approved AI algorithms in neuroradiology, including 47 for CT, 31 for MR, and 2 for PET [36]. Fasen et al. compared StrokeViewer's performance with radiologists, general radiologists, and senior residents, evaluating 474 patients with CT angiography, 75 of which had arterial occlusions. StrokeViewer demonstrated 77.3% sensitivity and 88.5% specificity, correctly identifying 40 of 42 Large Vessel Occlusions [38]. Despite the numerous AI algorithms developed for musculoskeletal radiology, only 22 have received FDA approval [39]. These algorithms primarily focus on fracture detection, osteoarthritis diagnosis, quantitative bone imaging, and image quality enhancement. Studies demonstrate promising results, with BoneView [40] outperforming radiologists in fracture detection on trauma X-rays and Rayvolve achieving a 96% sensitivity in fracture identification [41]. AI can improve radiology workflow by automating protocol creation, enhancing image quality, and reducing radiation dose. It can also optimize scanner usage and reduce patient waiting times. Additionally, AI can be used to triage patients by identifying exams with critical findings. However, the prioritization of studies from different modalities remains an unresolved issue [34].

4.7 Clinical Decision Support

AI has the potential to help healthcare providers in making better decisions, particularly in predicting the disease's method of treatment. In a study, patients' gene expression data was utilized in conjunction with machine learning techniques to accurately forecast chemotherapy response [42]. A study found that AI could predict the response with more than 80% accuracy for multiple drugs, which shows that AI can be useful in choosing the best treatment for each patient [43]. Another study used AI to analyze the health records of over 17,000 patients who took antidepressants and estimate how well they would respond to different types of antidepressants [44]. The study found that AI could accurately predict the response using real-world data, which suggests that AI can help doctors select the most effective antidepressant for each patient. These studies are examples of how AI can use different types of data, such as genes, health records, or images, to predict treatment outcomes.

4.8 Virtual Health Assistant

In the face of increasing healthcare demands and limited resources worldwide, finding solutions to overcome these challenges is crucial [45]. Virtual health assistants have emerged as an innovative technology that is transforming the healthcare industry by providing support to healthcare professionals. These assistants are designed to simulate human conversation and offer personalized patient care based on input from the patient [46]. Powered by AI applications, chatbots, audio, and user interfaces, these digital assistants can help patients with various tasks such as identifying underlying health issues based on symptoms, providing medical advice, medication reminders, scheduling doctor appointments, and monitoring vital signs. They can also collect daily health information and share reports with assigned physicians. By assuming some of these responsibilities, virtual assistants can alleviate the workload of human healthcare providers and enhance patient outcomes. Moreover, virtual health assistants offer the advantage of constant availability, making it easier for patients to access healthcare whenever needed [20]. Another valuable medical service provided by AI-driven mobile applications is triaging patients and assessing the urgency of their problems based on the previously mentioned symptoms. The National Health Service (NHS) has tested such an app in north London, and it is now being used by approximately 1.2 million people as an AI chatbot alternative to contacting the NHS non-emergency number [47]. Additionally, the introduction of intelligent speakers into the market holds significant benefits for elderly and chronically ill patients who may struggle with using smartphone apps efficiently. Overall, virtual health assistants have the potential to significantly enhance the quality, efficiency, and cost-effectiveness of healthcare delivery, while also increasing patient engagement and providing them with a better overall experience [20].

4.9 AI Assistant Model in Mental Health Care

AI-powered mental health applications can facilitate early detection and diagnosis of mental health conditions, providing tailored treatment and support [48, 49]. These applications also offer round-the-clock accessibility, alleviating the need for in-person appointments and minimizing waiting times. Moreover, digital tools can monitor patient progress and medication adherence, providing valuable insights into treatment efficacy. Depression stands out as the most extensively researched mental disorder in the context of AI applications for mental health. Additionally, AI-powered apps have demonstrated promising outcomes in supporting individuals with substance use disorders. A recent study evaluated the efficacy of a mental health digital app called Woebot in individuals with substance use disorders, revealing significant improvements in substance use, cravings, depression, and anxiety among those using Woebot [50]. Nevertheless, notable limitations warrant careful consideration within the domain of AI-powered mental health diagnosis. A significant limitation is the potential for bias inherent in the data and algorithms employed for AI-powered diagnosis. Improper representation of diverse populations in training data can lead to biased and inaccurate results. Additionally, AI-powered diagnosis may not adequately address the intricate nature of mental health conditions, which can manifest differently across individuals. Furthermore, there is a risk that AI-powered diagnosis may lack personalization and empathy, which are essential components of successful mental health treatment [51]. Therefore, despite its value as a tool in mental health care, AI-powered diagnosis should be utilized as an adjunct to professional diagnosis and treatment, not a replacement.

4.10 AI in Patient Education

Emerging applications of AI in patient education are gaining traction, particularly the utilization of AI-powered chatbots for various healthcare purposes, such as diet recommendations [52, 53], smoking cessation, and cognitive-behavioral therapy [54]. Patient education plays a pivotal role in healthcare, fostering comprehension of medical diagnoses, treatment options, and preventive measures [55]. Informed patients are more likely to adhere to treatment regimens and achieve favorable health outcomes [56]. AI's potential to significantly enhance patient education lies in its ability to provide personalized and interactive information and guidance to patients and their caregivers [57]. A notable example is the PROSCA (Prostate Cancer Communication Assistant) chatbot, which demonstrated a moderate to clear improvement in participants' knowledge about prostate cancer [58]. ChatGPT assists patients with diabetes by supporting their understanding of their diagnosis, treatment options, symptom monitoring, medication adherence, feedback, support, and query resolution [59]. AI technology can further be employed to adapt patient education materials to different reading levels, enabling patients to better grasp their diagnosis, treatment options, and self-care instructions [20]. However, addressing certain challenges and limitations associated with AI in patient education is crucial. These include ensuring the accuracy, reliability, and transparency of AI-provided information, safeguarding patient data privacy and confidentiality, and maintaining a human touch and empathetic communication approach [60]. While the application of AI in patient education is in its preliminary stages, its potential to revolutionize patient health knowledge acquisition is immense. As AI technology continues to evolve, we can anticipate even more innovative and effective approaches to leveraging AI for patient education.

4.11 Surgery Robot

Traditional mechanical robots have evolved to interpret the movements of a surgeon's hands, translating them into precise surgical actions through vibrating tools [61]. The next wave of surgical robots utilizes 3D digital segmentation and tailors human-initiated activities to customized surgical plans established before the procedure. This surge in intelligent robotic development results from the collaboration between surgical firms and tech giants, driven by the synergy of cloud computing, big data analytics, and AI. Deep learning, a pivotal force, has fueled the creation of intelligent surgical robots, necessitating detailed phases in processing and analyzing massive datasets [62].

The rise of robotic-assisted surgery, exemplified by a tenfold to fortyfold increase in typical general surgical procedures in 2017 compared to laparoscopic surgery, underscores the need for comprehensive training programs. Virtual simulation tools have facilitated training for robotic surgery, and this review delves into the validation processes for existing curricula and robotic-assisted surgery training simulators [63].

Robotic surgery has made significant progress in various medical specialties. In head and neck surgery, it capitalizes on specific anatomy and built-in entry points, revolutionizing minimally invasive concepts.

In pelvic organ prolapse treatment, robotic surgery has modified traditional approaches, offering minimally invasive solutions with reduced postoperative recovery. The advantages of robotic surgery are further exemplified in colorectal surgery, where it simplifies total mesorectal excision treatments for rectal cancer, maintaining the benefits of minimally invasive surgery. Laparoscopic surgery for colon cancer has gained recognition, but robotic surgery surpasses it in addressing anatomical restrictions, providing benefits like 3D imaging, dexterity, lack of vibration, and a quick learning curve. The adoption of robotic platforms continues to rise, with advancements expected to overcome technical restrictions [64, 65].

In thyroidectomy, robotic surgery exhibits advantages over open surgery in patient-reported outcomes, such as scar cosmesis and discomfort, despite drawbacks like extended operating times and costs. Robotic simulation and augmented reality contribute to surgical education, and robotic-assisted laparo-endoscopic single-site surgery is evolving to handle pathologic surgical anatomy [66, 67].

Robotic surgery's expansion into various medical fields, including pediatric urology and parathyroid surgery, showcases its versatility. However, challenges exist, necessitating the exploration of proposals for AI and machine learning to enhance robotic surgery's capabilities and address technical limitations [68, 69].

4.12 AI in Biomedical Research

4.12.1 Pharmaceutical Sciences and Drug Discovery

AI has a wide range of applications in the pharmaceutical industry, ranging from laboratory research to clinical applications. It facilitates rational drug design and assists in the identification of effective compounds by accelerating drug discovery methods. Its ability to analyze vast amounts of data and predict drug-protein interactions expedites drug development and reduces costs by integrating AI into medication manufacturing, allowing the pharmaceutical industry to achieve enhanced outcomes and reduce the high failure rate of clinical trials [70].

4.12.2 Pharmaceutical Industry

The pharmaceutical industry is rapidly adopting AI methods, particularly for analyzing biological data. The sheer volume of data generated in drug discovery and development is overwhelming, making AI-powered solutions highly sought after. Machine learning, deep learning, and AI in general, coupled with advancements in data processing and hardware capabilities, are poised to revolutionize drug discovery and development, moving towards algorithm-based approaches. Several biopharmaceutical companies have partnered with technology firms to develop platforms specifically designed for molecule discovery, with a focus on oncology and cardiovascular diseases. These collaborations allow pharmaceutical firms to leverage the expertise and innovative approaches of AI-powered drug discovery startups, accelerating the identification of promising therapeutic candidates [71].

The increasing complexity of pharmaceutical manufacturing necessitates the integration of human expertise with machine capabilities. AI tools hold immense potential to automate various pharmaceutical activities, not only addressing manufacturing challenges but also facilitating numerical simulations [71].

4.12.3 Quality Control and Quality Assurance

The adoption of AI in pharmaceutical quality assurance and control offers promising advancements over conventional band-based methods. The practical implementation of AI in all quality control checks may not yet be feasible due to the need for manual checks to ensure batch-to-batch homogeneity. An emerging solution, the Electronic Lab Notebook (ELN), combines automated data entry with AI algorithms, providing a promising platform for quality assurance. Embracing new technologies like intelligent quality control through expert systems, data mining, and knowledge discovery techniques could further revolutionize pharmaceutical quality assurance [71].

4.12.4 Clinical Trial Design

The clinical trial process, which aims to validate the safety and efficacy of a new drug, is a lengthy and costly undertaking, with only one in ten molecules successfully completing the process [72]. This high failure rate is attributed to various factors, including inadequate patient selection, technical limitations, and insufficient infrastructure. AI offers the potential to address these challenges and improve clinical trial outcomes. By analyzing patient-specific genome-exposome profiles, AI can identify individuals with the most relevant disease characteristics, facilitating the recruitment of a targeted patient population for Phase II and III trials [73]. Additionally, AI can be applied in the preclinical stage to predict the most promising lead compounds for clinical development, increasing the likelihood of success. AI-powered applications can effectively monitor patients' drug intake and provide personalized support, thereby enhancing adherence rates and contributing to trial success. This approach has been shown to increase adherence by 25%. In conclusion, the integration of AI into clinical trials holds significant promise for improving trial efficiency, reducing failure rates, and bringing promising new therapies to patients faster [71].

4.12.5 Precision and Genomic Medicine

Precision medicine is a movement towards changing healthcare through insights gained from DNA sequencing [74]. The healthcare landscape is transforming from community-oriented care (evidence-based medicine) to precision medicine. A big data analysis involves this transition, along with the evaluation of an individual genetics, environment, and lifestyle [75]. With the addition of medical history, behavioral factors, and environmental information, it is possible to accurately explain health conditions, disease states, and therapeutic options for affected individuals [76].

Precision medicine brings tangible advantages to healthcare, including early disease prediction and prevention [77]. Personalized treatment is possible by various data collection and analytics technologies. A novel application of high-throughput genotyping and the widespread use of electronic health records (EHRs) allows scientists to define new taxonomy from clinical and biomarker data [42]. The capacity to offer real-time recommendations relies on developing machine-learning algorithms that predict which patients may require a specific medication based on genomic information. The core of tailored treatments lies in genotyping those patients pre-emptively before the information becomes necessary [78].

Recently highlighted by Zou and colleagues, deep learning has been employed to integrate insights from scientific literature with sequencing results to suggest 3D protein configurations, pinpoint transcription start sites, simulate regulatory elements, and predict gene expression patterns from genotype data. These interpretations play a fundamental role in establishing connections between genomic variations and the manifestation of disease, the effectiveness of therapies, and prognosis [79].

The initial success of AI in image recognition has led to the emergence of radiogenomics. As an innovative field in precision medicine, radiogenomics aims to establish connections between cancer imaging characteristics and gene expression patterns. This effort aims to predict the likelihood of toxicity in patients undergoing radiotherapy. AI algorithms have been employed in various studies to analyze adverse events monitored at the bedside, along with other clinical parameters, aiming to predict organ dysfunction and failure [80]. Digital biomarkers are expected to enable remote disease monitoring beyond traditional hospital boundaries, supporting decentralized clinical trials.

4.12.6 AI for Public and Global Health

Many definitions of public health exist, but one that is commonly used is that it is "the science and art of preventing disease, prolonging life, and promoting health through the organized efforts and informed choices of society, organizations, public and private, communities, and individuals" [81]. Experiments with applicable AI solutions are now being conducted in a number of areas of public health. A few of these topics are covered in further detail below.

4.12.7 Health System Management

AI techniques provide comprehensive support for health services management, benefiting doctors, nurses, and administrators by offering continuous access to up-to-date medical information from various sources, including journals, textbooks, and clinical practices [82]. Particularly vital during the ongoing COVID-19 pandemic, these applications play a crucial role in ensuring a constant flow of information for effective global pandemic management. Beyond information access, AI applications contribute to coordinating patient information tools, facilitating health risk alerts, and predicting health outcomes [83].

In terms of healthcare efficiency, AI applications enable clinicians to access data immediately, ensuring enhanced patient safety during medication administration by nurses. Patients are also kept informed and engaged in their care through communication with medical teams during hospital stays. Additionally, AI plays a key role in optimizing logistics processes, such as the timely supply of drugs and equipment through just-in-time systems based on predictive algorithms. These applications also support the training of healthcare personnel, potentially bridging the gap between urban and rural health services [84, 85].

AI has a profound impact on public health and health management, particularly in the areas of early disease detection, predictive analysis, resource allocation, health monitoring, and health data management. In the next section, we will focus on the discussion of predictive analysis and risk assessment, which are key components of AI in healthcare.

4.12.8 Predictive Analysis

By analyzing real-time data, AI can accurately assess the probability of disease occurrence, progression, and response to the treatment. Large datasets are used to develop predictive models identifying individuals at increased risk of developing specific health conditions. These models, informed by machine learning algorithms and advanced data mining techniques, enable healthcare professionals to implement targeted preventive interventions, potentially delaying or preventing the onset of chronic diseases. By considering factors such as demographics, medical history, and lifestyle, predictive models can detect individuals with a higher likelihood of developing disease and enable targeted interventions for prevention and treatment [86, 87]. Another application of predictive analytics is in the prediction of hospital readmissions. Through the analysis of patient demographics, medical history, and social health factors, predictive models can identify patients at an elevated risk of readmission, allowing for targeted interventions to prevent such occurrences [88].

Additionally, AI tackles vaccine production and supply chain challenges, optimizing vaccine supply chains through robust algorithms and edge analytics, which detect anomalies and predicts events related to infectious diseases [89].

AI is revolutionizing the establishment of guidelines across various fields, including healthcare. Traditionally, creating guidelines involves a lengthy process, from identifying knowledge gaps to publishing and distributing the guidelines. AI can accelerate this process by identifying newly published data related to clinical trials and real-world patient outcomes in specific areas of interest. By mining this information, AI can assist in the initial stage of gathering relevant data. Subsequently, under the supervision of scientists and domain experts, AI algorithms can analyze vast amounts of data to uncover patterns and trends. This information can then inform the development of evidence-based guidelines in real-time, facilitating rapid information exchange with clinicians for critical evaluation of clinical and ethical implications [90, 91]. Professional organizations have developed frameworks to address unique concerns in developing, reporting, and validating AI in medicine. Regulatory oversight of AI in healthcare is still in its early stages. The US Food and Drug Administration (FDA) is working on guidelines to assess real-world applications of AI in medicine and has published a framework to guide the role of AI and machine learning in software as medical devices. The European Commission has led multidisciplinary efforts to enhance the credibility of AI, and the European Medicines Agency (EMA) has identified the regulation of AI as a strategic priority [92, 93]. These legislative efforts aim to shape the future of healthcare by ensuring the sector is well-prepared for technological advancements. In summary, AI plays a crucial role in establishing guidelines by providing data-driven insights and recommendations based on extensive information. This can lead to more efficient and effective decision-making, improved outcomes, and cost reduction. However, it is essential to ensure that AI-based guidelines are transparent, fair, unbiased, and informed by human expertise and ethical considerations [90].

4.12.9 Big Data in Healthcare

AI applications pose challenges and opportunities in handling patient data, particularly concerning the vast amount of medical big data generated from patients. These technologies assist medical researchers in managing data from clinical activities such as screening, diagnosis, and treatment assignment, enabling the identification of associations between subject features and outcomes of interest [94].

4.13 Challenges and Ethical Consideration of AI

Even though artificial intelligence is a tool that continually advances aspects of medical science, it also brings ethical challenges that must be acknowledged. While some of these challenges may improve over time as AI itself progresses, others necessitate examination and legislation by healthcare authorities.

AI in healthcare poses many ethical and social challenges, including responsibility for decisions, the possibility of errors, identity confirmation problems, and private information safety. The lack of enough information, particularly in minority groups such as patients with rare diseases in medical research, is a significant concern that needs to be mentioned. Keeping the safety, security, and transparency of patients' private data poses other challenges to the rational use of AI in health care.

The increasing integration of AI in healthcare necessitates a commitment to ethical accountability. Ensuring diversity and inclusivity in programming teams, along with regular algorithm audits, including their system implementation, is crucial. While AI cannot entirely replace clinical judgment, it serves as a valuable tool for clinicians in decision-making. As the implementation of AI in healthcare rises, effective governance becomes imperative to address regulatory, ethical, and trust issues. Governance at both the hospital and healthcare system levels is critical for patient safety, clinician confidence, and system accountability. The fast-paced development of AI necessitates national and international regulations to align with medical ethics principles. Initiatives such as the General Data Protection Regulation and the proposed Artificial Intelligence aim to regulate AI applications in healthcare, emphasizing pre-deployment conformity assessments and post-market analyses [95, 96].

Technical challenges in AI implementation involve simplicity in models for efficient healthcare professional operation (HCP). Issues such as lacking IT infrastructure, high costs, and data validity concerns pose hurdles. AI algorithms may suffer from inapplicability and bias. Overcoming these challenges requires healthcare providers to develop effective strategic plans addressing cost, technological infrastructure, and HCP utilization. Additionally, the need for more trust in AI-grounded clinical decision support systems necessitates the adoption of Explainable AI (XAI) solutions. Factors influencing clinicians' perceptions, including risks and trustworthiness of AI, workload, and willingness to receive AI training, should be considered [97, 98].

4.14 Data Accessibility

The Accessibility of a considerable quantity of data is the principle for the success of AI in healthcare systems. These data are applied to the system's subsequent training. Access to data from diverse database providers might lead to additional costs for a health institute or pharmaceutical company, and the data needs to be reliable and of high quality to ensure accurate result prediction. Several obstacles delay the complete integration of AI in health systems [99].

4.15 AI Accountability and User-Friendly Interface

One critical challenge is the black-box problem, where HCPs need more understanding of how AI arrives at results. Blaming healthcare practitioners for AI errors could delay the adoption. Therefore, policies and measures are recommended to protect doctors and AI healthcare systems. Ensuring a user-friendly AI interface, engaging all stakeholders, and understanding clinical needs are essential steps before the widespread adoption of AI in healthcare [100].

In conclusion, integrating AI in healthcare presents ethical, governance, and technical challenges that necessitate careful consideration and strategic planning. Ensuring adherence to medical-ethical principles, implementing effective governance structures, and addressing technical hurdles are vital. As AI continues to evolve, a collaborative approach involving healthcare professionals, regulators, and technology developers is crucial to control the full potential of AI while mitigating associated risks [101, 102].

4.16 Future Prospects and Development

Due to the recent technological advancements and remarkable progress in artificial intelligence and cutting-edge technologies, their utilization in the healthcare industry appears imperative. Beyond technical advancements, it is essential for companies, governments, and individuals involved in the field to study and optimize the use of these systems. Investing in this domain and facilitating the refinement of regulations governing the implementation of this technology would contribute significantly to the worthy progress of the healthcare industry. This perspective is particularly crucial in less developed and economically challenged countries, where the absence of suitable management structures and services makes it even more vital. Enabling the transfer of proposed models from developed countries to these regions using artificial intelligence tools and their algorithms holds particular significance. Potential funding sources, such as the European Next Generation EU program and national investment initiatives, could be explored to support this modernization [103].

A comparative analysis between countries actively participating in this research field and those not involved could help identify variables influencing the presence or absence of AI technologies in healthcare organizations [104].

A noticeable gap exists in quantitative analyses of the costs and profits associated with healthcare organizations implementing AI technologies in accounting, business, and management research. Research in this direction could enhance our understanding of the topic and increase accessibility to AI-based technologies for a broader spectrum of healthcare organizations. As discussed, interdisciplinary studies are crucial to reinforce the connection between AI, data quality management, and ethical considerations in healthcare [105].

As we envision the future of AI in medicine, collaboration among healthcare professionals, patients, computer scientists, pharmaceutical companies, and the public emerges as a cornerstone for the advancement and responsible deployment of this transformative technology. This collaborative synergy harnesses the expertise of healthcare providers, amplifying the impact of AI applications, particularly in specialized domains like radiology.

Policymakers and regulatory authorities play a pivotal role in shaping the trajectory of AI in healthcare by establishing comprehensive guidelines and regulations. By mandating the disclosure of methodologies and performance metrics, these entities promote accountability and ensure that AI applications are developed and deployed in a manner that upholds ethical principles. The allocation of resources towards AI innovation and the resolution of fairness and equity issues contributes to the evolution of a healthcare system that is more inclusive, minimizing biases and leveraging AI to improve the well-being of all patients.

4.17 Conclusion

The healthcare landscape increasingly recognizes the significance of AI-driven tools in advancing next-generation healthcare technology. It is widely acknowledged that AI has the potential to enhance various aspects of healthcare operations and service delivery. The incorporation of AI into virtual health and mental health support has shown promise in terms of improving patient care. Notably, one compelling factor driving the integration of AI applications is the expected cost savings within the healthcare system. As we are at the convergence of AI and medicine, the future promises a healthcare environment with remarkable precision, efficiency, and creativity. Collaboration among healthcare professionals, technologists, and politicians is critical to realizing this promise and assuring ethical considerations and equal access to the benefits of AI in medicine.

References

1. Xu Y, Liu X, Cao X, Huang C, Liu E, Qian S, et al. Artificial intelligence: a powerful paradigm for scientific research. Innovations. 2021;2(4):100179.
2. Shukla Shubhendu S, Vijay J. Applicability of artificial intelligence in different fields of life. Int J Sci Eng Res. 2013;1(1):28–35.
3. McCarthy J. What is artificial intelligence? 2004.
4. Lupton M. Some ethical and legal consequences of the application of artificial intelligence in the field of medicine. Trends Med. 2018;18(4):100147.
5. Alpaydin E. Introduction to machine learning. Cambridge: MIT Press; 2020.
6. Erickson BJ, Korfiatis P, Akkus Z, Kline TL. Machine learning for medical imaging. Radiographics. 2017;37(2):505–15.
7. Davenport T, Kalakota R. The potential for artificial intelligence in healthcare. Future Healthc J. 2019;6(2):94.
8. Misawa M, Kudo S-e, Mori Y, Cho T, Kataoka S, Yamauchi A, et al. Artificial intelligence-assisted polyp detection for colonoscopy: initial experience. Gastroenterology. 2018;154(8):2027–9.
9. Murdoch TB, Detsky AS. The inevitable application of big data to health care. JAMA. 2013;309(13):1351–2.
10. Wang F, Preininger A. AI in health: state of the art, challenges, and future directions. Yearb Med Inform. 2019;28(1):016–26.
11. Beneke F, Mackenrodt M-O. Artificial intelligence and collusion. Int Rev Intell Property Compet Law. 2019;50:109–34.
12. Steels L, Brooks R. The artificial life route to artificial intelligence: building embodied, situated agents. London: Routledge; 2018.
13. Bielecki A, Bielecki A. Foundations of artificial neural networks. In: Models of neurons and perceptrons: selected problems and challenges. Cham: Springer; 2019. p. 15–28.
14. Danilevsky M, Qian K, Aharonov R, Katsis Y, Kawas B, Sen P. A survey of the state of explainable AI for natural language processing. arXiv preprint arXiv:201000711. 2020.
15. Kakani V, Nguyen VH, Kumar BP, Kim H, Pasupuleti VR. A critical review on computer vision and artificial intelligence in food industry. J Agric Food Res. 2020;2:100033.
16. Bini SA. Artificial intelligence, machine learning, deep learning, and cognitive computing: what do these terms mean and how will they impact health care? J Arthroplast. 2018;33(8):2358–61.
17. Roski J, Maier EJ, Vigilante K, Kane EA, Matheny ME. Enhancing trust in AI through industry self-governance. J Am Med Inform Assoc. 2021;28(7):1582–90.
18. Gómez-González E, Gomez E, Márquez-Rivas J, Guerrero-Claro M, Fernández-Lizaranzu I, Relimpio-López MI, et al. Artificial intelligence in medicine and healthcare: a review and classification of current and near-future applications and their ethical and social Impact. arXiv preprint arXiv:200109778. 2020.
19. Topol EJ. High-performance medicine: the convergence of human and artificial intelligence. Nat Med. 2019;25(1):44–56.
20. Alowais SA, Alghamdi SS, Alsuhebany N, Alqahtani T, Alshaya AI, Almohareb SN, et al. Revolutionizing healthcare: the role of artificial intelligence in clinical practice. BMC Med Educ. 2023;23(1):689.
21. McKinney SM, Sieniek M, Godbole V, Godwin J, Antropova N, Ashrafian H, et al. International evaluation of an AI system for breast cancer screening. Nature. 2020;577(7788):89–94.
22. Haenssle HA, Fink C, Schneiderbauer R, Toberer F, Buhl T, Blum A, et al. Man against machine: diagnostic performance of a deep learning convolutional neural network for dermoscopic melanoma recognition in comparison to 58 dermatologists. Ann Oncol. 2018;29(8):1836–42.
23. Han SS, Park I, Chang SE, Lim W, Kim MS, Park GH, et al. Augmented intelligence dermatology: deep neural networks empower medical professionals in diagnosing skin cancer and predicting treatment options for 134 skin disorders. J Invest Dermatol. 2020;140(9):1753–61.
24. Li S, Zhao R, Zou H. Artificial intelligence for diabetic retinopathy. Chin Med J. 2022;135(3):253–60.
25. Becker J, Decker JA, Römmele C, Kahn M, Messmann H, Wehler M, et al. Artificial intelligence-based detection of pneumonia in chest radiographs. Diagnostics. 2022;12(6):1465.
26. Raghunath S, Pfeifer JM, Ulloa-Cerna AE, Nemani A, Carbonati T, Jing L, et al. Deep neural networks can predict new-onset atrial fibrillation from the 12-lead ECG and help identify those at risk of atrial fibrillation–related stroke. Circulation. 2021;143(13):1287–98.
27. Mijwil MM, Aggarwal K. A diagnostic testing for people with appendicitis using machine learning techniques. Multimed Tools Appl. 2022;81(5):7011–23.
28. Smith KP, Kang AD, Kirby JE. Automated interpretation of blood culture gram stains by use of a deep convolutional neural network. J Clin Microbiol. 2018;56(3):e01521.
29. Peiffer-Smadja N, Dellière S, Rodriguez C, Birgand G, Lescure F-X, Fourati S, Ruppé E. Machine learning in the clinical microbiology laboratory: has the time come for routine practice? Clin Microbiol Infect. 2020;26(10):1300–9.
30. Smith KP, Kirby JE. Image analysis and artificial intelligence in infectious disease diagnostics. Clin Microbiol Infect. 2020;26(10):1318–23.
31. Matheny ME, Whicher D, Israni ST. Artificial intelligence in health care: a report from the National Academy of Medicine. JAMA. 2020;323(6):509–10.
32. Berlyand Y, Raja AS, Dorner SC, Prabhakar AM, Sonis JD, Gottumukkala RV, et al. How artificial

intelligence could transform emergency department operations. Am J Emerg Med. 2018;36(8):1515–7.
33. Hautz WE, Kämmer JE, Hautz SC, Sauter TC, Zwaan L, Exadaktylos AK, et al. Diagnostic error increases mortality and length of hospital stay in patients presenting through the emergency room. Scand J Trauma Resusc Emerg Med. 2019;27(1):1–12.
34. Mello-Thoms C, Mello CA. Clinical applications of artificial intelligence in radiology. Br J Radiol. 2023;96:20221031.
35. Tadavarthi Y, Vey B, Krupinski E, Prater A, Gichoya J, Safdar N, Trivedi H. The state of radiology AI: considerations for purchase decisions and current market offerings. Radiol Artif Intell. 2020;2(6):e200004.
36. Allen B, Agarwal S, Coombs L, Wald C, Dreyer K. 2020 ACR data science institute artificial intelligence survey. J Am Coll Radiol. 2021;18(8):1153–9.
37. Bakradze E, Liberman AL. Diagnostic error in stroke—reasons and proposed solutions. Curr Atheroscler Rep. 2018;20:1–13.
38. Fasen BA, Berendsen RC, Kwee RM. Artificial intelligence software for diagnosing intracranial arterial occlusion in patients with acute ischemic stroke. Neuroradiology. 2022;64(8):1579–83.
39. Allen B, Dreyer K, Stibolt R Jr, Agarwal S, Coombs L, Treml C, et al. Evaluation and real-world performance monitoring of artificial intelligence models in clinical practice: try it, buy it, check it. J Am Coll Radiol. 2021;18(11):1489–96.
40. Regnard N-E, Lanseur B, Ventre J, Ducarouge A, Clovis L, Lassalle L, et al. Assessment of performances of a deep learning algorithm for the detection of limbs and pelvic fractures, dislocations, focal bone lesions, and elbow effusions on trauma X-rays. Eur J Radiol. 2022;154:110447.
41. Reichert G, Bellamine A, Fontaine M, Naipeanu B, Altar A, Mejean E, et al. How can a deep learning algorithm improve fracture detection on X-rays in the emergency room? J Imaging. 2021;7(7):105.
42. Johnson KB, Wei WQ, Weeraratne D, Frisse ME, Misulis K, Rhee K, et al. Precision medicine, AI, and the future of personalized health care. Clin Transl Sci. 2021;14(1):86–93.
43. Huang C, Clayton EA, Matyunina LV, McDonald LD, Benigno BB, Vannberg F, McDonald JF. Machine learning predicts individual cancer patient responses to therapeutic drugs with high accuracy. Sci Rep. 2018;8(1):16444.
44. Sheu Y-h, Magdamo C, Miller M, Das S, Blacker D, Smoller JW. AI-assisted prediction of differential response to antidepressant classes using electronic health records. NPJ Dig Med. 2023;6(1):73.
45. Buch VH, Ahmed I, Maruthappu M. Artificial intelligence in medicine: current trends and future possibilities. Br J Gen Pract. 2018;68(668):143–4.
46. Curtis RG, Bartel B, Ferguson T, Blake HT, Northcott C, Virgara R, Maher CA. Improving user experience of virtual health assistants: scoping review. J Med Internet Res. 2021;23(12):e31737.
47. Burgess M. The NHS is trialling an AI chatbot to answer your medical questions. 2017. http://www.wired.co.uk/article/babylon-nhs-chatbot-app.
48. Luxton DD. Artificial intelligence in psychological practice: current and future applications and implications. Prof Psychol Res Pract. 2014;45(5):332.
49. Graham S, Depp C, Lee EE, Nebeker C, Tu X, Kim H-C, Jeste DV. Artificial intelligence for mental health and mental illnesses: an overview. Curr Psychiatry Rep. 2019;21:1–18.
50. Prochaska JJ, Vogel EA, Chieng A, Kendra M, Baiocchi M, Pajarito S, Robinson A. A therapeutic relational agent for reducing problematic substance use (Woebot): development and usability study. J Med Internet Res. 2021;23(3):e24850.
51. Lee EE, Torous J, De Choudhury M, Depp CA, Graham SA, Kim H-C, et al. Artificial intelligence for mental health care: clinical applications, barriers, facilitators, and artificial wisdom. Biol Psychiatry. 2021;6(9):856–64.
52. Zhang J, Oh YJ, Lange P, Yu Z, Fukuoka Y. Artificial intelligence chatbot behavior change model for designing artificial intelligence chatbots to promote physical activity and a healthy diet. J Med Internet Res. 2020;22(9):e22845.
53. Chew HSJ. The use of artificial intelligence–based conversational agents (Chatbots) for weight loss: scoping review and practical recommendations. JMIR Med Inform. 2022;10(4):e32578.
54. Wang H, Zhang Q, Ip M, Lau JTF. Social media–based conversational agents for health management and interventions. Computer. 2018;51(8):26–33.
55. Bombard Y, Baker GR, Orlando E, Fancott C, Bhatia P, Casalino S, et al. Engaging patients to improve quality of care: a systematic review. Implement Sci. 2018;13:1–22.
56. Wong CK, Yeung DY, Ho HC, Tse K-P, Lam C-Y. Chinese older adults' internet use for health information. J Appl Gerontol. 2014;33(3):316–35.
57. Aggarwal A, Tam CC, Wu D, Li X, Qiao S. Artificial intelligence–based chatbots for promoting health behavioral changes: systematic review. J Med Internet Res. 2023;25:e40789.
58. Görtz M, Baumgärtner K, Schmid T, Muschko M, Woessner P, Gerlach A, et al. An artificial intelligence-based chatbot for prostate cancer education: design and patient evaluation study. Dig Health. 2023;9:20552076231173304.
59. Nakhleh A, Spitzer S, Shehadeh N. ChatGPT's response to the diabetes knowledge questionnaire: implications for diabetes education. Diabetes Technol Therap. 2023;25(8):571–3.
60. Lee D, Yoon SN. Application of artificial intelligence-based technologies in the healthcare industry: opportunities and challenges. Int J Environ Res Public Health. 2021;18(1):271.
61. Bramhe S, Pathak SS. Robotic surgery: a narrative review. Cureus. 2022;14(9):e29179.

62. Bhandari M, Zeffiro T, Reddiboina M. Artificial intelligence and robotic surgery: current perspective and future directions. Curr Opin Urol. 2020;30(1):48–54.
63. Chen R, Rodrigues Armijo P, Krause C, Force SRT, Siu K-C, Oleynikov D. A comprehensive review of robotic surgery curriculum and training for residents, fellows, and postgraduate surgical education. Surg Endosc. 2020;34:361–7.
64. Nozawa H, Watanabe T. Robotic surgery for rectal cancer. Asian J Endosc Surg. 2017;10(4):364–71.
65. Pai A, Marecik S, Park J, Prasad L. Robotic colorectal surgery for neoplasia. Surg Clin. 2017;97(3):561–72.
66. Gettman M, Rivera M. Innovations in robotic surgery. Curr Opin Urol. 2016;26(3):271–6.
67. Aidan P, Arora A, Lorincz B, Tolley N, Garas G. Robotic thyroid surgery: current perspectives and future considerations. ORL. 2018;80(3-4):186–94.
68. Mizuno K, Kojima Y, Nishio H, Hoshi S, Sato Y, Hayashi Y. Robotic surgery in pediatric urology: current status. Asian J Endosc Surg. 2018;11(4):308–17.
69. O'Sullivan S, Nevejans N, Allen C, Blyth A, Leonard S, Pagallo U, et al. Legal, regulatory, and ethical frameworks for development of standards in artificial intelligence (AI) and autonomous robotic surgery. Int J Med Robot Comp Assisted Surg. 2019;15(1):e1968.
70. Paul D, Sanap G, Shenoy S, Kalyane D, Kalia K, Tekade RK. Artificial intelligence in drug discovery and development. Drug Discov Today. 2021;26(1):80–93.
71. Mak K-K, Pichika MR. Artificial intelligence in drug development: present status and future prospects. Drug Discov Today. 2019;24(3):773–80.
72. Hay M, Thomas DW, Craighead JL, Economides C, Rosenthal J. Clinical development success rates for investigational drugs. Nat Biotechnol. 2014;32(1):40–51.
73. Harrer S, Shah P, Antony B, Hu J. Artificial intelligence for clinical trial design. Trends Pharmacol Sci. 2019;40(8):577–91.
74. Council NR. Toward precision medicine: building a knowledge network for biomedical research and a new taxonomy of disease. Washington: National Academies Press; 2011.
75. Mesko B. The role of artificial intelligence in precision medicine. Expert Rev Precis Med Drug Dev. 2017;2(5):239–41.
76. Ziegelstein RC. Personomics and precision medicine. Trans Am Clin Climatol Assoc. 2017;128:160.
77. Hartmaier RJ, Albacker LA, Chmielecki J, Bailey M, He J, Goldberg ME, et al. High-throughput genomic profiling of adult solid tumors reveals novel insights into cancer pathogenesis. Cancer Res. 2017;77(9):2464–75.
78. Schildcrout JS, Shi Y, Danciu I, Bowton E, Field JR, Pulley JM, et al. A prognostic model based on readily available clinical data enriched a pre-emptive pharmacogenetic testing program. J Clin Epidemiol. 2016;72:107–15.
79. Zou J, Huss M, Abid A, Mohammadi P, Torkamani A, Telenti A. A primer on deep learning in genomics. Nat Genet. 2019;51(1):12–8.
80. Qiu Q, Nian Y-j, Guo Y, Tang L, Lu N, Wen L-z, et al. Development and validation of three machine-learning models for predicting multiple organ failure in moderately severe and severe acute pancreatitis. BMC Gastroenterol. 2019;19:1–9.
81. Griffiths S, Jewell T, Donnelly P. Public health in practice: the three domains of public health. Public Health. 2005;119(10):907–13.
82. Tran BX, Vu GT, Ha GH, Vuong Q-H, Ho M-T, Vuong T-T, et al. Global evolution of research in artificial intelligence in health and medicine: a bibliometric study. J Clin Med. 2019;8(3):360.
83. Agrawal A, Gans JS, Goldfarb A. Exploring the impact of artificial intelligence: prediction versus judgment. Inf Econ Policy. 2019;47:1–6.
84. Guo J, Li B. The application of medical artificial intelligence technology in rural areas of developing countries. Health Equity. 2018;2(1):174–81.
85. Chakradhar S. Predictable response: finding optimal drugs and doses using artificial intelligence. Nat Med. 2017;23(11):1244–8.
86. Nelson KM, Chang ET, Zulman DM, Rubenstein LV, Kirkland FD, Fihn SD. Using predictive analytics to guide patient care and research in a national health system. J Gen Intern Med. 2019;34:1379–80.
87. Amarasingham R, Patzer RE, Huesch M, Nguyen NQ, Xie B. Implementing electronic health care predictive analytics: considerations and challenges. Health Aff. 2014;33(7):1148–54.
88. Donzé J, Aujesky D, Williams D, Schnipper JL. Potentially avoidable 30-day hospital readmissions in medical patients: derivation and validation of a prediction model. JAMA Intern Med. 2013;173(8):632–8.
89. Alotaibi S, Mehmood R, Katib I, Rana O, Albeshri A. Sehaa: a big data analytics tool for healthcare symptoms and diseases detection using Twitter, Apache Spark, and Machine Learning. Appl Sci. 2020;10(4):1398.
90. Rivera S, Liu X, Chan A, Denniston A, Calvert M. Guidelines for clinical trial protocols for interventions involving artificial intelligence: the spirit-ai extension. BMJ. 2020;370:3210.
91. Administration. Artificial intelligence and machine learning in software as a medical device. 2021. https://www.fda.gov/medical-devices/software-medical-device-samd/artificial-intelligence-and-machine-learning-software-medical-device.
92. Radanliev P, De Roure D. Disease X vaccine production and supply chains: risk assessing healthcare systems operating with artificial intelligence and industry 4.0. Heal Technol. 2023;13(1):11–5.

93. Nikolinakos NT. A European approach to excellence and trust: the 2020 white paper on artificial intelligence. In: EU policy and legal framework for artificial intelligence, robotics and related technologies-the AI act. Cham: Springer; 2023. p. 211–80.
94. Jiang F, Jiang Y, Zhi H, Dong Y, Li H, Ma S, et al. Artificial intelligence in healthcare: past, present and future. Stroke Vasc Neurol. 2017;2(4):101.
95. Guidance W. Ethics and governance of artificial intelligence for health. Geneva: World Health Organization; 2021.
96. Reddy S, Allan S, Coghlan S, Cooper P. A governance model for the application of AI in health care. J Am Med Inform Assoc. 2020;27(3):491–7.
97. Kelly CJ, Karthikesalingam A, Suleyman M, Corrado G, King D. Key challenges for delivering clinical impact with artificial intelligence. BMC Med. 2019;17:1–9.
98. Tachkov K, Zemplenyi A, Kamusheva M, Dimitrova M, Siirtola P, Pontén J, et al. Barriers to use artificial intelligence methodologies in health technology assessment in central and east European countries. Front Public Health. 2022;10:921226.
99. Lamberti MJ, Wilkinson M, Donzanti BA, Wohlhieter GE, Parikh S, Wilkins RG, Getz K. A study on the application and use of artificial intelligence to support drug development. Clin Ther. 2019;41(8):1414–26.
100. Al Kuwaiti A, Nazer K, Al-Reedy A, Al-Shehri S, Al-Muhanna A, Subbarayalu AV, et al. A review of the role of artificial intelligence in healthcare. J Personal Med. 2023;13(6):951.
101. Tiwari R, Explainable AI. (XAI) and its applications in building trust and understanding in AI decision making. Int J Sci Res Eng Manage. 2023;7:1–13.
102. Clement T, Kemmerzell N, Abdelaal M, Amberg M. XAIR: a systematic metareview of explainable AI (XAI) aligned to the software development process. Mach Learn Knowl Extract. 2023;5(1):78–108.
103. Picek O. Spillover effects from next generation EU. Intereconomics. 2020;55(5):325–31.
104. Secinaro S, Calandra D, Secinaro A, Muthurangu V, Biancone P. The role of artificial intelligence in healthcare: a structured literature review. BMC Med Inform Decis Mak. 2021;21:1–23.
105. Sousa M, Dal Mas F, Pesqueira A, Lemos C, Verde JM, Cobianchi L. The potential of AI in health higher education to increase the students' learning outcomes. TEM J. 2021;2:488–97.

Genetic Health

5

Marius Geanta, Adriana Boata, Bianca Cucos, Cosmina Cioroboiu, and Ramona Popescu

What Will You Learn in This Chapter?
In this chapter we discuss the transformative role of genomic technologies in understanding health and disease. It begins by tracing the evolution of genetic technologies, from the discovery of DNA's structure to the completion of the Human Genome Project and beyond, highlighting how these advances have reshaped our understanding of what constitutes health and disease. It expands on how personalized medicine, powered by genomic insights, is redefining healthcare practices. We explore the multifaceted impact of genomics on disease prevention, diagnosis, and treatment, emphasizing the advent of new screening technologies, the importance of personalized prevention, and the revolutionary potential of consumer-directed genetic tests. Additionally, the chapter addresses the societal revolution spurred by the integration of genomics into public health, underscoring the emergence of new disciplines that blends genetic science with public health strategies.

Rationale and Importance
The rationale behind this chapter is to provide a comprehensive understanding of how genomic information is being integrated into healthcare and public health, marking a shift towards more personalized, predictive, and preventive medicine. The importance of this chapter lies in its ability to illuminate the profound implications of genomic technologies for individual and population health. It emphasizes the need for a new ethical and regulatory framework to navigate the challenges posed by this integration, including issues of privacy, consent, and equity. By exploring the potential of genomics to revolutionize health, the chapter aims to inform and inspire a broad audience about the present and future of

M. Geanta (✉)
Centre for Innovation in Medicine, Bucharest, Romania

KOL Medical Media, Bucharest, Romania

United Nations University-Maastricht Economic and Social Research Institute on Innovation and Technology, Maastricht, The Netherlands
e-mail: marius.geanta@ino-med.ro

A. Boata
Centre for Innovation in Medicine, Bucharest, Romania

Hesy Biotech Consulting, Bucharest, Romania

B. Cucos
Centre for Innovation in Medicine, Bucharest, Romania

Genomed Consulting, Bucharest, Romania

C. Cioroboiu
Centre for Innovation in Medicine, Bucharest, Romania

KOL Medical Media, Bucharest, Romania

R. Popescu
Centre for Innovation in Medicine, Bucharest, Romania

M. Hasanzad (ed.), *A Glimpse at Medicine in the Future*,
https://doi.org/10.1007/978-981-97-4974-4_5

medicine and public health, highlighting the opportunities and responsibilities of these technological advances.

5.1 Introduction: Unraveling the Code of Life

In the history of scientific discovery, few moments stand as pivotal as the unveiling of the structure of DNA. This revelation, which occurred in 1953 by James Watson and Francis Crick, marked the beginning of a new era in biology and medicine. DNA, or deoxyribonucleic acid, is the molecule that carries the genetic instructions used in the growth, development, functioning, and reproduction of all known living organisms. With its elegant yet simple structure, Watson and Crick's discovery of the DNA double helix provided the first clues into the complex world of genetics and heredity. This foundational understanding set the stage for decades of research that would further unravel the genetic code—the sequence of nucleotides that spells out the instructions for life [1].

The genetic code is essentially a biological language with just four letters: A (adenine), T (thymine), C (cytosine), and G (guanine). These letters form words, or codons, each consisting of three nucleotides. Every group of three nucleotides in a DNA sequence specifies a particular amino acid (the building block of proteins) in a way that's remarkably consistent across organisms. This universality underscores a profound truth about the interconnectedness of life on Earth. The elucidation of the genetic code in the 1960s was a monumental achievement, revealing how sequences of DNA are translated into the proteins that perform critical functions in our cells [2].

However, understanding the letters and words was just the beginning. The true complexity and dynamism of genetics were laid bare by the Human Genome Project (HGP), an international scientific research project with the goal of determining the sequence of the human genome and identifying the genes that it contains. Completed in 2003, the HGP provided a comprehensive catalog of the human genome's approximately 3 billion base pairs. This monumental achievement has often been compared to landing on the moon in its scope and impact on humanity [3].

The HGP has reshaped our understanding of health and disease at the most fundamental level. It has revealed that humans are 99.9% identical at the genetic level, yet the 0.1% variation is profoundly significant. This small genetic variation contributes to the diverse tapestry of human phenotypes, influencing everything from eye color and height to predispositions for diseases like cancer, diabetes, and heart disease [4].

The findings from the HGP have also illuminated the concept of polygenic traits—characteristics influenced by multiple genes—and the intricate ways in which genes interact with environmental factors. Diseases once thought to be caused by single genetic mutations are now understood to be far more complex, often involving multiple genetic and environmental factors. This has profound implications for medicine and public health, shifting the paradigm from a one-size-fits-all approach to a more personalized strategy. Understanding an individual's genetic makeup can inform more precise diagnoses, predict susceptibilities to diseases, and tailor treatments to the unique genetic profile of each patient, a practice known as personalized or precision medicine [5].

Moreover, the HGP has spurred the development of new technologies and fields of study. High-throughput sequencing technologies, which allow scientists to sequence DNA at unprecedented speed and cost-effectiveness, are a direct outcome of the HGP. These technologies have democratized genetic research, making it possible to conduct large-scale studies that link genetic variations with diseases and traits. They have also enabled the study of genomics in more diverse populations, which is critical for understanding the full spectrum of human genetic diversity and for ensuring that the benefits of genomic medicine are accessible to all [6].

Genetics and genomics are closely related fields that are often used interchangeably, but they have distinct differences. Genetics is the study of individual genes and their roles in inheritance—how traits or conditions are passed down from generation to generation. It focuses on the function and composition of a single gene and how the gene's hereditary information is encoded, mutated, and expressed. Genomics, on the other hand, is a more recent field that involves the comprehensive study of the complete genetic makeup of organisms—the entire genome. Genomics examines the interactions, functions, and structure of all the genes in the genome collectively and how they interact with each other and with the environment. While genetics aims to understand the impact of a single gene on a trait or disease, genomics looks at all genetic information to provide a holistic view of how complex biological systems interact and influence an organism's growth, development, and health [7].

The evolution of genomics has significantly revolutionized the landscape of modern medicine, particularly in fields such as cancer care and cardiovascular disease, offering new strategies to reduce the burden of chronic diseases. In cancer care, genomics has paved the way for personalized medicine, allowing treatments to be specifically tailored to individual genetic profiles, thus enhancing their effectiveness and minimizing side effects. This approach is exemplified in the development of targeted therapies, like trastuzumab for certain breast cancers, which specifically counteract the effects of genetic mutations found in cancer cells. Furthermore, genomics facilitates early detection and prevention, enabling healthcare providers to identify and intervene in high-risk individuals before the onset of disease symptoms.

In the realm of cardiovascular disease, genomics has been instrumental in creating genetic risk profiles that identify individuals with a heightened risk, leading to early and more targeted preventive measures [8]. It has also deepened the understanding of the mechanisms behind cardiovascular conditions, such as hypertension and hyperlipidemia, fostering the development of more precise treatments. Pharmacogenomics, a subset of genomics, has further refined the approach to cardiovascular disease by tailoring drug therapy based on an individual's genetic makeup, thus optimizing treatment efficacy and safety.

Beyond individual patient care, the impact of genomics extends to public health strategies. Identifying genetic risk factors at the population level enables the development of targeted public health campaigns and policies aimed at reducing the incidence of chronic diseases. The predictive power of genomic information is a cornerstone in this effort, allowing for early identification and intervention in individuals predisposed to chronic diseases, potentially preventing their onset and easing the overall healthcare burden [9].

Overall, the integration of genomics into medical practice signifies a shift towards more personalized, predictive, and preventive healthcare. This shift not only enhances the management of chronic diseases like cancer and cardiovascular disease but also promises to transform the future of healthcare, making it more precise, effective, and efficient. As genomic technologies continue to evolve, they hold the potential for even greater advancements in medical science and public health [10].

5.2 The New Frontier of Screening Technologies

The advent of genomic integration into medical screening technologies has marked a turning point in the early detection and management of diseases [11].

The advent of new genetic technologies has ushered in a transformative era in disease screening, offering remarkable advantages over traditional methods. These advancements are not only refining current screening programs but also paving the way for innovative screening strategies. Central to this revolution is the growing field of whole genome sequencing (WGS), a technology that has the potential to redefine our approach to health and disease management. By decoding an individual's entire genetic makeup, WGS provides a comprehensive view of the genetic factors that may contribute to disease risk. This holistic approach contrasts with earlier methods that typically targeted specific genes or markers associated with particular conditions. The ability to analyze the entire genome from birth presents a groundbreaking opportunity: diseases can be predicted and potentially prevented before they manifest, enabling a shift from reactive to proactive healthcare [12].

The implementation of whole genome sequencing and similar initiatives marks the dawn of a modern screening era. This new paradigm of genetic screening offers the prospect of more personalized, precise, and predictive healthcare. For instance, newborn screening programs, traditionally reliant on biochemical markers for a limited number of conditions, could be significantly expanded with WGS, encompassing a broader spectrum of genetic diseases. This approach not only facilitates early intervention but also tailors preventive measures and treatments to the individual's genetic profile. In adults, similar sequencing initiatives could revolutionize cancer screening, allowing for the detection of genetic predispositions to various cancers and enabling earlier and more effective interventions. The impact of these technologies extends beyond individual patient care; they promise to enhance public health strategies by enabling the early detection of diseases at a population level. As these genomic technologies continue to advance and become more accessible, they are poised to fundamentally alter the landscape of disease screening and prevention, heralding a new era of personalized and precision medicine [5, 10].

The applications of genetic screening are diverse and transformative across various medical fields. In prenatal care, noninvasive prenatal testing (NIPT) analyzes cell-free fetal DNA in a pregnant woman's blood to detect chromosomal abnormalities such as Down syndrome. This method offers a safer alternative to invasive tests like amniocentesis, significantly reducing the risk to the fetus [13].

Recently reported in the New England Journal of Medicine, desNIPT is a next-generation blood test for prenatal genetic screening, developed by the University of Southern Denmark, as a noninvasive method that allows for the analysis of the exome (the part of DNA that encodes proteins) with an accuracy as high as genetic testing of samples obtained through amniocentesis or chorionic villus sampling (CVS). The study reports a 100% concordance with the results obtained through invasive methods. The results show that the test allows for a comprehensive analysis of fetal genes using a blood sample from the mother [14].

In oncology, among the most groundbreaking advancements are pan-cancer screening tests, which epitomize the potential of genomics to revolutionize cancer diagnosis and treatment. Pan-cancer screening tests, such as those developed by Grail, are designed to detect a wide array of cancers at an early stage, often before symptoms appear. These tests utilize sophisticated genomic analysis techniques to identify tiny fragments of cancer DNA circulating in the blood. Grail's tests, for instance, leverage deep sequencing methods to analyze cell-free DNA for specific patterns indicative of cancer. This approach represents a paradigm shift in cancer screening, moving from a focus on specific cancer types to a broader, more holistic view that can potentially detect multiple cancer types simultaneously. The implications of this are profound, offering the possibility of one test that screens for a multitude of cancers, thereby simplifying screening protocols and potentially increasing early detection rates [15].

When it comes to infectious diseases, genomic screening technologies are being used to rapidly identify pathogens, understand their resistance profiles, and track disease outbreaks. This was notably demonstrated during the COVID-19 pandemic, where genomic sequencing played a crucial role in identifying and monitoring the spread of the virus and its variants [16, 17].

Genetic screening is also instrumental in pharmacogenomics, which predicts an individual's response to drugs. This approach can identify the most effective medications while minimizing adverse drug reactions, leading to safer and more effective treatments [18].

Cardiovascular diseases, too, benefit from genomic screening. Genetic tests can identify individuals at high risk for conditions like familial hypercholesterolemia, enabling early intervention and management to prevent more serious cardiac events [8].

The impact of these new screening technologies extends beyond individual patient care to public health at large. By enabling earlier detection and more precise targeting of interventions, they promise to reduce the overall disease burden. Moreover, as these technologies become more accessible and cost-effective, they have the potential to democratize health care, making early detection and personalized medicine a reality for a broader segment of the population [19].

However, with these advances come new challenges and responsibilities. The ethical implications of widespread genetic screening, issues of data privacy, and the potential for genetic discrimination are areas that require careful consideration and regulation. The integration of genomics into screening technologies is not just a scientific or medical advancement; it's a societal shift that necessitates a comprehensive and multidisciplinary approach to ensure its benefits are realized responsibly and equitably [10].

5.3 Tailored for You: The Rise of Personalized Prevention

The urgent need to address the rising burden of non-communicable diseases like cardiovascular disease and cancer has brought personalized prevention to the forefront. Traditional disease prevention strategies, which are largely based on "one size fits all" recommendations, have been re-evaluated in light of the complex and heterogeneous nature of these diseases [20]. The risk factors for such conditions are not uniform but are a tapestry woven from clinical, genetic, and environmental threads unique to each individual.

Recent advances in genetics and multi-omics technologies have been pivotal in this shift toward personalized prevention. These technologies have provided new opportunities to stratify disease risks at an individual level, allowing for a more nuanced approach. By integrating individual biological data (such as genomic, transcriptomic, and metabolomic profiles) with sociodemographic, clinical, and environmental factors, a more detailed and personalized assessment of potential disease risks is now possible [21]

Personalized prevention is not just about predicting disease occurrence; it's also about implementing strategies that are targeted to an individual's unique profile of genomic data and other biomarkers. This approach not only improves the quality of care and life for patients but also reduces healthcare time, effort, and costs. By focusing on high-risk individuals for specific diseases based on risk calculations, personalized preventive interventions can lead to more cost-effective outcomes, particularly by avoiding over-diagnosis and redundant treatments [10, 21].

In the realm of personalized cancer prevention, the application of genetics plays a pivotal role in both primary and secondary prevention strategies. These approaches are tailored to an individual's unique genetic profile, enabling more effective and targeted prevention measures.

5.4 Cancer Personalized Prevention

For primary prevention, genetic risk assessment stands as a cornerstone. This involves using genetic testing to identify individuals with an increased risk due to familial or hereditary factors. For example, the presence of BRCA1 or BRCA2 gene mutations significantly raises the risk for breast and ovarian cancers. Knowing one's genetic risk can guide individuals in making informed decisions about preventive measures. Alongside this, targeted lifestyle changes are recommended based on genetic predispositions. Certain genetic markers can indicate a higher susceptibility to cancers associated with factors like obesity, smoking, or alcohol consumption. Personalized advice on diet and physical activity is thus provided, aligning with these genetic insights [22]. Additionally, chemoprevention becomes a viable strategy for individuals with specific genetic risks. Women with a genetic predisposition to breast cancer, for instance, might be advised to take preventive medications such as tamoxifen. Moreover, vaccination strategies, like the HPV vaccine, are implemented more assertively in individuals whose genetic makeup increases their susceptibility to certain cancers. In some cases, prophylactic surgical interventions are considered for those with a very high genetic risk, such as individuals with BRCA mutations. Prophylactic surgeries, like mastectomy or oophorectomy, might be an option to significantly reduce the risk of developing cancer. Additionally, personalized radiological imaging, such as the use of MRI screenings alongside traditional methods, is employed for those with a genetic predisposition to specific cancers.

Personalized prevention of colorectal cancer (CRC) is another example of how genetic information can be leveraged to tailor prevention strategies to individual risk profiles. In individuals with a family history or genetic predispositions, such as those with Lynch syndrome or familial adenomatous polyposis (FAP), the risk of developing CRC is significantly higher [23]. For these high-risk groups, genetic testing plays a crucial role in early identification, allowing for the implementation of targeted surveillance and preventive measures [24]. Enhanced screening protocols, such as more frequent colonoscopies starting at an earlier age, are recommended for these individuals. Beyond genetic factors, lifestyle modifications, including diet, physical activity, and smoking cessation, are advised and tailored to each person's genetic and environmental background. In some cases, chemoprevention, using drugs like aspirin, may be considered to reduce the risk of polyp formation and CRC development [25]. For individuals with extremely high genetic risk, prophylactic surgical options, such as colectomy, might be discussed as a preventive measure [26].

Secondary prevention strategies in cancer care, deeply rooted in genetics, focus on early detection and intervention. Reimagining cancer secondary prevention through the lens of genomics involves leveraging specific biomarkers and comprehensive tests to tailor follow-up and treatment strategies. This approach significantly enhances the ability to monitor for cancer recurrence, assess treatment efficacy, and manage patient care more effectively. One of the most promising biomarkers in cancer secondary prevention is circulating tumor DNA (ctDNA). ctDNA tests involve detecting fragments of DNA shed by cancer cells into the bloodstream. By monitoring ctDNA levels, clinicians can detect minimal residual disease or early signs of recurrence, often before they are clinically evident through imaging or physical symptoms. For example, in colorectal cancer, elevated ctDNA levels post-surgery can indicate a high risk of recurrence, guiding decisions regarding adjuvant chemotherapy [27]. In certain cancers, specific gene mutations serve as crucial biomarkers for secondary prevention. In lung cancer, mutations in genes like EGFR or ALK are not only targets for treatment but also markers for monitoring treatment response and disease progression. Periodic testing for these mutations can help adjust therapy regimens or identify resistance to targeted therapies [28].

CGP tests analyze a wide array of cancer-related genes simultaneously. These tests provide a detailed genetic profile of a tumor, which can be crucial for secondary prevention. For example, in melanoma, CGP can identify mutations in the BRAF gene, which not only guide initial treatment choices but also inform ongoing monitoring and subsequent treatment decisions if the cancer recurs [29].

These strategies highlight the shift towards a more personalized approach in cancer prevention, leveraging genetic information to tailor prevention to the unique risks and needs of each individual. This genetic focus not only enhances the effectiveness of prevention strategies but also underscores the importance of early detection and intervention in reducing the overall burden of cancer.

5.5 CVD Personalized Prevention

In the field of cardiovascular disease (CVD), personalized prevention strategies are increasingly adopted in practice. This shift towards a more individualized method is particularly critical given the complex interplay of genetic, lifestyle, and environmental factors in the development of CVD. Personalized prevention of CVD involves a comprehensive approach, encompassing genetic screening, tailored lifestyle modifications, and specific medical interventions, all aimed at reducing individual risk factors and preventing the onset of heart-related illnesses [8].

A prime example of this approach is seen in the management of Familial Hypercholesterolemia (FH), a genetic disorder characterized by significantly elevated LDL-cholesterol levels. Individuals with FH are identified through genetic testing and thorough family history assessments, which are crucial for early detection. Given the genetic basis of FH, lifestyle modifications such as dietary changes to lower LDL-C levels, increased physical activity, and smoking cessation are personalized based on each individual's genetic profile and overall health. In addition to lifestyle changes, pharmacological interventions play a key role. Statins are commonly used, but due to the severity of FH, more aggressive treatments are often necessary. This is where the new generations of therapies, like PCSK9 inhibitors, come into play. These advanced treatments target specific pathways involved in cholesterol metabolism and are particularly beneficial for those who have not achieved desired LDL-C levels through conventional statin therapy or who are intolerant to them [30]. Regular monitoring of lipid profiles and cardiovascular risk factors is essential to tailor these interventions effectively and adjust them over time. Additionally, the genetic nature of FH necessitates a family-based approach to screening and prevention, ensuring that first-degree relatives who might be at risk are identified and treated early [31].

The development of a new class of drugs based on small interfering RNA (siRNA) represents a significant advancement in medical science, particularly in the field of cardiovascular disease prevention. These drugs, part of the broader category of nucleic acid therapeutics, operate by selectively silencing specific genes at the RNA level. This method allows for targeted intervention in various biological processes contributing to disease [32].

Inclisiran is a prime example of this innovative approach. As an siRNA-based therapy, it specifically targets and silences the gene responsible for the production of the PCSK9 protein, a key regulator of cholesterol metabolism. By inhibiting PCSK9, inclisiran effectively lowers levels of low-density lipoprotein cholesterol (LDL-C) in the bloodstream. Its unique mechanism of action distinguishes it from traditional cholesterol-lowering treatments like statins and provides an alternative for patients who have not achieved optimal results with existing therapies [33].

The introduction of inclisiran into national health programs, such as those in the UK, marks a significant step in the adoption of siRNA-based therapies for public health. The National Health Service (NHS) in the UK has integrated inclisiran into its cardiovascular disease prevention strategies, targeting patients with high cholesterol levels, including those with Familial Hypercholesterolemia or a history of cardiovascular events. The key advantage of inclisiran, apart from its efficacy, is the dosing schedule—an injection only twice a year, enhancing patient compliance.

The use of inclisiran in these programs is not just a treatment strategy but also part of a larger public health initiative aimed at reducing the incidence of heart attacks and strokes. This approach highlights the potential of siRNA-based therapies in transforming the landscape of disease prevention and management, offering new hope for patients and a novel direction for healthcare systems worldwide [34].

However, the full realization of personalized prevention's potential necessitates further research, particularly in the realm of behavioral science. Sustainable integration of personalized preventive interventions and behaviors is crucial for the long-term success of this approach. Additionally, more cost-effective calculations, especially for omics-testing technologies, are needed to ensure that the benefits of these advancements are realized in a practical, scalable manner [10].

5.6 All Eyes on Public Health: Embedding Genetic Information to Promote Population Health

The integration of genomic information into public health strategies is reshaping our approach to healthcare on a population level. This field, a dynamic intersection of genomics and public health, focuses on applying genetic and molecular insights to prevent disease and promote health across communities [35].

The new discipline of public health genomics is based on the understanding that genetic factors play a significant role in the onset of many common diseases. By identifying these genetic components, public health initiatives can be more effectively tailored to address the specific needs of different populations. This approach goes beyond the traditional one-size-fits-all public health strategies, enabling more precise interventions.

A key aspect of public health genomics is the use of large-scale genomic data to identify population-specific risk factors for diseases. Such data helps understand the prevalence and impact of certain genetic variants within different communities. This knowledge can inform targeted screening programs, lifestyle modification initiatives, and community-based interventions, improving the overall effectiveness of public health efforts.

Moreover, public health genomics plays a crucial role in understanding the environmental and social determinants of health [36]. By analyzing how genetic factors interact with environmental and lifestyle variables, it's possible to develop comprehensive strategies considering the multifaceted nature of health and disease. For instance, understanding the genetic predisposition to conditions like hypertension or diabetes can guide the development of community-level programs focused on diet, exercise, and education [37].

An example of public health genomics intervention is the application of the "founder effect" in specific populations, such as the Ashkenazi Jewish community. This group exhibits a higher prevalence of certain genetic disorders like Tay-Sachs disease and Canavan disease due to the founder effect [38].

In the United States, public health genomics has significantly advanced in identifying and managing Tier 1 genomic diseases—conditions with a strong genetic component where effective intervention can lead to substantial public health benefits. Three such diseases, Familial Hypercholesterolemia (FH), Hereditary Breast and Ovarian Cancer Syndrome (HBOC), and Lynch Syndrome, exemplify the impact of genetics in disease prevention and management. Their inclusion as priority diseases in public health strategies showcases the integration of genomics into healthcare and highlights the evolving nature of disease prevention and management [39].

The "All of Us" program is another example of an ambitious international initiative in the field of public health genomics. This program, part of the Precision Medicine Initiative, aims to gather health data from one million or more people living in the U.S. to accelerate research and improve health. The uniqueness of this program lies in its inclusivity—it seeks to gather data from a diverse range of participants, especially those who have been underrepresented in biomedical research. By doing so, "All of Us" aims to develop a more detailed understanding of individual differences in lifestyle, environment, and biology, including genetics [10].

The data collected through this initiative will be used to study a wide range of health conditions, with the goal of developing more precise and effective healthcare tailored to individual differences. This approach to research promises to usher in new ways to predict disease risks, identify optimal treatments, and prevent diseases before they occur.

The field also extends to infectious disease control. Genomic sequencing of pathogens provides critical insights into disease transmission, drug resistance, and outbreak dynamics. This information is invaluable for developing effective control measures, vaccines, and treatment strategies [17]. One of the key benefits of this genomic surveillance was the identification of new variants of the virus. As SARS-CoV-2 mutated, some variants showed increased transmissibility, while others demonstrated potential resistance to initial vaccine formulations. By rapidly identifying and characterizing these variants, health authorities could adjust public health strategies accordingly. This included implementing targeted lockdowns, updating testing protocols, and informing vaccine development. In addition to informing immediate response strategies, genomic surveillance provided valuable data for future pandemic preparedness. Insights gained from the surveillance of SARS-CoV-2 contribute to the development of more effective vaccines and therapeutics, for not just for COVID-19 but also other potential viral threats [40].

One of the most groundbreaking developments in public health genomics is the potential for personalized medicine at the population level. This involves tailoring public health policies and programs to address the specific genetic makeup of a population, leading to more effective prevention and treatment strategies [41].

The availability of genomic data for public health, coupled with the advent of sophisticated data analytics systems, marks a significant stride in disease surveillance and management. The vast reservoirs of genomic data generated globally are invaluable for public health initiatives, but their true potential is unlocked through advanced analytics. These data analytics systems harness machine learning, artificial intelligence, and computational biology to parse complex genomic datasets, identifying patterns and correlations that might elude traditional analysis. By analyzing genomic data alongside environmental and lifestyle factors, public health professionals can develop more targeted intervention strategies, improving the efficacy of disease prevention and health promotion programs. This integration of genomic data with advanced analytics heralds a new era in public health, where decisions are informed by a deep understanding of the genetic basis of health and disease, leading to more effective, efficient, and tailored healthcare solutions [42].

However, the implementation of public health genomics also raises ethical, legal, and social issues. These include concerns about genetic privacy, potential discrimination, and equitable access to genomic technologies. Addressing these challenges requires a collaborative approach involving policymakers, healthcare providers, and the public to ensure that the benefits of public health genomics are realized responsibly and equitably.

5.7 The Elephant in the Room: Ethical, Legal, and Policy Frameworks for Genomic Data

As we step into the era of personalized medicine, propelled by advances in genetic and genomic technologies, we encounter a landscape rich in potential but challenged with ethical, legal, and social challenges. The field of public health genomics not only offers unprecedented opportunities for disease prevention and tailored treatments but also raises crucial questions about the responsible use and governance of genetic information.

Ethical considerations in genomic medicine revolve around the sensitivity of individual privacy and the right to confidentiality. As genomic data can divulge intimate details about an individual's health predispositions and potential future health trajectories, safeguarding this information against unauthorized access and misuse is paramount. The complexity of genetic data also intensifies the process of informed consent. Patients and research participants must be thoroughly informed about what genetic testing entails, the scope of information it can reveal, and the potential implications of this knowledge—not just for the individual but also for family members who may share genetic traits [43].

Furthermore, the specter of genetic discrimination looms large. How do we ensure that individuals are not marginalized or discriminated against based on their genetic makeup? Laws like the Genetic Information Nondiscrimination Act in the U.S. represent significant steps in protecting individuals from genetic discrimination in health insurance and employment. However, as the field of genomics expands, so does the need for ongoing legal vigilance and adaptation to new contexts and challenges [44].

The rise of direct-to-consumer genetic testing kits has democratized access to genetic information, allowing individuals to gain insights into their genetic makeup without necessarily involving healthcare professionals. While empowering on one front, this trend raises concerns about the accuracy of these tests, the potential for misinterpretation of results, and the adequacy of post-test counseling and support. These concerns highlight the need for regulatory frameworks to ensure that these tests meet certain standards of reliability and validity and that consumers are adequately informed and supported in their use [45].

In parallel, the global nature of genetic research and the transnational flow of genetic data call for international collaboration in policy development. Harmonizing regulations across borders is essential in addressing the challenges of data sharing, privacy protection, and ethical research practices. This global approach is not just about creating uniform standards but also about respecting cultural diversity and ethical pluralism in the application of genomics [46].

With genetics increasingly becoming a part of everyday health decisions, enhancing public understanding of genetics is crucial. Education initiatives aimed at both the public and healthcare professionals are needed to navigate the complexities of genomics. For healthcare providers, this includes training in the interpretation of genetic tests, understanding the nuances of genetic risk, and communicating this information effectively to patients [47].

In conclusion, the integration of genomics into healthcare is a journey marked by remarkable scientific achievements and complex ethical and legal considerations. As we harness the power of genomic information to improve health outcomes, we must also commit to developing robust, ethical, and responsive frameworks that protect individuals and uphold the values of equity, privacy, and informed choice. This journey is not just about scientific advancement but about shaping a future where genomics is used responsibly and equitably for the betterment of all.

5.8 Empowered Health: Citizen Engagement and Social Implications

Recent evidence suggests that as genomic data becomes increasingly accessible, it is not only transforming healthcare but also redefining how individuals perceive and interact with their health.

5.8.1 Societal Transformation Through Genomic Data

The widespread availability of genomic information fosters a more informed and proactive citizen engagement. People are now more involved in their health decisions, armed with insights about their genetic predispositions to certain diseases. This empowerment, however, comes with its challenges. Knowing one's genomic information from birth can lead to a heightened sense of vulnerability or deterministic thinking about health and disease. There is also the risk of genetic information being misinterpreted or causing unnecessary anxiety, emphasizing the need for proper genetic counseling and education [48].

The ethical implications of accessing one's genomic information from birth are profound. It raises questions about consent, especially for children who are too young to make their own health decisions. There is also the potential for genetic discrimination in areas like insurance and employment, necessitating robust legal protections. Furthermore, the psychological impact of knowing one's genetic risks from an early age is an area requiring careful consideration and support.

5.8.2 Predicting the Future: Scenarios in a Genetically Informed World

5.8.2.1 Positive Scenarios

A Proactive Health Paradigm: In a future where genetic data is easily accessible, healthcare shifts from being reactive to proactive. Regular genomic screenings become a part of routine health check-ups, allowing for early intervention and personalized disease prevention strategies. This approach significantly reduces the incidence and impact of genetic diseases.

A Connected Health Ecosystem: Genomic data integration into electronic health records leads to a more connected health ecosystem. Physicians, genetic counselors, and patients collaboratively use this information to make informed decisions. AI-driven tools analyze genomic data, offering insights and recommendations for health optimization.

Societal Shifts in Health Responsibility: As individuals gain more insights into their genetic predispositions, there is a societal shift towards personal responsibility for health. Lifestyle changes and preventive measures are more widely adopted, driven by a better understanding of personal health risks.

5.8.2.2 Negative Scenarios

Exacerbated Healthcare Disparities: In a future scenario where genomic data is integral to healthcare, a significant risk emerges: the exacerbation of healthcare disparities. Despite the advancements in genetic testing and personalized medicine, these cutting-edge services remain inaccessible to lower socio-economic groups. Wealthier segments of society have the means to benefit from personalized health insights and tailored treatments, leading to a significant improvement in their health outcomes. In contrast, underprivileged communities, lacking access to these technologies, continue to rely on traditional, less effective healthcare methods.

This divide is not just in terms of access but also in health literacy. The affluent, better informed about the potential and limitations of genomic data, are able to make more informed health decisions. Meanwhile, those in lower socio-economic groups, with limited understanding and resources, are unable to leverage the benefits of genomic medicine. This disparity leads to a two-tiered healthcare system, where your health prospects are increasingly dictated by your economic status, deepening existing health inequalities and creating a societal rift.

Ethical Dilemmas in Genetic Editing: Another future scenario unfolds around the ethical dilemmas posed by genetic editing technologies, like CRISPR-Cas9. In this world, gene editing has progressed beyond disease prevention to include enhancements and modifications for non-medical reasons. Parents are now faced with the option of not just screening for genetic diseases but also selecting desirable traits in their children, such as increased intelligence, physical ability, or appearance. This capability ignites a host of ethical, moral, and societal questions.

The practice of genetic editing for enhancement purposes creates a societal divide, with some advocating for the freedom to use these technologies as they see fit, while others decry it as unnatural and ethically problematic. Debates intensify over what constitutes a "normal" human trait versus an enhancement. Governments and regulatory bodies struggle to keep pace with these advancements, grappling with how to regulate such practices. The possibility of a genetically modified elite class emerges, leading to new forms of discrimination and societal stratification based on genetic modifications. This scenario presents a profound ethical quandary about the extent of human intervention in natural genetic processes and the societal implications of such choices [10, 49].

In conclusion, the integration of genomic data into society is a double-edged sword, offering immense potential for improved health outcomes and citizen empowerment but also presenting significant ethical, legal, and psychological challenges. Navigating this landscape requires a concerted effort from policymakers, healthcare providers, and society at large to ensure that the benefits of genomic data are realized responsibly and equitably.

References

1. Trent RJ. History of molecular medicine. In: Trent RJ, editor. Molecular medicine. 3rd ed. Burlington: Academic Press; 2005. p. 1–15. https://www.sciencedirect.com/science/article/pii/B9780126990577500011.

2. 1953: DNA double helix [Internet]. [cited 2024 Jan 22]. https://www.genome.gov/25520255/online-education-kit-1953-dna-double-helix.
3. Hull MA, Rees CJ, Sharp L, Koo S. A risk-stratified approach to colorectal cancer prevention and diagnosis. Nat Rev Gastroenterol Hepatol. 2020;17(12):773–80.
4. Laestadius LI, Rich JR, Auer PL. All your data (effectively) belong to us: data practices among direct-to-consumer genetic testing firms. Genet Med. 2017;19(5):513–20.
5. An overview of pharmacogenomic testing for psychiatric disorders - NCBI Bookshelf [Internet]. [cited 2024 Jan 22]. https://www.ncbi.nlm.nih.gov/books/NBK595332/.
6. Guo CG, Ma W, Drew DA, Cao Y, Nguyen LH, Joshi AD, et al. Aspirin use and risk of colorectal cancer among older adults. JAMA Oncol. 2021;7(3):428–35.
7. ChatGPT [Internet]. [cited 2024 Jan 22]. https://chat.openai.com.
8. Jaskulski S, Nuszbaum C, Michels KB. Components, prospects and challenges of personalized prevention. Front Public Health. 2023;11:1075076.
9. Comprehensive genomic profiling: a required tool for precision oncology | ASCO Daily News [Internet]. [cited 2024 Jan 22]. https://dailynews.ascopubs.org/do/comprehensive-genomic-profiling-required-tool-precision-oncology.
10. Miceikaitė I, Hao Q, Brasch-Andersen C, Fagerberg CR, Torring PM, Kristiansen BS, et al. Comprehensive noninvasive fetal screening by deep trio-exome sequencing. N Engl J Med. 2023;389(21):2017–9.
11. Alvarellos M, Sheppard HE, Knarston I, Davison C, Raine N, Seeger T, et al. Democratizing clinical-genomic data: how federated platforms can promote benefits sharing in genomics. Front Genet. 2023;13:1045450. https://www.frontiersin.org/articles/10.3389/fgene.2022.1045450.
12. Determinants of health [Internet]. [cited 2024 Jan 22]. https://www.who.int/news-room/questions-and-answers/item/determinants-of-health.
13. Education [Internet]. Genomics education programme. [cited 2024 Jan 22]. https://www.genomicseducation.hee.nhs.uk/education/.
14. ESMO. ESMO clinical practice guidelines: gastrointestinal cancers [Internet]. [cited 2024 Jan 22]. https://www.esmo.org/guidelines/guidelines-by-topic/esmo-clinical-practice-guidelines-gastrointestinal-cancers.
15. Horton R, Lucassen A. Ethical considerations in research with genomic data. N Bioethics. 2023;29(1):37–51.
16. Bittla P, Kaur S, Sojitra V, Zahra A, Hutchinson J, Folawemi O, et al. Exploring circulating tumor DNA (CtDNA) and its role in early detection of cancer: a systematic review. Cureus. 2023;15(9):e45784.
17. WHO global genomic surveillance strategy for pathogens with pandemic and epidemic potential 2022-2032 [Internet]. [cited 2024 Jan 22]. https://www.who.int/initiatives/genomic-surveillance-strategy.
18. Alonso R, Perez de Isla L, Muñiz-Grijalvo O, Diaz-Diaz JL, Mata P. Familial hypercholesterolaemia diagnosis and management. Eur Cardiol. 2018;13(1):14–20.
19. US EEOC [Internet]. [cited 2024 Jan 22]. Genetic information discrimination. https://www.eeoc.gov/genetic-information-discrimination.
20. Steinke-Lange V, Holinski-Feder E. Genetic screening and personalized prevention in colorectal cancer. Vis Med. 2019;35(4):226–30.
21. Medeiros AM, Bourbon M. Genetic testing in familial hypercholesterolemia: is it for everyone? Curr Atheroscler Rep. 2023;25(4):127–32.
22. Hajar R. Genetics in cardiovascular disease. Heart Views. 2020;21(1):55–6.
23. Genetics vs. Genomics Fact Sheet [Internet]. [cited 2024 Jan 22]. https://www.genome.gov/about-genomics/fact-sheets/Genetics-vs-Genomics.
24. Genomic Medicine [Internet]. [cited 2024 Jan 22]. https://www.genome.gov/genetics-glossary/Genomic-Medicine.
25. The Lancet. Genomic newborn screening: current concerns and challenges. Lancet. 2023;402(10398):265. https://www.thelancet.com/journals/lancet/article/PIIS0140-6736(23)01513-1/fulltext.
26. Chen Z, Azman AS, Chen X, Zou J, Tian Y, Sun R, et al. Global landscape of SARS-CoV-2 genomic surveillance and data sharing. Nat Genet. 2022;54(4):499–507.
27. Middleton A, Milne R, Almarri MA, Anwer S, Atutornu J, Baranova EE, et al. Global public perceptions of genomic data sharing: what shapes the willingness to donate DNA and health data? Am J Hum Genet. 2020;107(4):743–52.
28. American Association for Cancer Research. Healthy lifestyle may help mitigate high genetic risk of cancer. 2024. https://www.aacr.org/about-the-aacr/newsroom/news-releases/healthy-lifestyle-may-help-mitigate-high-genetic-risk-of-cancer/.
29. Albosta MS, Grant JK, Taub P, Blumenthal RS, Martin SS, Michos ED. Inclisiran: a new strategy for LDL-C lowering and prevention of atherosclerotic cardiovascular disease. Vasc Health Risk Manag. 2023;19:421–31.
30. Villalobos P, Wistuba II. Lung cancer biomarkers. Hematol Oncol Clin North Am. 2017;31(1):13–29.
31. Budreviciute A, Damiati S, Sabir DK, Onder K, Schuller-Goetzburg P, Plakys G, et al. Management and prevention strategies for non-communicable diseases (NCDs) and their risk factors. Front Public Health. 2020;8:574111. https://www.frontiersin.org/articles/10.3389/fpubh.2020.574111.
32. Noninvasive prenatal testing: current perspectives and future challenges - PMC [Internet]. [cited 2024 Jan 22]. https://www.ncbi.nlm.nih.gov/pmc/articles/PMC7824607/.
33. Holdenrieder S, Ungerer V, Oberhofer A, Bronkhorst AJ. Pan-cancer screening by circulating tumor DNA (ctDNA) – recent breakthroughs and chronic pitfalls. J Lab Med. 2022;46(4):247–53.

34. Annas GJ. Personalized medicine or public health? Bioethics, human rights, and choice. Rev Port Saúde Públ. 2014;32(2):158–63.
35. Carmi S, Hui KY, Kochav E, Liu X, Xue J, Grady F, et al. Sequencing an Ashkenazi reference panel supports population-targeted personal genomics and illuminates Jewish and European origins. Nat Commun. 2014;5(1):4835.
36. Borry P, Bentzen HB, Budin-Ljøsne I, Cornel MC, Howard HC, Feeney O, et al. The challenges of the expanded availability of genomic information: an agenda-setting paper. J Community Genet. 2018;9(2):103–16.
37. Walton NA, Hafen B, Graceffo S, Sutherland N, Emmerson M, Palmquist R, et al. The development of an infrastructure to facilitate the use of whole genome sequencing for population health. J Pers Med. 2022;12(11):1867.
38. Molster CM, Bowman FL, Bilkey GA, Cho AS, Burns BL, Nowak KJ, et al. The evolution of public health genomics: exploring its past, present, and future. Front Public Health. 2018;6:247. https://www.frontiersin.org/articles/10.3389/fpubh.2018.00247.
39. Tamura K. The genetic code: Francis Crick's legacy and beyond. Life. 2016;6(3):36.
40. Hood L, Rowen L. The human genome project: big science transforms biology and medicine. Genome Med. 2013;5(9):79.
41. Damrongchai N, Johnston R, Tegart G. The impact of foresight studies on human healthcare in the post-genomic era. Int J Foresight Innov Policy. 2010;6:151–65.
42. Brand A, Brand H, Schulte Bäumen T. The impact of genetics and genomics on public health. Eur J Hum Genet. 2008;16(1):5–13.
43. Foss KS, O'Daniel JM, Berg JS, Powell SN, Cadigan RJ, Kuczynski KJ, et al. The rise of population genomic screening: characteristics of current programs and the need for evidence regarding optimal implementation. J Pers Med. 2022;12(5):692.
44. DePhillips C. The role of RNA interference therapeutics in hypercholesterolemia and implications for practice. J Nurse Pract. 2023;19(6):104619.
45. Ostwald DA, Schmitt M, Peristeris P, Gerritzen T, Durand A. The societal impact of Inclisiran in England: evidence from a population health approach. Value Health. 2023;26(9):1353–62.
46. Trends in tier 1 genomic applications 2013-2022 | Blogs | CDC [Internet]. 2022 [cited 2024 Jan 22]. https://blogs.cdc.gov/genomics/2022/09/19/trends-in-tier-1/.
47. Reay WR, Clarke E, Eslick S, Riveros C, Holliday EG, McEvoy MA, et al. Using genetics to inform interventions related to sodium and potassium in hypertension. Circulation. 2024;149(13):1019–32.
48. Reay WR, et al. Using genetics to inform interventions related to sodium and potassium in hypertension. Circulation. 2024;149(13):1019–32. https://doi.org/10.1161/CIRCULATIONAHA.123.065394.
49. What is personalised medicine? - ICPerMed [Internet]. [cited 2024 Jan 22]. https://www.icpermed.eu/en/icpermed-medicine.php.

Regenerative Medicine and Bioprinting

6

Zahra Ghezel-Ayagh, Andisheh Qashqaie, Yaser Tahamtani, Mandana Hasanzad, and Hamid Reza Aghayan

What Will You Learn in This Chapter?
In this chapter, we'll explore the exciting world of regenerative medicine and bioprinting. Regenerative medicine involves harnessing the body's natural healing processes to repair or replace damaged tissues and organs. Bioprinting takes this concept a step further by using 3D printing technology to create living tissue constructs that can be implanted into the body. Together, these cutting-edge fields hold incredible promise for revolutionizing healthcare and improving the quality of life for patients around the world. We'll learn about new ways scientists are using technology to help the body heal itself. From different types of stem cells and growing tissues in labs to printing organs, we'll see how these advances could change how we treat diseases and injuries.

Rationale and Importance
Regenerative medicine and bioprinting represent a paradigm shift in healthcare, offering innovative solutions to address critical medical challenges. State-of-the-art technologies could completely change how we think about replacing organs and regenerating tissue. They provide hope to individuals with serious injuries, chronic diseases, and organ failure by offering alternatives to standard transplants and invasive treatments. Moreover, regenerative medicine and bioprinting not only aim to restore function but also enhance quality of life.

6.1 Introduction

The available pharmacological therapies can cure most pathological disorders. However, as the human body has a low regenerative potential, they have limited therapeutic value in certain dis-

Z. Ghezel-Ayagh · Y. Tahamtani
Department of Basic and Population Based Studies in NCD, Reproductive Epidemiology Research Center, Rayan Institute, ACECR, Tehran, Iran

Department of Stem Cells and Developmental Biology, Cell Science Research Center, Royan Institute for Stem Cell Biology and Technology, ACECR, Tehran, Iran

A. Qashqaie
Oncology Hematology and Cell Therapy Research Institute, Shariati Hospital, Tehran University of Medical Sciences, Tehran, Iran

M. Hasanzad
Personalized Medicine Research Center, Endocrinology and Metabolism Clinical Sciences Institute, Tehran University of Medical Sciences, Tehran, Iran

H. R. Aghayan (✉)
Cell Therapy and Regenerative Medicine Research Center, Endocrinology and Metabolism Molecular-Cellular Sciences Institute, Tehran University of Medical Sciences, Tehran, Iran
e-mail: shraghayan@tums.ac.ir

M. Hasanzad (ed.), *A Glimpse at Medicine in the Future*,
https://doi.org/10.1007/978-981-97-4974-4_6

orders. The congenital defects, trauma, and specific diseases such as myocardial infarction, peripheral vascular disease, and renal failure damage the organs and tissues, leading to organ loss [1]. Traditionally, organ and tissue transplantation has been suggested as a therapeutic option for organ failure or tissue damage. In addition to the total cost that society bears in terms of considering patients with devastating diseases or failed organs, the supply of donated tissues and organs cannot encounter the waiting list of people requesting organ transplantation. Therefore, scientists employed both engineering and biological principles to develop sources of tissues and organs to promote the regeneration of damaged organs or the creation of new tissues [2]. In the last few decades, a promising modality of therapy for various degenerative diseases, known as "Regenerative Medicine," has been established that aims to regenerate, repair, or replace damaged tissues and organs in patients who have limited or no other therapeutic options and is now considered as an alternative or adjuvant to traditional treatments [1, 3]. According to the Alliance for Regenerative Medicine definition, "regenerative medicine includes gene therapies, cell therapies, and tissue-engineered therapies intended to augment, repair, replace, or regenerate organs, tissues, cells, genes, and metabolic processes in the body." The ability to regenerate or replace a variety of tissues and organs and return normal function to them is the ultimate target of regenerative medicine [4, 5]. This new source of treatment has shown promising results in skin, heart, kidney, joint, and liver diseases. The main concept of regenerative medicine is to obscure tissue or organ regeneration with the use of different kinds of technologies, including cell transplantation, tissue engineering, cell-derivates application, and gene editing [6–8]. Recent advances in technology and interdisciplinary sciences have enabled a revolutionary improvement in this branch of medicine. Serious, uncurable health problems such as diabetes, spinal cord injury, heart failure, and cancer can be potentially cured by regenerative medicine. Studies in regenerative medicine and tissue engineering have made major advances and innovations in replacing damaged human organs through the application of several broad-spectrum technologies [1, 2]. The different approaches are organ transplantation, gene editing and miRNA treatment, stem cell therapy, organoid development, and also constructing artificial tissue and organ bioprinting by the application of different scaffolds, nanomedicine, and three-dimensional (3D) bioprinters. Artificial intelligence, wireless data exchange, soft intelligence biomaterials, living robotics, and soft nanorobotics can be the next revolutionary approaches in regenerative medicine. In other words, multidisciplinary fields include biomechanics, material science, cell biology, genetics, and computer science [9, 10]. The objective of this chapter is to present a summary of the potential of regenerative medicine in the future, focusing on stem cells, organoids, and bioprinting applications.

6.2 Stem Cells as Tools for Regenerative Medicine

Since the discovery of stem cells and the identification of their unique properties, these cells have been widely considered as proper candidates for regenerative medicine to repair organs or tissues [11]. To replace the damaged tissue so that it can be functional or it can be able to stimulate the original tissue's function, a combination of different and appropriate materials such as scaffolds, growth factors, and stem cells must be used [12, 13]. Due to the tremendous therapeutic potential of different stem cells, which are inflammation inhibition, apoptosis inhibition, cell recruitment, angiogenesis stimulation, and differentiation potential, they can be good candidates for regenerative therapies [3]. Different adult, fetal, or embryonic-derived stem cells (Fig. 6.1) have been successfully applied in clinical studies. Some cells like mesenchymal stem/stromal cells (MSCs), hematopoietic stem/progenitor cells, fibroblasts, chondrocytes, and keratinocytes have more popularity in clinical applications [14, 15]. The stem cells must be employed in emergencies and their robustness must be retained during adult life. Adult stem cells or progenitor cells can

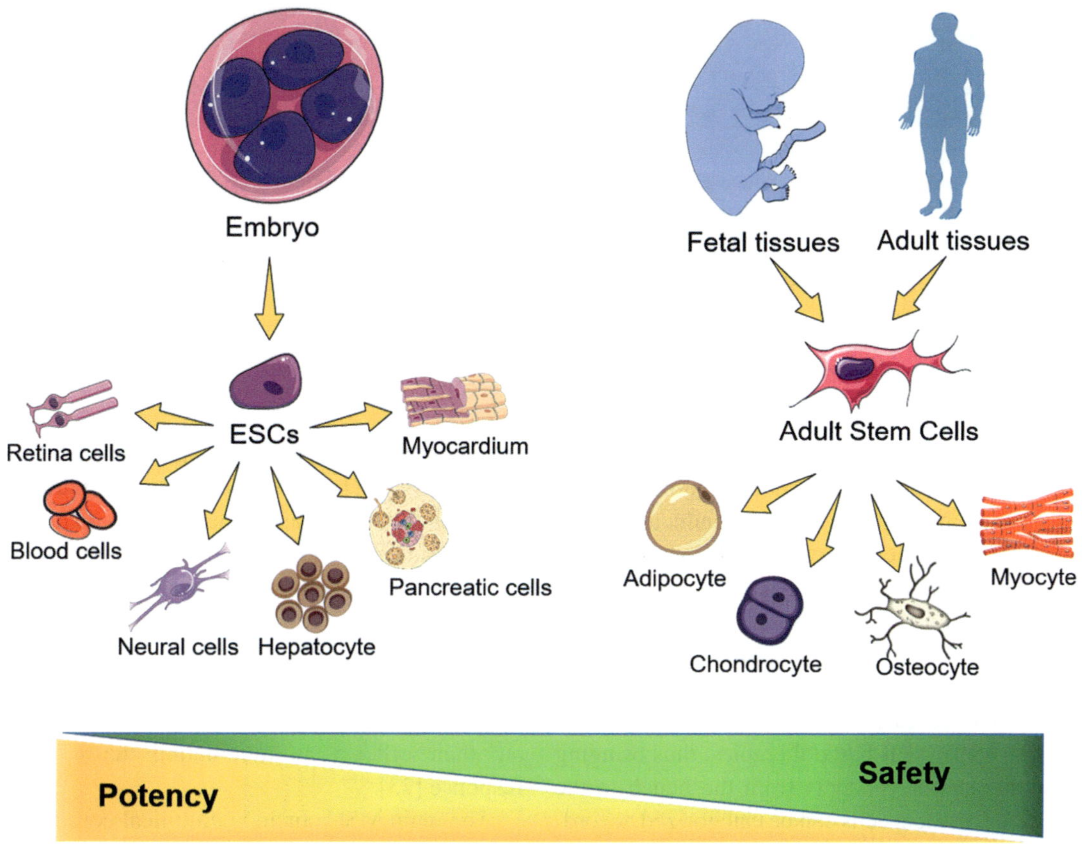

Fig. 6.1 Different sources of stem cells

be injected into damaged organs after isolation from adult tissues, resulting in tissue structural regeneration and functional improvement [16]. The stem cells used can have an autologous or allogeneic origin from an individual human. In addition, regenerative medicine strategies have also focused on xenogenic cells from animals such as pigs or cows [2]. In addition, the biologically active molecules of stem cells which are secreted as extracellular microvesicles or known as their secrotomes can also be employed in therapy [17]. Figure 6.2 demonstrates the different aspects of regenerative therapies.

In recent years, the number of stem-cell-based companies that introduce clinical applications has increased. Also, many clinical trials using stem cells for the treatment of various diseases have been conducted. Therefore, it may be pos-

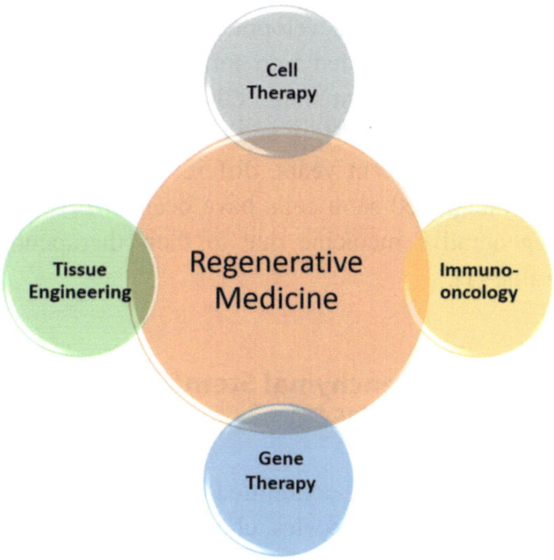

Fig. 6.2 Therapeutic aspects of regenerative medicine

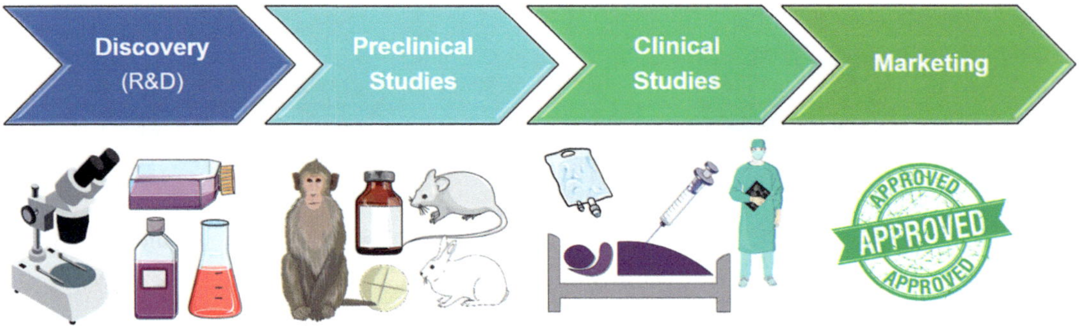

Fig. 6.3 Commercialization pathway of regenerative medicine products

sible to find more stem-cell-based products in the clinical market soon. The translation of research ideas into the clinical field is conducted under well-controlled development processes including proof-of-concept studies, research and development, preclinical studies, and clinical trials (Fig. 6.3). The competent regulatory authorities provide highly defined regulations and directives by which researchers can achieve safe and effective stem-cell-based therapies, thus bringing new therapeutic products from the bench to the bedside [3]. Stem cells can be categorized according to different aspects of these cells. Considering their developmental stage, stem cells can be divided into embryonic stem cells (ESCs), fetal stem cells (FSCs), and adult stem cells (ASCs). Considering their developmental potential, stem cells can be divided into totipotent stem cells (TSCs), pluripotent stem cells (PSCs), multipotent stem cells (MSCs), and unipotent stem cells (USCs). In recent years, different categories of the mentioned stem cells have been involved in regenerative medicine due to their therapeutic value [1].

6.2.1 Mesenchymal Stem/Stromal Cells (MSCs)

MSCs are lineage-committed cells that can differentiate into osteocytes, chondrocytes, and adipocytes [18]. The MSCs are well known for their immunosuppressive functions, preventing activated T cell proliferation via immunosuppressive cytokine secretion. They also suppress apoptosis and programmed cell death signaling [19, 20]. MSCs possess characteristics including multiple immunomodulatory and regenerative capacities, differentiation potential, the immune suppression effect on stimulated lymphocytes, and low immunogenicity under certain biological conditions [21, 22]. Most clinical trials have used different sources of adult MSCs to understand the mechanisms of protection provided by MSCs and evaluate their efficacy in modulating the immune response [23].

The main MSCs include umbilical cord blood (UCB)-derived MSCs, adipose (AD)-derived MSCs, amniotic fluid-derived MSCs, Wharton's jelly-derived MSCs, placenta-derived MSCs, and bone marrow (BM)-derived MSCs. MSCs can also be extracted from the retina, liver, gastric mucosa, tendon, cartilage, peripheral blood, as well as different embryonic and fetal organs [24–29]. Based on the origin of the MSCs, they have different characteristics, including self-renewal and differentiation capacity, senescence, long-term cultivation, and immunomodulatory action [30–32]. BM-MSCs can be used in autologous and allogenic contexts and have a longer doubling time but have a lower efficiency of collection compared to other MSCs [33]. AD-MSCs are not only easy to collect and have a shorter doubling time but are also approximately 100–500 times more efficient to harvest [34]. However, they have a strong characteristic for adipogenic differentiation and the number of cytokines secreted is significantly lower [35]. UC-MSCs can be easily collected compared to other stem cells, and ethical or donor-age issues can be avoided. They have higher proliferation and differentiation capabilities and a short doubling time [30, 36].

From preclinical and clinical research data, it seems that the therapeutic transplantation of MSCs may represent a safe and effective treatment for different diseases [21, 22]. Studies have focused on transplanting MSC-derived insulin-producing cells to overcome the underlying T cell-mediated autoimmune dysfunction, restore immune homeostasis, preserve residual β cells, and replace damaged β cells. However, clear evidence supporting this hypothesis is currently lacking [21, 22]. Two clinical trials have reported positive outcomes of MSC-derived pancreatic progenitor transplantation; however, these studies lack clear evidence for mature functioning MSC-derived β cells [23]. MSCs have been used to examine their treating potential for Parkinson's and Alzheimer's diseases [18, 37], musculoskeletal disorders and joint injuries [38, 39], immune-related inflammation disease [3], renal disorders [40], cardiovascular disorders, and peripheral vascular diseases [41]. In the latter, they have enhanced therapeutic angiogenesis and vascular regeneration [41]. They can stop cell aging by preventing telomere shortening because of their protective properties against oxidative stress [1].

6.2.2 Hematopoietic Stem Cells (HSCs)

HSCs are multipotent and give origin to all the cells of the hematopoietic system, namely myeloid cells and lymphocytes. They can be extracted from peripheral blood, bone marrow, placenta, and cord blood. They can cause immunological problems, such as transplant rejection, their main limitation when used for regenerative medicine. Yet, till now, they have been effective in leukemia, malignant lymphoma, regenerative anemia, congenital metabolism and immunodeficiency, nonresponsive autoimmune disease, and solid cancer treatments. Additionally, HSCs have been used in different clinical studies to regenerate liver, heart, and neural tissues [42]. These new indications are under investigation and regulatory bodies have only approved HSCs for treating hematological disorders. Recently, transplantation of gene-edited HSCs has been authorized for treating genetic disorders like hemophilia and thalassemia [43].

6.2.3 Embryonic Stem Cells (ESCs)

ESCs are pluripotent stem cells that can be theoretically differentiated into all embryonic cell types by inducing them with external stimulating factors during in vitro culture. ESCs also can be genetically modified. As they are self-renewal so far, different cell lines have been established and maintained through in vitro culture [1]. ESCs-derived progenitor cells have been clinically applied in different conditions like diabetes, ischemic heart disease, age-related macular degeneration, and interstitial cystitis. All of these clinical studies are in the early phase and more studies are needed to prove the safety and efficacy of ESCs-derived cells in humans [44–47].

6.2.4 Induced Pluripotent Stem Cells (iPSCs)

iPSCs, introduced in 2006, are made by reprogramming fibroblast cells or other adult somatic cells and therefore are known as artificially created stem cells. Self-renewability, differentiation capacity, and malformed species performance are iPSCs characteristics common with ESCs. However, scientific evidence relating to changes in cell-specific regulatory pathways, gene expression, and epigenetic regulation is still little compared to ESCs. Not understanding these issues poses a risk of cell dysfunction or tissue chimerism [48]. Because of serious concerns for their genomic instability and safety, the number of clinical studies using iPSCs-derived cells is minimal. A report by Kim et al. demonstrated that only 19 clinical trials have been registered in clinical trial registry databases. The highest number of these trials (10 of 19) were registered by Japanese researchers. The main areas of research were Parkinson's disease, age-related macular degeneration, Aplastic anemia, Retinitis pigmentosa, heart failure, GVHD, COVID-19 ARDS, and myocardial ischemia [49].

6.3 Organoid-Based Regenerative Medicine

Organoids are 3D self-organized constructions extracted or derived from embryonic or adult stem/progenitor cells that contain fully differentiated functioning cell types. These miniature organ-like structures that are grown in vitro resemble the structure and function of the native organs [50]. The spatiotemporal process of organ development during embryogenesis can be monitored dynamically by these organoid structures. Organoids offer competent models for studying developmental and degenerative disorders such as Alzheimer's disease [51]. They launch new horizons in personalized medicine, cell therapy, and drug screening and replace the growing need for organ transplantation [52, 53]. Multistep pluripotent stem cell differentiation protocols aligned with the co-culture of various cell types may be required for organoid generation [54]. Gene-modified immunotherapeutic organoids could be generated by merging stem cell-derived organoids with the genome engineering/gene editing tools via CRISPR-Cas9 technology [55–57]. The immunotherapeutic organoids provide a new prospect in drug screening and cell-based therapy for precision medicine [58]. The first study highlighting organoids' potential for regenerative medicine was reported in 2012. It was an intestinal organoid showing therapeutic effects after transplantation into an animal model of inflammatory bowel disease [59]. Tissue-specific adult stem/progenitor cell-derived organoids can be collected through minimally invasive procedures. They are capable of long-term mass propagation and large quantities of cultivation, facilitating their use as a therapeutic agent. These organoids can differentiate into specific target tissue cells or help to regenerate the target tissue upon transplantation [60]. Considering the importance of this new technology in the future of medicine, more details are presented.

6.3.1 Organoids for Ectodermal Derivatives

Brain organoids may be a future therapeutic option for treating cerebral diseases. As these organoids have shown satisfactory functionality, they can provide new opportunities for testing new drugs [52]. For studying neurovascular diseases such as brain ischemia and infectious diseases such as microcephaly in newborn babies induced by the ZIKA virus, brain organoids have been generated from human pluripotent stem cells and neural stem cells [61]. When transplanted into the mouse's medial prefrontal cortex, human brain organoids are differentiated into various brain regions, including the cerebral cortex, cerebellum, and midbrain. In addition, various studies have been conducted on mini-brains generated from single brain organoids [52, 61]. Retinal organoids have been produced and transplanted in the subretinal space of mice. These organoids have integrated into the host retina and have developed photoreceptors, enhancing color vision during daylight [62, 63]. However, scientists have further maturated their differentiation protocols to obtain improved visual functions and light sensitivity in these organoids. Moreover, physical contact between the implanted organoids and the host inner retinal neurons must be improved [64]. For the first time, the development of functional eye models in 3D configuration has been achieved by organoid technology [65]. Otic and inner ear organoids have been generated from PSCs containing otic progenitor cells, hair cells, and neurons. These structures provide in vitro models of the inner ear that could aid in identifying new regenerative drugs, cell-based therapies, and gene therapies for inner ear diseases [66]. Implantation of 3D heterotypic cell mixtures of otic organoids in the ex-ovo chick embryo chorioallantoic membrane stimulated the maturation of the inner ear hair-like cells by supplying a vascularization network [67]. Several studies have been conducted to treat ophthalmologic diseases such as dry eye and Sjogren's syndrome through lacrimal gland organoids. These organoids were integrated into lacrimal gland tissues, secreted tear components, and regenerated damaged tissues when transplanted into mouse inflammation models. Lacrimal tissue stem cells can also be cultured in large quantities for extended periods [68, 69]. Moreover, skin-like structures containing skin progenitor cells and hair follicles are observed in skin organoids resembling fetal skin [70, 71].

6.3.2 Organoids for Endodermal Derivatives

Gastrointestinal and airway organoids such as intestinal, thyroid, stomach, liver, pancreas, biliary tract, salivary gland, and lung organoids have been studied [60].

Gastrointestinal organoids are built by epithelial and mesenchymal cells derived from PSCs [72]. Mamoru Watanabe et al. were the first to study intestinal organoid therapeutics and to start the development of organoid-based regenerative medicine clinical trials. It seems that organoid-based regenerative therapeutic agents can be used as a fundamental treatment for intestinal Behçet's disease and radiation proctitis. Intestinal organoid transplantation to a mouse model of radiation proctitis has successfully regenerated the intestinal tissue. Moreover, a clinical trial for chronic bowel disease patients with ulcerative colitis was started in 2020 [73]. Additionally, intestinal organoid-based regenerative medicine has also demonstrated therapeutic efficacy in the bowel disease mouse model [74]. The transplantation of fetal intestine- or adult small intestine stem cell-derived organoids to the colonic ulcers of an animal model successfully repaired the damaged tissues and reconstructed the crypt-villus structures. Self-organoids generated from intact crypts, when engrafted to the adjacent damaged areas, overcome rejection of the graft, providing for the first time an efficient therapeutic approach for personalized regenerative medicine for these diseases [75, 76]. Thyroid organoids [77] and stomach organoids [78] have also been studied as gastrointestinal organoids. Developing functional in vitro mini-liver due to the increase in chronic liver diseases such as liver cirrhosis and hepatocarcinoma is essential [79]. The treatment of inherited metabolic diseases is being well studied after the first report of liver organoid establishment by Hans Clevers et al. The transplantation of liver organoids into a tyrosinemia animal model generated normal hepatocytes and enhanced the survival time [80]. Hepatic organoid technology has obtained differentiated hepatic cells and vascular networks [81, 82]. Hepatic organoids have been successfully transplanted in mice suffering from acute liver failure, resulting in functional hepatic tissue development and mouse recovery [83]. Additionally, organoids have the potential to be used as ex vivo gene therapeutic agents. Normal liver organoids were generated by Copper Metabolism Domain Containing 1 (COMMD1) gene transfection after the establishment of liver organoids from COMMD1-deficient dogs with inherited copper metabolic insufficiency. These organoids were able to express the COMMD1 gene for a prolonged period and improve the signs of copper accumulation disease after being transplanted to the body through the portal vein [84]. Pancreas organoid technology has enabled studying unknown pancreatic signaling pathways [85], has enhanced the development of novel therapeutic approaches for type I and II diabetes and metabolic disorders [86], and has provided a preclinical model for studying pancreatic ductal adenocarcinoma [87–89]. After 5 weeks of transplanting pancreatic organoids in the peritoneal cavity of immunodeficient mice, a great neovascularization network was built. These organoids were further matured into different cells of islets of Langerhans, especially beta cells [90–93]. Recently, the transplantation of biliary tract cells derived-organoids into a biliary tract injured animal model has demonstrated a regenerative effect [94]. A salivary gland organoid-based regenerative medicine is actively being conducted to target radiation treatment- and Sjogren's syndrome-related xerostomia, and a clinical trial is ongoing. These organoids are established from salivary gland tissues, and when transplanted into a mouse model of radiation-induced salivary gland dysfunction, they showed salivation improvement [95]. Lung diseases, including asthma and cancer, can be potentially treated by developing lung organoids in the future. Although multiple microporous scaffold usage gained advantages in lung organoid production, cell maturation, morphogenesis, and mesenchymal nature are still challenges for the development of mini-lungs [96, 97].

6.3.3 Organoids for Mesodermal Derivatives

Studies have focused on the production of the kidney, heart, and different reproductive tract organoids as mesoderm-derived organoids [60]. The hPSC-derived kidney or renal organoids were maturated after transplantation under the kidney capsule of immunodeficient mice, leading to podocyte development, the establishment of glomerular perfusion, and the connection of the developed glomerular vascularization with the pre-existing vascular networks [98, 99]. The differentiation of 3D iPSC-derived cardiomyocytes has identical characteristics to their native phenotype [100]. Moreover, after 24 h of transplanting hiPSC-derived cardiac organoids into the internal abdominal muscle of mice, cardiac progenitor cells, MSCs, and endothelial cells were observed; and only after 1 month of transplantation, a widespread neovascularization network was built, and physiological functions were seen [101]. Prostate organoids can be used to validate preclinical anti-cancer drugs to predict their therapeutic efficiency [102]. At present, new therapeutic concepts for female reproductive system diseases, such as endometriosis and endometrial, ovarian, and cervical carcinoma [103] can be provided by endometrial organoids [104, 105], cervical organoids [106], vaginal organoids [107], and fallopian tube organoids [108]. Moreover, this represents a huge step in studying infertility and disorders related to pregnancy, and miscarriage can be mimicked by the development of mini-placenta organoids in the future [109]. Additionally, these organoids can be used as therapeutic approaches for regenerative medicine to target a wider range of diseases in the future [60].

6.4 Limitations and Challenges of Cell-Based Regenerative Medicine

Different challenges such as safety, efficacy, scale-up, dosing, and final product cost should be considered when applying regenerative medicine to treat different diseases [110]. Determining the best types of cells in the culture, developing an easily operated and low-cost protocol for cell differentiation, isolating and purifying the right cells from a multi-type cell mixture, defining the appropriate quality attributes, and selecting the suitable therapeutic approach are the crucial challenges in the therapeutic application of stem cell therapy [111]. Stem cells are derived from various sources with different characteristics, capabilities, potential, and efficiency. Therefore, selecting the accurate source of stem cells for the target can be effective in ensuring treatment efficiency [1]. Adult stem cells have proven their safe use in regenerative medicine [112] and the regulatory agencies have approved a few of them for clinical applications. Although pluripotent stem cells give rise to cells from all three germ layers and can serve as unlimited cell sources, regenerative medicine is still searching for ethically acceptable, safe, and efficient ideal stem cells for therapeutic applications [113]. Moreover, as these cells might be incompletely differentiated, they might increase the risk of teratoma formation due to their oncogenic potential [55, 114]. Moreover, they exhibit several other issues, such as genomic instability and possible rejection [113]. Different applications have been improved to resolve the limitations of stem cell therapy, such as combining cells with scaffolds, using suitable biochemical properties in cell cultures, and immunomodulation and gene editing of stem cells [115–118]. However, preclinical studies must be applied due to their status as newer technologies [119]. In addition, although regenerative medicine has a promising future, the development and regulation for the potential application of stem cell therapies are in diverse states in different countries [113]. Table 6.1 shows a list of the marketed regenerative medicine products in

Table 6.1 Samples of marketed regenerative medicine products (Adapted from: https://alliancerm.org/available-products)

Company	Product name	Cell type/product specification	Indication	Country/region of marketing authorization
Cell therapy products				
TIGENIX	Alofisel	Allogenic ADMSC	Perianal Crohn's fistulas	EU
NIPRO CORP	Stemirac	Autologous BMMSC	Spinal cord injury	Japan
TEGO SCIENCE	Rosmir	Autologous fibroblast	Under eye wrinkles	South Korea
STEMPEUTICS PVT	Stempeucel	Allogenic BMMSC	Critical limb ischemia	India
CO.DON AG	Spherox	Autologous chondrocytes	Cartilage defects	EU
JCR PHARMACEUTICALS	TEMCELL	Allogenic BMMSC	Refractory GVHD	Japan
AURION BIOTECH	Vyznova	Corneal endothelial cell	Bullous keratopathy	Japan
CHIESI FARMACEUTICI	Holoclar	Autologous limbal SC	Limbal stem cell deficiency	EU
BIOSOLUTIONS	KeraHeal-Allo	Allogenic keratinocyte	Second-degree burns	South Korea
CORESTEM	Neuronata-r	Allogenic BMMSC	ALS	South Korea
MEDIPOST	Cartistem	Allogenic UCMSCs	Knee cartilage defects	South Korea
ANTEROGEN	Cupistem	Autologous ADMSC	Knee cartilage defects	South Korea
FCB PHARMICELL	Cellgram-AMI	Autologous BMMSC	Acute MI	South Korea
FIBROCELL TECHNOLOGIES	LaViv (Azficel-T)	Autologous fibroblast	Nasolabial fold wrinkles	USA
ANTEROGEN	Queencell	Autologous SVF	Connective tissue disorders	South Korea
S. BIOMEDICS	CureSkin	Autologous fibroblast	Depressed acne scars	South Korea
TEGO SCIENCES	Kaloderm	Allogenic keratinocyte	Diabetic foot ulcers	South Korea
TEGO SCIENCES	Kaloderm	Allogenic keratinocyte	Deep second-degree burn	South Korea
BIOSOLUTIONS	KeraHeal	Autologous keratinocyte	Second-degree burns	South Korea
Tissue-engineered products				
NUO THERAPEUTICS	Aurix	PRP gel	Wound treatment	USA
ORGANOGENESIS, INC. & NOVARTIS	Apligraf	Bi-layered living skin substitute	Chronic venous leg ulcers and diabetic foot	USA
ADMEDUS	CardioCel	Cardiovascular scaffold	Cardiovascular abnormalities	USA, EU, Canada
ORGANOGENESIS	Dermagraft	Allogenic fibroblasts /scaffold	Chronic foot ulcers	USA
VERICEL	Epicel	Autologous keratinocytes (CEA)	Full thickness burns	EU

(continued)

Table 6.1 (continued)

Company	Product name	Cell type/product specification	Indication	Country/region of marketing authorization
TEGO SCIENCES	Holoderm	Autologous keratinocytes	Burns, vitiligo, nevi, and scars	South Korea
CHA BIO	Hyalograft 3D	Autologous fibroblasts in 3D scaffolds	Diabetic foot ulcers	South Korea
J-TEC	JACC	Autologous cultured chondrocytes and collagen gel	Traumatic cartilage defect or osteochondritis dissecans of the knee	Japan
J-TEC	J-TEC	Autologous keratinocytes	Deep dermal and full-thickness burns	Japan
VERICEL	MACI	Autologous chondrocyte on a scaffold	Cartilage defects of the knee	USA
AESCULAP BIOLOGICS	Novocart 3D	3D scaffolds with embedded autologous chondrocytes	Articular cartilage repair	EU
INTEGRA	Omnigraft	Bilayer dermal substitute	Diabetic foot ulcers	USA
ORTHOCELL	Ortho-ACI	Autologous chondrocytes on scaffold	Cartilage defects	Australia
SEWON CELLONTECH CO.	RMS Ossron™	Cultured autologous osteoblasts	Bone defects	South Korea
ENZYVANT	RETHYMIC®	Allogeneic processed thymus tissue	Congenital athymia	USA
MALLINCKRODT	Stratagraft	Allogeneic keratinocytes and fibroblast in collagen	Deep partial-thickness burns	USA
ORGANOGENESIS	Transcyte	Skin substitute	Epidermolysis Bullosa	USA
COLLPLANT	Vergenix-STR	Collagen type I matrices	Connective tissue disorders	EU
COLLPLANT	Vergenix FG	Collagen type I matrices	Wound treatment	EU
Gene therapy products				
CSL Behring LLC	HEMGENIX	Adeno-associated virus	Hemophilia B	USA, EU
BLUEBIRD BIO	Abecma	(CAR) T-cell	Multiple myeloma	USA, EU, Canada, Japan
LEGEND BIOTECH & JANSSEN	Carvykti	(CAR) T-cell	Multiple myeloma	USA, EU
BRISTOL MYERS SQUIBB	Breyanzi	(CAR) T-cell	B-cell lymphoma	USA, EU
JW THERAPEUTICS	Carteyva	(CAR) T-cell	B-cell lymphoma	China
CASI PHARMACEUTICALS INC.	Inaticabtagene Autoleucel	(CAR) T-cell	B-cell lymphoma	China

(continued)

Table 6.1 (continued)

Company	Product name	Cell type/product specification	Indication	Country/region of marketing authorization
NOVARTIS	Kymriah	(CAR) T-cell	B-cell lymphoma	USA, EU, Canada, Japan, Singapore
KITE Pharma	Yescarta	(CAR) T-cell	B cell malignancies	USA, EU, Canada, Japan, China
KITE Pharma	Tecartus	(CAR) T-cell	Mantle cell lymphoma	USA, EU
VERTEX PHARMACEUTICALS	Casgevy	CRISPR gene editing	Sickle-cell disease and β-thalassaemia	United Kingdom, Bahrain
SAREPTA THERAPEUTICS	Elevidys	Adeno-associated virus	Duchenne muscular dystrophy	USA
SHENZHEN SIBIONO GENETECH	Gendicine	Engineered adenovirus	Head and neck squamous cell cancer	China
AMGEN	Imlygic	Herpes simplex virus-1	Recurrent melanoma	USA, EU, Australia
ORCHARD THERAPEUTICS	Libmeldy	Lentiviral vector	Metachromatic leukodystrophy	EU
SPARK THERAPEUTICS	Luxturna	Adeno-associated virus	Retinal dystrophy	USA, EU, Canada
BIOMARIN	Roctavian	Adeno-associated virus	Hemophilia A	USA, EU
BLUEBIRD BIO	Skysona	Gene-modified CD34+ cell	Active cerebral adrenoleukodystrophy	USA
GSK	Strimvelis	Gene-modified CD34+ cell	Adenosine deaminase severe combined immune deficiency	EU
PTC THERAPEUTICS	Upstaza	Adeno-associated virus	Aromatic L-amino acid decarboxylase deficiency	EU
KRYSTAL BIOTECH	VYJUVEK	Herpes simplex virus-1	Dystrophic epidermolysis bullosa	USA
NOVARTIS GENE THERAPIES	Zolgensma	Adeno-associated virus	Spinal muscular atrophy	USA, EU, Canada, Australia, South Korea
BLUEBIRD BIO	Zynteglo	Gene-modified CD34+ cell	ß-thalassemia	USA

three categories: cell therapy products, tissue-engineered products, and gene therapy products.

On the other hand, despite the recent advances in organoid generation, they cannot completely mimic the organs regarding cell types and maturation. Body organs grow in a 3D structure and contain cells belonging to different germ layers. Therefore, in-vitro organoids should recapitulate the exact organogenesis, giving rise to a 3D functional tissue/organ composed of cells from different germ layers [113]. In addition, technical challenges facing in-vitro organoid cultures are difficulties in the growth factor delivery responsible for organoid maturation, poor reproducibility for massive production, the unsuitable size of generated organoids, and limited maturity and cell diversity [1]. The nanoparticles and bioengineering materials being used in therapeutic protocols also face serious challenges such as safety hazards, their therapeutic properties, the lack of understanding of the in vivo metabolization, and the mechanisms of their interaction with different stem cells [110, 111]. These technical obstacles and challenges limiting the application of different regenerative medicine assays are currently under investigation worldwide [1].

6.5 Bioprinting

Organ failure is a big health issue with limited solutions. For decades, organ transplantation has been considered an ultimate therapeutic intervention for end-organ damage. Organ shortage, life-long immunosuppression, complex surgical procedures and postoperative care, and high costs are major limitations of this therapeutic option. These challenges and new achievements in regenerative medicine have driven biomedical scientists to produce functional organs in vitro [120]. The tissue engineering approach has shown progress in areas like in vitro manufacturing of skin substitutes and cartilage (Table S1) but still faces big challenges in producing more complex organs. Adding new technologies such as 3D-bioprinting to the conventional tissue engineering methods may accelerate achieving laboratory-grown functional organ [121].

6.5.1 Historical Background

Pioneers like Vladimir Mironov, Gabor Forgas, and Thomas Boland have combined technologies like cell patterning and commercial inkjet printing to create lab-grown human organs for transplantation. The first Extrusion-based bioprinting technology was reported in 2002, followed by the development of the first inkjet bioprinter in 2003. Mironov et al. introduced the concepts of layer-by-layer cell printing on a thermo-reversible gel for 3D organ bioprinting. In 2006, electro hydrodynamic was applied to deposit living cells, and scaffold-free vascular tissue was engineered in 2009. In 2015, coaxial technology was adopted for tubular structure fabrication, and rapid continuous optical 3D printing based on digital light processing (DLP) was applied in 2016. In 2019, a perfusable scale-down heart was manufactured, and collagen-based human hearts were bioprinted at various scales using freedom reversible embedding of suspended hydrogels (FRESH) technology [122–124].

6.5.2 Main Parameters

Among different bioprinting methods, extrusion-based bioprinting is the most widely used method for organ and tissue engineering. With this method, a wide range of viscosities can be applied which is in favor of using high cell density bio-inks. However, setting up extrusion-based bioprinters is quite simple, and commercial plotters or desktop 3D printers can easily be converted into extrusion-based bioprinters, making them suitable for low-cost customized services. Their main limitations are the low volume of bio-ink and strict criteria for bio-ink selection [120]. To select a suitable material for bio-ink, its printability, mechanical properties, biocompatibility, and similarity to extracellular matrices should be considered. Typical bio-inks include hydrogels, decellularized matrix components, microcarriers, tissue spheroids and strands, cell pellets, nanocomposite, and supramolecular bio-inks. Hydrogel is a popular material for bio-ink preparation because of its capability to provide a viable microenvironment for cell adhesion, growth, and proliferation. Some natural or synthetic materials such as alginate, fibrinogen, gelatin, collagen, silk fibroin, chitosan, agarose, pluronics, hyaluronic acid, Polyethylene glycol (PEG), Gelatin methacryloyl (GelMA), and polyethylene oxide (PEO) have been used for hydrogel formation. As aforementioned, the ideal bio-ink should be able to mimic the natural ECM of the targeted tissue or organ. For example, the addition of hyaluronic acid to chondrocyte bio-ink could be a good choice for cartilage bioprinting. Another factor affecting the functionality of a bioprinted organ or tissue is its vascularization. Efficient blood vessel formation is crucial for the growth, viability, differentiation, and engraftment of bioprinted constructs [121, 124, 125].

6.5.3 Artificial Intelligence and Bioprinting

Recently, the importance of artificial intelligence (AI) in the bioprinting of human organs/tissues is increasingly recognized, as it has become a powerful tool in this field. AI enhances the efficiency, accuracy, and overall success of bioprinting processes in several ways. One crucial aspect of AI contribution is designing and optimizing organ structures. Creating complex organ structures with intricate vascular networks is challenging, but AI algorithms can analyze vast data to generate optimized designs that mimic natural organs. By considering factors such as tissue compatibility, mechanical properties, and functionality requirements, AI helps researchers create more efficient and functional organ models. AI also enables real-time monitoring and control during the bioprinting process. Another significant contribution of AI is its ability to analyze large datasets related to cell behavior and tissue development. By processing this information, AI algorithms can identify patterns and predict how different cell types interact within a printed organ structure. This predictive capability helps researchers optimize cell placement strategies for enhanced tissue integration and functionality. Furthermore, AI plays a vital role in quality control during the bioprinting process. It can analyze images or scans of printed organs to detect any structural defects or inconsistencies that may affect their functionality or viability. This automated inspection saves time compared to manual examination while ensuring higher accuracy. In addition to these technical aspects, AI facilitates collaboration among researchers by enabling data-sharing platforms and fostering knowledge exchange. Machine learning algorithms can leverage collective insights gained from various research projects to further improve bioprinting techniques. Overall, artificial intelligence has become indispensable in advancing the field of bioprinting human organs. Continued advancements in both AI and bioprinting technologies hold promise for overcoming organ shortages and transforming healthcare as we know it [126, 127].

6.5.4 Future Directions

Despite the promising future of 3D bioprinting in filling unmet medical needs, its current applications are limited to drug screening, toxicology studies, disease modeling, and the production of personalized implants. The first commercialized 3D-printed tissue, a hepatic organoid, was introduced by Organovo Inc. (USA) for medical research, disease modeling, and drug discovery [128, 129]. If regulatory agencies accept the use of bioprinted organs/tissue for preclinical and translational studies, the market for this new technology will be considerably expanded. However, future research still needs to address limitations, such as improving bioprinting techniques, optimizing cellular sources, and selecting a suitable biomaterial compound. Tissues and organs have different sizes, rigidities, and cell types, making accurate bioprinting challenging. To overcome this complexity, multi-scale, multi-material, and multi-cell bioprinting is a priority for its future technology development. There is also a critical need to develop better assays, which can analyze cell viability and functionality in 3D constructs. Finally, to use the bioprinted constructs in clinical and translational studies, it is necessary to integrate functional vasculature to ensure long-term graft survival [130].

6.6 Conclusion

Despite considerable progress in regenerative therapies, manufacturing a laboratory-grown functional organ is challenging and far from the market. The current application of regenerative medicine is repairing and regenerating damaged tissues and organs using cell-based therapies, tissue-engineered products, and gene manipulation. A combination of new advancements in regenerative therapies with new technologies such as artificial intelligence and 3D-bioprinting could significantly accelerate reaching the valuable objective of organ replacement.

References

1. Altyar AE, El-Sayed A, Abdeen A, Piscopo M, Mousa SA, Najda A, Abdel-Daim MM. Future regenerative medicine developments and their therapeutic applications. Biomed Pharmacother. 2023;158:114131.
2. Dzobo K, Thomford NE, Senthebane DA, Shipanga H, Rowe A, Dandara C, et al. Advances in regenerative medicine and tissue engineering: innovation and transformation of medicine. Stem Cells Int. 2018;2018:2495848.
3. Mousaei Ghasroldasht M, Seok J, Park HS, Liakath Ali FB, Al-Hendy A. Stem cell therapy: from idea to clinical practice. Int J Mol Sci. 2022;23(5):2850.
4. Atala A. Advances in tissue and organ replacement. Curr Stem Cell Res Ther. 2008;3(1):21–31.
5. Mendelson A, Frenette PS. Hematopoietic stem cell niche maintenance during homeostasis and regeneration. Nat Med. 2014;20(8):833–46.
6. Bailey AM, Mendicino M, Au P. An FDA perspective on preclinical development of cell-based regenerative medicine products. Nat Biotechnol. 2014;32(8):721–3.
7. Knoepfler PS. From bench to FDA to bedside: US regulatory trends for new stem cell therapies. Adv Drug Deliv Rev. 2015;82-83:192–6.
8. Tang XL, Li Q, Rokosh G, Sanganalmath SK, Chen N, Ou Q, et al. Long-term outcome of administration of c-kit(POS) cardiac progenitor cells after acute myocardial infarction: transplanted cells do not become cardiomyocytes, but structural and functional improvement and proliferation of endogenous cells persist for at least one year. Circ Res. 2016;118(7):1091–105.
9. Polak JM, Mantalaris S. Stem cells bioprocessing: an important milestone to move regenerative medicine research into the clinical arena. Pediatr Res. 2008;63(5):461–6.
10. Chen FM, Zhao YM, Jin Y, Shi S. Prospects for translational regenerative medicine. Biotechnol Adv. 2012;30(3):658–72.
11. Rajabzadeh N, Fathi E, Farahzadi R. Stem cell-based regenerative medicine. Stem Cell Invest. 2019;6:19.
12. Guan X, Avci-Adali M, Alarçin E, Cheng H, Kashaf SS, Li Y, et al. Development of hydrogels for regenerative engineering. Biotechnol J. 2017;12(5):600394.
13. Gao G, Cui X. Three-dimensional bioprinting in tissue engineering and regenerative medicine. Biotechnol Lett. 2016;38(2):203–11.
14. Kraeutler MJ, Belk JW, Purcell JM, McCarty EC. Microfracture versus autologous chondrocyte implantation for articular cartilage lesions in the knee: a systematic review of 5-year outcomes. Am J Sports Med. 2018;46(4):995–9.
15. Mistry H, Connock M, Pink J, Shyangdan D, Clar C, Royle P, et al. Autologous chondrocyte implantation in the knee: systematic review and economic evaluation. Health Technol Assess. 2017;21(6):1–294.
16. Ratajczak MZ, Bujko K, Wojakowski W. Stem cells and clinical practice: new advances and challenges at the time of emerging problems with induced pluripotent stem cell therapies. Pol Arch Med Wewn. 2016;126(11):879–90.
17. Ratajczak MZ, Ratajczak J. Extracellular microvesicles as game changers in better understanding the complexity of cellular interactions-from bench to clinical applications. Am J Med Sci. 2017;354(5):449–52.
18. Feng Z, Gao F. Stem cell challenges in the treatment of neurodegenerative disease. CNS Neurosci Ther. 2012;18(2):142–8.
19. Lee JM, Jung J, Lee HJ, Jeong SJ, Cho KJ, Hwang SG, Kim GJ. Comparison of immunomodulatory effects of placenta mesenchymal stem cells with bone marrow and adipose mesenchymal stem cells. Int Immunopharmacol. 2012;13(2):219–24.
20. Zhu XY, Klomjit N, Conley SM, Ostlie MM, Jordan KL, Lerman A, Lerman LO. Impaired immunomodulatory capacity in adipose tissue-derived mesenchymal stem/stromal cells isolated from obese patients. J Cell Mol Med. 2021;25(18):9051–9.
21. Stiner R, Alexander M, Liu G, Liao W, Liu Y, Yu J, et al. Transplantation of stem cells from umbilical cord blood as therapy for type I diabetes. Cell Tissue Res. 2019;378:155–62.
22. Pavathuparambil Abdul Manaph N, Sivanathan KN, Nitschke J, Zhou X-F, Coates PT, Drogemuller CJ. An overview on small molecule-induced differentiation of mesenchymal stem cells into beta cells for diabetic therapy. Stem Cell Res Ther. 2019;10(1):1–18.
23. De Klerk E, Hebrok M. Stem cell-based clinical trials for diabetes mellitus. Front Endocrinol. 2021;12:631463.
24. Wang Y, Yu X, Chen E, Li L. Liver-derived human mesenchymal stem cells: a novel therapeutic source for liver diseases. Stem Cell Res Ther. 2016;7(1):71.
25. Secunda R, Vennila R, Mohanashankar AM, Rajasundari M, Jeswanth S, Surendran R. Isolation, expansion and characterisation of mesenchymal stem cells from human bone marrow, adipose tissue, umbilical cord blood and matrix: a comparative study. Cytotechnology. 2015;67(5):793–807.
26. Rama P, Matuska S, Paganoni G, Spinelli A, De Luca M, Pellegrini G. Limbal stem-cell therapy and long-term corneal regeneration. N Engl J Med. 2010;363(2):147–55.
27. Jabbarpour Z, Aghayan S, Arjmand B, Fallahzadeh K, Alavi-Moghadam S, Larijani B, Aghayan HR. Xeno-free protocol for GMP-compliant manufacturing of human fetal pancreas-derived mesenchymal stem cells. Stem Cell Res Ther. 2022;13(1):268.
28. Larijani B, Aghayan H-R, Goodarzi P, Arjmand B. GMP-grade human fetal liver-derived mesenchymal stem cells for clinical transplantation. In: Turksen K, editor. Stem cells and good manufacturing practices: methods, protocols, and regulations. New York: Springer; 2015. p. 123–36.

29. Larijani B, Arjmand B, Ahmadbeigi N, Falahzadeh K, Soleimani M, Sayahpour FA, Aghayan HR. A simple and cost-effective method for isolation and expansion of human fetal pancreas derived mesenchymal stem cells. Arch Iran Med. 2015;18(11):770–5.
30. Hass R, Kasper C, Böhm S, Jacobs R. Different populations and sources of human mesenchymal stem cells (MSC): a comparison of adult and neonatal tissue-derived MSC. Cell Commun Signal. 2011;9:12.
31. Mastrolia I, Foppiani EM, Murgia A, Candini O, Samarelli AV, Grisendi G, et al. Challenges in clinical development of mesenchymal stromal/stem cells: concise review. Stem Cells Transl Med. 2019;8(11):1135–48.
32. Saeedi P, Halabian R, Imani Fooladi AA. A revealing review of mesenchymal stem cells therapy, clinical perspectives and modification strategies. Stem Cell Invest. 2019;6:34.
33. Peng L, Jia Z, Yin X, Zhang X, Liu Y, Chen P, et al. Comparative analysis of mesenchymal stem cells from bone marrow, cartilage, and adipose tissue. Stem Cells Dev. 2008;17(4):761–73.
34. Bourin P, Bunnell BA, Casteilla L, Dominici M, Katz AJ, March KL, et al. Stromal cells from the adipose tissue-derived stromal vascular fraction and culture expanded adipose tissue-derived stromal/stem cells: a joint statement of the International Federation for Adipose Therapeutics and Science (IFATS) and the International Society for Cellular Therapy (ISCT). Cytotherapy. 2013;15(6):641–8.
35. Kuhbier JW, Weyand B, Radtke C, Vogt PM, Kasper C, Reimers K. Isolation, characterization, differentiation, and application of adipose-derived stem cells. Adv Biochem Eng Biotechnol. 2010;123:55–105.
36. Choudhery MS, Badowski M, Muise A, Harris DT. Utility of cryopreserved umbilical cord tissue for regenerative medicine. Curr Stem Cell Res Ther. 2013;8(5):370–80.
37. Nabetani M, Shintaku H, Hamazaki T. Future perspectives of cell therapy for neonatal hypoxic-ischemic encephalopathy. Pediatr Res. 2018;83(1-2):356–63.
38. Barry F, Murphy M. Mesenchymal stem cells in joint disease and repair. Nat Rev Rheumatol. 2013;9(10):584–94.
39. Rivera-Izquierdo M, Cabeza L, Láinez-Ramos-Bossini A, Quesada R, Perazzoli G, Alvarez P, et al. An updated review of adipose derived-mesenchymal stem cells and their applications in musculoskeletal disorders. Expert Opin Biol Ther. 2019;19(3):233–48.
40. Reinders ME, Fibbe WE, Rabelink TJ. Multipotent mesenchymal stromal cell therapy in renal disease and kidney transplantation. Nephrol Dial Transplant. 2010;25(1):17–24.
41. Sun Y, Lu Y, Yin L, Liu Z. The roles of nanoparticles in stem cell-based therapy for cardiovascular disease. Front Bioeng Biotechnol. 2020;8:947.
42. Müller AM, Huppertz S, Henschler R. Hematopoietic stem cells in regenerative medicine: astray or on the path? Transfus Med Hemother. 2016;43(4):247–54.
43. Sheridan C. For hemophilia and thalassemia, a new era of 'one-and-done' gene therapies has arrived. Nat Biotechnol. 2022;40(11):1531–3.
44. Shin JH, Ryu C-M, Yu HY, Park J, Kang AR, Shin JM, et al. Safety of human embryonic stem cell-derived mesenchymal stem cells for treating interstitial cystitis: a phase I study. Stem Cells Transl Med. 2022;11(10):1010–20.
45. Menasché P, Vanneaux V, Hagège A, Bel A, Cholley B, Parouchev A, et al. Transplantation of human embryonic stem cell–derived cardiovascular progenitors for severe ischemic left ventricular dysfunction. J Am Coll Cardiol. 2018;71(4):429–38.
46. da Cruz L, Fynes K, Georgiadis O, Kerby J, Luo YH, Ahmado A, et al. Phase 1 clinical study of an embryonic stem cell–derived retinal pigment epithelium patch in age-related macular degeneration. Nat Biotechnol. 2018;36(4):328–37.
47. Shapiro AJ, Thompson D, Donner TW, Bellin MD, Hsueh W, Pettus J, et al. Insulin expression and C-peptide in type 1 diabetes subjects implanted with stem cell-derived pancreatic endoderm cells in an encapsulation device. Cell Rep Med. 2021;2(12):100466.
48. Sharkis SJ, Jones RJ, Civin C, Jang YY. Pluripotent stem cell-based cancer therapy: promise and challenges. Sci Transl Med. 2012;4:127–9.
49. Kim JY, Nam Y, Rim YA, Ju JH. Review of the current trends in clinical trials involving induced pluripotent stem cells. Stem Cell Rev Rep. 2022;18(1):142–54.
50. Hofer M, Lutolf MP. Engineering organoids. Nat Rev Mater. 2021;6(5):402–20.
51. Sun N, Meng X, Liu Y, Song D, Jiang C, Cai J. Applications of brain organoids in neurodevelopment and neurological diseases. J Biomed Sci. 2021;28(1):30.
52. Corrò C, Novellasdemunt L, Li VSW. A brief history of organoids. Am J Physiol Cell Physiol. 2020;319(1):151–65.
53. Vives J, Batlle-Morera L. The challenge of developing human 3D organoids into medicines. Stem Cell Res Ther. 2020;11(1):72.
54. Kim J, Koo B-K, Knoblich JA. Human organoids: model systems for human biology and medicine. Nat Rev Mol Cell Biol. 2020;21(10):571–84.
55. Lee H, Son MY. Current challenges associated with the use of human induced pluripotent stem cell-derived organoids in regenerative medicine. Int J Stem Cells. 2021;14(1):9–20.
56. Li Y, Tang P, Cai S, Peng J, Hua G. Organoid based personalized medicine: from bench to bedside. Cell Regen. 2020;9(1):21.

57. Schwank G, Koo BK, Sasselli V, Dekkers JF, Heo I, Demircan T, et al. Functional repair of CFTR by CRISPR/Cas9 in intestinal stem cell organoids of cystic fibrosis patients. Cell Stem Cell. 2013;13(6):653–8.
58. Homicsko K. Organoid technology and applications in cancer immunotherapy and precision medicine. Curr Opin Biotechnol. 2020;65:242–7.
59. Yui S, Nakamura T, Sato T, Nemoto Y, Mizutani T, Zheng X, et al. Functional engraftment of colon epithelium expanded in vitro from a single adult $Lgr5^+$ stem cell. Nat Med. 2012;18(4):618–23.
60. Choi WH, Bae DH, Yoo J. Current status and prospects of organoid-based regenerative medicine. BMB Rep. 2023;56(1):10–4.
61. Kim MS, Kim DH, Kang HK, Kook MG, Choi SW, Kang KS. Modeling of hypoxic brain injury through 3D human neural organoids. Cells. 2021;10(2):234.
62. Kruczek K, Gonzalez-Cordero A, Goh D, Naeem A, Jonikas M, Blackford SJI, et al. Differentiation and transplantation of embryonic stem cell-derived cone photoreceptors into a mouse model of end-stage retinal degeneration. Stem Cell Rep. 2017;8(6):1659–74.
63. Santos-Ferreira T, Völkner M, Borsch O, Haas J, Cimalla P, Vasudevan P, et al. Stem cell-derived photoreceptor transplants differentially integrate into mouse models of cone-rod dystrophy. Invest Ophthalmol Vis Sci. 2016;57(7):3509–20.
64. McLelland BT, Lin B, Mathur A, Aramant RB, Thomas BB, Nistor G, et al. Transplanted hESC-derived retina organoid sheets differentiate, integrate, and improve visual function in retinal degenerate rats. Invest Ophthalmol Vis Sci. 2018;59(6):2586–603.
65. O'Hara-Wright M, Gonzalez-Cordero A. Retinal organoids: a window into human retinal development. Development. 2020;147(24):189746.
66. van der Valk WH, Steinhart MR, Zhang J, Koehler KR. Building inner ears: recent advances and future challenges for in vitro organoid systems. Cell Death Differ. 2021;28(1):24–34.
67. Moeinvaziri F, Shojaei A, Haghparast N, Yakhkeshi S, Nemati S, Hassani S-N, Baharvand H. Towards maturation of human otic hair cell–like cells in pluripotent stem cell–derived organoid transplants. Cell Tissue Res. 2021;386(2):321–33.
68. Bannier-Hélaouët M, Post Y, Korving J, Trani Bustos M, Gehart H, Begthel H, et al. Exploring the human lacrimal gland using organoids and single-cell sequencing. Cell Stem Cell. 2021;28(7):1221–32.
69. Jeong SY, Choi WH, Jeon SG, Lee S, Park JM, Park M, et al. Establishment of functional epithelial organoids from human lacrimal glands. Stem Cell Res Ther. 2021;12(1):247.
70. Sun H, Zhang YX, Li YM. Generation of skin organoids: potential opportunities and challenges. Front Cell Dev Biol. 2021;9:709824.
71. Lee J, Rabbani CC, Gao H, Steinhart MR, Woodruff BM, Pflum ZE, et al. Hair-bearing human skin generated entirely from pluripotent stem cells. Nature. 2020;582(7812):399–404.
72. Múnera JO, Wells JM. Generation of gastrointestinal organoids from human pluripotent stem cells. Methods Mol Biol. 2017;1597:167–77.
73. Jee J, Park JH, Im JH, Kim MS, Park E, Lim T, et al. Functional recovery by colon organoid transplantation in a mouse model of radiation proctitis. Biomaterials. 2021;275:120925.
74. Sugimoto S, Kobayashi E, Fujii M, Ohta Y, Arai K, Matano M, et al. An organoid-based organ-repurposing approach to treat short bowel syndrome. Nature. 2021;592(7852):99–104.
75. Okamoto R, Shimizu H, Suzuki K, Kawamoto A, Takahashi J, Kawai M, et al. Organoid-based regenerative medicine for inflammatory bowel disease. Regener Ther. 2020;13:1–6.
76. Fawzy A, Prince A, Hassan A, Fayed A, Zschöck M, Naga M, et al. Epidemiological studies on Johne's disease in ruminants and Crohn's disease in humans in Egypt. Amsterdam: Elsevier; 2013. p. 79–86.
77. Ogundipe VML, Groen AH, Hosper N, Nagle PW, Hess J, Faber H, et al. Generation and differentiation of adult tissue-derived human thyroid organoids. Stem Cell Rep. 2021;16(4):913–25.
78. Bartfeld S, Bayram T, van de Wetering M, Huch M, Begthel H, Kujala P, et al. In vitro expansion of human gastric epithelial stem cells and their responses to bacterial infection. Gastroenterology. 2015;148(1):126–36.
79. Nuciforo S, Heim MH. Organoids to model liver disease. JHEP Rep. 2021;3(1):100198.
80. Huch M, Dorrell C, Boj SF, van Es JH, Li VS, van de Wetering M, et al. In vitro expansion of single Lgr5+ liver stem cells induced by Wnt-driven regeneration. Nature. 2013;494(7436):247–50.
81. Peng WC, Logan CY, Fish M, Anbarchian T, Aguisanda F, Álvarez-Varela A, et al. Inflammatory cytokine TNFα promotes the long-term expansion of primary hepatocytes in 3D culture. Cell. 2018;175(6):1607–19.
82. Takebe T, Sekine K, Enomura M, Koike H, Kimura M, Ogaeri T, et al. Vascularized and functional human liver from an iPSC-derived organ bud transplant. Nature. 2013;499(7459):481–4.
83. Nie Y-Z, Zheng Y-W, Ogawa M, Miyagi E, Taniguchi H. Human liver organoids generated with single donor-derived multiple cells rescue mice from acute liver failure. Stem Cell Res Ther. 2018;9(1):5.
84. Kruitwagen HS, Oosterhoff LA, van Wolferen ME, Chen C, Nantasanti Assawarachan S, Schneeberger K, et al. Long-term survival of transplanted autologous canine liver organoids in a COMMD1-deficient dog model of metabolic liver disease. Cells. 2020;9(2):410.
85. Ghezelayagh Z, Zabihi M, Kazemi Ashtiani M, Ghezelayagh Z, Lynn FC, Tahamtani Y. Recapitulating pancreatic cell–cell interactions

through bioengineering approaches: the momentous role of non-epithelial cells for diabetes cell therapy. Sciences. 2021;78(23):7107–32.

86. Zhang X, Ma Z, Song E, Xu T. Islet organoid as a promising model for diabetes. Protein Cell. 2022;13(4):239–57.
87. Bengtsson A, Andersson R, Rahm J, Ganganna K, Andersson B, Ansari D. Organoid technology for personalized pancreatic cancer therapy. Cell Oncol. 2021;44(2):251–60.
88. Petersen MBK, Gonçalves CAC, Kim YH, Grapin-Botton A. Recapitulating and deciphering human pancreas development from human pluripotent stem cells in a dish. Curr Top Dev Biol. 2018;129:143–90.
89. Yu XX, Xu CR. Understanding generation and regeneration of pancreatic β cells from a single-cell perspective. Development. 2020;147(7):179051.
90. Wang D, Wang J, Bai L, Pan H, Feng H, Clevers H, Zeng YA. Long-term expansion of pancreatic islet organoids from resident Procr(+) progenitors. Cell. 2020;180(6):1198–211.e19.
91. Soltanian A, Ghezelayagh Z, Mazidi Z, Halvaei M, Mardpour S, Ashtiani MK, et al. Generation of functional human pancreatic organoids by transplants of embryonic stem cell derivatives in a 3D-printed tissue trapper. J Cell Physiol. 2019;234(6):9564–76.
92. Lebreton F, Lavallard V, Bellofatto K, Bonnet R, Wassmer CH, Perez L, et al. Insulin-producing organoids engineered from islet and amniotic epithelial cells to treat diabetes. Nat Commun. 2019;10(1):4491.
93. Ghezelayagh Z, Zabihi M, Zarkesh I, Gonçalves CAC, Larsen M, Hagh-Parast N, et al. Improved differentiation of hESC-derived pancreatic progenitors by using human fetal pancreatic mesenchymal cells in a micro-scalable three-dimensional co-culture system. Stem Cell Rev Rep. 2022;18(1):360–77.
94. Sampaziotis F, Muraro D, Tysoe OC, Sawiak S, Beach TE, Godfrey EM, et al. Cholangiocyte organoids can repair bile ducts after transplantation in the human liver. Science. 2021;371(6531):839–46.
95. Pringle S, Maimets M, van der Zwaag M, Stokman MA, van Gosliga D, Zwart E, et al. Human salivary gland stem cells functionally restore radiation damaged salivary glands. Stem Cells. 2016;34(3):640–52.
96. Dye BR, Youngblood RL, Oakes RS, Kasputis T, Clough DW, Spence JR, Shea LD. Human lung organoids develop into adult airway-like structures directed by physico-chemical biomaterial properties. Biomaterials. 2020;234:119757.
97. Weiner AI, Jackson SR, Zhao G, Quansah KK, Farshchian JN, Neupauer KM, et al. Mesenchyme-free expansion and transplantation of adult alveolar progenitor cells: steps toward cell-based regenerative therapies. NPJ Regen Med. 2019;4:17.
98. Garreta E, Prado P, Tarantino C, Oria R, Fanlo L, Martí E, et al. Fine tuning the extracellular environment accelerates the derivation of kidney organoids from human pluripotent stem cells. Nat Mater. 2019;18(4):397–405.
99. van den Berg CW, Ritsma L, Avramut MC, Wiersma LE, van den Berg BM, Leuning DG, et al. Renal subcapsular transplantation of PSC-derived kidney organoids induces neo-vasculogenesis and significant glomerular and tubular maturation in vivo. Stem Cell Rep. 2018;10(3):751–65.
100. Filippo Buono M, von Boehmer L, Strang J, Hoerstrup SP, Emmert MY, Nugraha B. Human cardiac organoids for modeling genetic cardiomyopathy. Cells. 2020;9(7):1733.
101. Varzideh F, Pahlavan S, Ansari H, Halvaei M, Kostin S, Feiz MS, et al. Human cardiomyocytes undergo enhanced maturation in embryonic stem cell-derived organoid transplants. Biomaterials. 2019;192:537–50.
102. Pamarthy S, Sabaawy HE. Patient derived organoids in prostate cancer: improving therapeutic efficacy in precision medicine. Mol Cancer. 2021;20(1):125.
103. Chumduri C, Turco MY. Organoids of the female reproductive tract. J Mol Med. 2021;99(4):531–53.
104. Turco MY, Gardner L, Hughes J, Cindrova-Davies T, Gomez MJ, Farrell L, et al. Long-term, hormone-responsive organoid cultures of human endometrium in a chemically defined medium. Nat Cell Biol. 2017;19(5):568–77.
105. Syed SM, Kumar M, Ghosh A, Tomasetig F, Ali A, Whan RM, et al. Endometrial Axin2(+) cells drive epithelial homeostasis, regeneration, and cancer following oncogenic transformation. Cell Stem Cell. 2020;26(1):64–80.
106. Chumduri C, Gurumurthy RK, Berger H, Dietrich O, Kumar N, Koster S, et al. Opposing Wnt signals regulate cervical squamocolumnar homeostasis and emergence of metaplasia. Nat Cell Biol. 2021;23(2):184–97.
107. Ali A, Syed SM, Jamaluddin MFB, Colino-Sanguino Y, Gallego-Ortega D, Tanwar PS. Cell lineage tracing identifies hormone-regulated and Wnt-responsive vaginal epithelial stem cells. Cell Rep. 2020;30(5):1463–77.
108. Kessler M, Hoffmann K, Brinkmann V, Thieck O, Jackisch S, Toelle B, et al. The Notch and Wnt pathways regulate stemness and differentiation in human fallopian tube organoids. Nat Commun. 2015;6:8989.
109. Turco MY, Gardner L, Kay RG, Hamilton RS, Prater M, Hollinshead MS, et al. Trophoblast organoids as a model for maternal-fetal interactions during human placentation. Nature. 2018;564(7735):263–7.
110. Kim DS, Bae S. Impact and challenges of enactment for advanced regenerative medicine in South Korea. Front Bioeng Biotechnol. 2022;10:972865.
111. Wang Z, Ruan J, Cui D. Advances and prospect of nanotechnology in stem cells. Nanoscale Res Lett. 2009;4(7):593–605.

112. Liu S, Zhou J, Zhang X, Liu Y, Chen J, Hu B, et al. Strategies to optimize adult stem cell therapy for tissue regeneration. Int J Mol Sci. 2016;17(6):982.
113. Suman S, Domingues A, Ratajczak J, Ratajczak MZ. Potential clinical applications of stem cells in regenerative medicine. Adv Exp Med Biol. 2019;1201:1–22.
114. He J, Zhang X, Xia X, Han M, Li F, Li C, et al. Organoid technology for tissue engineering. J Mol Cell Biol. 2020;12(8):569–79.
115. Goessler UR, Riedel K, Hormann K, Riedel F. Perspectives of gene therapy in stem cell tissue engineering. Cells Tissues Organs. 2006;183(4):169–79.
116. Howard D, Buttery LD, Shakesheff KM, Roberts SJ. Tissue engineering: strategies, stem cells and scaffolds. J Anat. 2008;213(1):66–72.
117. Boehler RM, Graham JG, Shea LD. Tissue engineering tools for modulation of the immune response. BioTechniques. 2011;51(4):239–40.
118. Andorko JI, Jewell CM. Designing biomaterials with immunomodulatory properties for tissue engineering and regenerative medicine. Bioeng Transl Med. 2017;2(2):139–55.
119. De Pieri A, Rochev Y, Zeugolis DI. Scaffold-free cell-based tissue engineering therapies: advances, shortfalls and forecast. NPJ Regen Med. 2021;6(1):18.
120. Gu Z, Fu J, Lin H, He Y. Development of 3D bioprinting: from printing methods to biomedical applications. Asian J Pharm Sci. 2020;15(5):529–57.
121. Karzyński K, Kosowska K, Ambrożkiewicz F, Berman A, Cichoń J, Klak M, et al. Use of 3D bioprinting in biomedical engineering for clinical application. Med Stud. 2018;34(1):93–7.
122. Mendoza-Cerezo L, Jesús MR, Macías-García A, Marcos-Romero AC, Díaz-Parralejo A. Evolution of bioprinting and current applications. Int J Bioprint. 2023;9(4):742.
123. Gupta D, Negi NP. 3D bioprinting: printing the future and recent advances. Bioprinting. 2022;27:e00211.
124. Grigoryan B, Sazer DW, Avila A, Albritton JL, Padhye A, Ta AH, et al. Development, characterization, and applications of multi-material stereolithography bioprinting. Sci Rep. 2021;11(1):3171.
125. Agarwal S, Saha S, Balla VK, Pal A, Barui A, Bodhak S. Current developments in 3D bioprinting for tissue and organ regeneration–a review. Front Mech Eng. 2020;6:589171.
126. Raj B, Gupta BB, Yamaguchi S, Gill SS. AI for big data-based engineering applications from security perspectives. Boca Raton: CRC Press; 2023.
127. Reddy VS, Ramasubramanian B, Telrandhe VM, Ramakrishna S. Contemporary standpoint and Future of 3D bioprinting in tissue/organs printing. Curr Opin Biomed Eng. 2023;2023:100461.
128. Javaid M, Haleem A, Singh RP, Suman R. 3D printing applications for healthcare research and development. Glob Health J. 2022;6:217–26.
129. Yaneva A, Shopova D, Bakova D, Mihaylova A, Kasnakova P, Hristozova M, Semerdjieva M. The progress in bioprinting and its potential impact on health-related quality of life. Bioengineering. 2023;10(8):910.
130. Datta P, Barui A, Wu Y, Ozbolat V, Moncal KK, Ozbolat IT. Essential steps in bioprinting: from pre-to post-bioprinting. Biotechnol Adv. 2018;36(5):1481–504.

Personalized Medicine and Therapeutics

7

Negar Sarhangi, George P. Patrinos, Hamid Reza Aghaei Meybodi, Shekoufeh Nikfar, Behnaz Sarrami, Pantea Kiani, Bagher Larijani, Nazli Khodayari, and Mandana Hasanzad

What Will You Learn in This Chapter?
In this chapter, you will gain a comprehensive understanding of personalized medicine and its transformative impact on clinical practice. The exploration begins by elucidating the fundamental principles and concepts underpinning personalized medicine. The chapter unfolds to elucidate how advancements in molecular biology and technology have paved the way for a paradigm shift in healthcare, allowing for tailored and precise treatment strategies. As you progress through the chapter, you will learn about the role of genomic information in stratifying patient populations, identifying susceptibility to diseases, and predicting treatment responses through a pharmacogenomics approach. Furthermore, the chapter navigates through the challenges and ethical considerations associated with implementing

N. Sarhangi
Medical Genomics Research Center, Tehran Medical Sciences, Islamic Azad University, Tehran, Iran

Endocrinology and Metabolism Research Center, Endocrinology and Metabolism Clinical Sciences Institute, Tehran University of Medical Sciences, Tehran, Iran

G. P. Patrinos
Department of Pharmacy, School of Health Sciences, University of Patras, Patras, Greece

Department of Genetics and Genomics, College of Medicine and Health Sciences, United Arab Emirates University, Al-Ain, Abu Dhabi, UAE

Zayed Center for Health Sciences, United Arab Emirates University, Al-Ain, Abu Dhabi, UAE

Faculty of Medicine and Health Sciences, Department of Pathology, Clinical Bioinformatics Unit, Erasmus University Medical Center, Rotterdam, The Netherlands

H. R. Aghaei Meybodi · B. Larijani
Endocrinology and Metabolism Research Center, Endocrinology and Metabolism Clinical Sciences Institute, Tehran University of Medical Sciences, Tehran, Iran

S. Nikfar
Faculty of Pharmacy, Department of Pharmacoeconomics and Pharmaceutical Administration, Tehran University of Medical Sciences, Tehran, Iran

B. Sarrami
Missouri Pharmacogenomics Consulting, St. Louis, MO, USA

P. Kiani
Division of Pharmacology, Utrecht Institute for Pharmaceutical Sciences, Utrecht University, Utrecht, The Netherlands

PanGenix, Maassluis, The Netherlands

N. Khodayari
Division of Pulmonary, Critical Care and Sleep Medicine, Department of Medicine, University of Florida, Gainesville, FL, USA

M. Hasanzad (✉)
Personalized Medicine Research Center, Endocrinology and Metabolism Clinical Sciences Institute, Tehran University of Medical Sciences, Tehran, Iran

M. Hasanzad (ed.), *A Glimpse at Medicine in the Future*,
https://doi.org/10.1007/978-981-97-4974-4_7

personalized medicine, addressing issues related to privacy, equity, and accessibility. By the conclusion of this chapter, you will have acquired a holistic perspective on personalized medicine, appreciating its multifaceted impacts on clinical decision-making, patient care, and the trajectory of healthcare as a whole. This knowledge will empower you to comprehend the ongoing developments, controversies, and potential future directions in the field, enabling you to critically evaluate the role of personalized medicine in shaping the future of clinical practice.

Rationale and Importance
Personalized medicine, also known as precision medicine, is a novel approach to medical treatment and healthcare that takes individual differences in patients' genes, environments, and lifestyles into consideration. Several essential aspects can be used to understand personalized medicine's rationale and importance, including targeted therapy, optimizing drug selection and dosages, reduced adverse drug reactions, and predictive and preventive strategies.

7.1 Principles of Personalized Medicine

The concept of personalized medicine is not new and can be traced back to Hippocrates. He believes in the concept of disease individuality. He said, "It is even more important to know what kind of individual suffers from a disease than knowing from what disease an individual suffers" [1].

The traditional medicine model used to be a "one size fits all" where everyone diagnosed with a particular disease would get treated the same way. However, problems started to emerge due to patients not responding or developing adverse effects; therefore, a different approach was needed. Inter-individual differences such as age, race, genetics, environmental factors, and comorbid conditions led to a more personalized model, which is now called Precision Medicine [2]. Precision Medicine takes into account individual differences to tailor both the prevention and treatment of diseases.

The principal concepts of P4 medicine—participatory, personalized (treatment), predictive, and preventive medicine—have emerged within the past 20 years [3]. P4 medicine is thus primarily focused on using huge amounts of data, mainly biological (omics data). This is especially true for personalized treatment, which is one of the pillars of P4 medicine [4]. In some references, precision medicine has seven pillars: Diagnostics, prediction, prevention, prognostics, personalized treatment (Pharmacogenomics), monitoring, and participatory.

The personalized medicine approach is regarded as an art of medicine. According to the US National Human Genome Research Institute. "Precision medicine (generally considered equivalent to personalized medicine or individualized medicine) is a novel approach which employs information about an individual's genomic, environmental, and lifestyle data to guide decisions related to their medical management."

According to the National Cancer Institute (NCI), personalized medicine is defined as follows: "A form of medicine that uses information about a person's genes or proteins to prevent, diagnose, or treat disease."

Personalized treatment, as a main component of personalized medicine, provides two approaches for targeting germline and somatic mutation, which are called pharmacogenomics and targeted therapy, respectively. Pharmacogenomics (PGx) focuses on the relationship between genetic variations and drug response. The concept of targeted therapy has had a significant impact on one field of medicine known as oncology [5, 6].

Precision medicine is defined as using cutting-edge molecular profiling that helps develop accurate diagnostic, prognostic, and therapeutic approaches tailored to a patient's needs [7].

Precision medicine appears to be delivering promising results, including the effectiveness of care, early detection of genetically caused diseases, and new taxonomy for disease.

In practice, precision medicine remains focused on the use of huge amounts of data, both in terms of the individuals involved and the omics (for example, genomics, transcriptomics, epigenomics, proteomics, metabolomics, and pharmacogenomics) data used.

According to the second international consensus report on gaps and opportunities for the clinical translation of precision diabetes medicine, precision medicine "focuses on minimizing errors and improving accuracy in medical decisions and health recommendations. It seeks to maximize efficacy, cost-effectiveness, safety, access for those in need, and compliance compared with contemporary evidence-based medicine. Precision medicine emphasizes tailoring diagnostics or therapeutics (prevention or treatment) to subgroups of populations sharing similar characteristics." Personalized and individualized medicine is "the use of a person's own data to objectively gauge the efficacy, safety, and tolerability of therapeutics, and, subjectively, to tailor health recommendations and/or medical decisions to the individual's preferences, circumstances, and capabilities."

Recent technological developments in genomics and many other OMIC sciences (transcriptomics, proteomics, etc.) have reshaped current and future clinical medicine practice. Many believe that genomics will revolutionize medical practice and healthcare through its potential for more precise early diagnosis and management of diseases for each patient.

Genomic information has been used in healthcare delivery, and genetic markers that can influence the prevention, treatment, and survival of various diseases have been identified (Fig. 7.1).

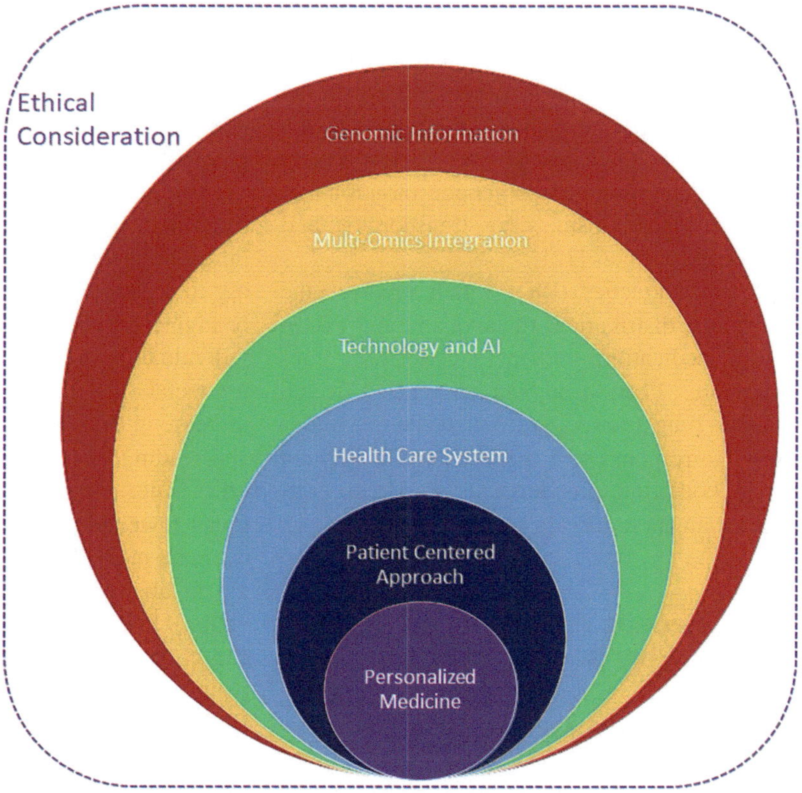

Fig. 7.1 Components of personalized medicine. This figure illustrates the interconnected components of personalized medicine, which orchestrate a harmonic combination of patient-centered care, a dynamic healthcare system, cutting-edge technology and AI, multiomics integration, and genetic information. These aspects work together to create a precise symphony, with the goal of transforming healthcare into a personalized, proactive, and patient-centered approach

7.2 Pharmacogenomics

Inter-individual variability in response to foods and medicines has been observed for millennia—Pythagoras (570 BC to 495 BC) reported that some people suffered from red blood cell hemolysis after consuming fava beans. This hemolytic response is now known to be caused by mutations in the *G6PD* gene, which results in glucose-6-phosphate dehydrogenase deficiency, the most common human enzyme deficiency, affecting 400 million people worldwide [8].

Pharmacogenomics (PGx), being the most important pillar of precision medicine, can predict response to medications, providing information such as the likelihood of side effects and drug efficacy [9]. PGx is being used in multiple disease states, such as cardiology, endocrinology, gastroenterology, infectious diseases, neurology, oncology, psychiatry, and more [10]. The human genome contains about 21,000 protein-coding genes [11]. Single-base change in an individual's DNA called single-nucleotide polymorphism (SNP) is the most common genetic variation that codes for metabolism, drug transporters, and drug-target proteins [12].

Pharmacokinetics (PK), which involves the absorption, distribution, metabolism, and excretion of medication, influences target drug concentration. Pharmacodynamics (PD), which involves drug transporters and receptors, influences the response to a medication [2]. PK and PD are both vital to understand in order to optimize treatment outcomes effectively [13].

7.3 Genetic Variation in Pharmacogenomics

Variations in genetics can sometimes lead to a difference in individual response to medications, such as how susceptible they are to adverse events or therapeutic efficacy [14]. Interestingly, the human DNA is about 99.9% identical, and SNPs can create those genetic variations [15]. SNPs are very common and can occur about once every 300 base pairs of sequence, and they code for both PK and PD pathways discussed earlier [16].

One group of extensively studied SNPs are found in the superfamily of Cytochrome P450 isoenzymes or CYP450. These metabolize most drugs; therefore, any variation can cause a decrease in efficacy or increase the risk of toxicity with certain medications. The most commonly studied CYPs are CYP2D6, CYP2C19, CYP2C9, and CYP3A4. The frequency of variations in the activity of these enzymes can also vary among different ethnic groups [17].

7.4 Pharmacokinetics and Pharmacodynamics

Pharmacokinetics is the study of how the body interacts with medication from when it is ingested to its excretion. Metabolism is an important part of PK, which occurs throughout the body, but many interactions occur in the liver. Metabolizing enzymes play a crucial role in the PK of drugs for both exogenous and endogenous substances. Metabolism, through the CYP family, enhances the water solubility of a drug to make elimination from the body easier [18]. Genetic variations can influence the activity level of the CYPs correlating to a genotype-phenotype, resulting in lower or higher plasma levels of a drug. For example, individuals with a lack of CYP2D6 enzyme activity metabolize drugs more slowly, causing an increase in the plasma concentration of that drug and leading to potential adverse events. On the other side, increased activity of CYP2D6 can mean lower plasma concentration, leading to a decrease in the drug's efficacy, and the intermediate metabolizer of cyp2d6 has an effect on medication [19]. These two variations of phenotype can contribute to either adverse events or therapeutic failure of a drug. In addition, because CYP2D6 is the predominant metabolism pathway for up to 25% of medications and has over 100 allelic variants, it is the single most important factor in individual variability among the population [20].

Neurons throughout our body communicate through chemicals or neurotransmitters. These chemicals signal messages from one neuron (pre-synaptic) to a receiving one (post-synaptic) through a gap between the two called synapses. Once the message is received, it has various downstream effects. The neurotransmitter is recycled and removed from the synapse through a transporter and stored for future use. The transporter called serotonin transporter is encoded by the *SLC6A4* gene [17]. Depending on the transporter status of SCL6A4 (long or short form), remission rates can vary amongst individuals using SSRIs [17].

On the post-synaptic neuron is the gene serotonin 2A receptor (HTR2A), which is one the most abundant serotonin receptors in the brain and is associated with multiple mental health disorders in regard to their susceptibility or treatment outcome. Two SNPs in the *HTR2A* gene that have been well studied can lower receptor expression, reducing excitation at the post-synaptic neuron and affecting antidepressant response, particularly serotonin reuptake inhibitors (SSRIs). This makes HTR2A an important factor in the pathogenesis of Major Depressive Disorder (MDD) and a biomarker for the treatment of antidepressants [21, 22].

7.5 Pharmacogenomics in Pain Management

Pain is very difficult to treat since it is subjective, and many do not respond the same way to the same treatment [23]. Chronic pain management has put a strain on the healthcare system, causing more than $600 billion annually [24]. Pain has a gene × environment interaction, which means there is a genetic disposition to the development and severity of chronic pain, in addition to environmental factors that play a role. This creates not only a great deal of variability in treatment options and response but also a great deal of much-needed understanding of the related pathways [25]. Personalized medicine in this disease state is also an important approach to consider, as it can tailor drug therapy to everyone's unique genetic makeup.

Multiple enzymes are involved in the metabolism, efficacy, and toxicity of opioids. The majority of metabolism and efficacy are seen in variations in *CYP2D6* and *OPRM1* gene [24]. CYP2D6 is highly polymorphic and accounts for 25% of all drugs metabolized, including opioids such as codeine, tramadol, and oxycodone. Analgesic effects of codeine are achieved by morphine, which occurs when codeine is converted to morphine by CYP2D6. However, individuals can have different activity levels of their CYP2D6. These can be categorized into four different groups: ultrarapid metabolizer (UM), which is increased enzyme activity; normal metabolizer (NM); intermediate metabolizer (IM); and poor metabolizer (PM), which is decreased enzyme activity [23]. About 10% of Caucasians are PM for CYP2D6, which means codeine activation to morphine is impacted in those individuals, and they are less likely to achieve adequate pain relief. On the other hand, if an individual is a UM for CYP2D6, they can run the risk of toxicity levels due to rapid conversion to morphine that is seen in 3.1% of the European population that found in DPWG guideline [23]. There are also well-known documents on the death of neonates whose breastfeeding mothers have the UM variation gene of CYP2D6. This not only confirms the impact of CYP2D6 UM but also that morphine is transferred from mother to babies during breastfeeding [26]. In the case of tramadol PM individuals cannot activate the drug and cosequently do not get any relief from drug and UM patients experience side efects.

The PD gene, μ-opioid receptor (*OPRM1*), is another major target for analgesics [25]. The reduction in the activity of OPRM1 can lead to a decrease in pain relief or opioid tolerance and hyperalgesia [27]. Therefore, our understanding of the pharmacogenomics impact of OPRM1 helps predict the most effective therapy for a patient suffering from chronic pain. OPRM1 is also a target of interest due to its regulation of the rewarding effects of many drugs leading to addiction [28].

7.6 Pharmacogenomics in Psychiatry

In addition, the worldwide cost of treating mental illness will rise to $6 trillion by the year 2039 [29, 30]. The challenge rising among both clinicians and patients has been the trial-and-error process of mental health disorder treatment [29]. SSRIs are the most common and first-line medications used for MDD [31]. One of the largest depressions trials to date, STAR*D, shows a response rate of about 50%, remission of 25% for first-line therapy, and a decrease in subsequent medication trials [32]. Pharmacogenomics reduces this trial-and-error process time by identifying the relevant genes and the different variations that can increase efficacy and minimize side effects [31].

The two major PK genes for psychiatry are *CYP2D6* and *CYP2C19*, which metabolize most drugs [33]. The metabolizer status of the two CYP genes has been shown to be associated with a potential increased risk for side effects, which can ultimately reduce patient adherence. Some drugs already have FDA genetic testing information within their package insert [34]. Citalopram, a relatively safe and well-tolerated drug for MDD, has an FDA warning for a decrease in dose if an individual is a PM for CYP2C19 due to QT prolongation [35].

7.7 Cardiovascular Pharmacogenomics

Cardiovascular pharmacogenomics focuses on genetic variations in an individual's genome that influence their response to cardiovascular medications and drug-related adverse events [36]. According to the American Heart Association, between 2017 and 2020, there were 127.9 million US adults that had some form of cardiovascular disease (CVD). The cost associated with CVD from 2018 to 2019 was $407.3 billion [37]. Therefore, it becomes crucial to understand the variabilities in genetics that can contribute to improving cardiovascular health and optimizing care [38].

Clopidogrel (Plavix) is the most commonly used antiplatelet drug in both the US and Canada. Clopidogrel is a prodrug (inactive) that requires activation by CYP2C19. However, CYP2C19 is highly polymorphic and has over 2000 genetic variants. If a genetic variant causes loss of function of the protein (PM), it can result in an increased risk for cardiovascular events, which clopidogrel was intended to decrease as a result of taking the medication in the first place. The FDA has, therefore, issued a warning in the package insert of the drug to warn providers to consider an alternative option for patients who are CYP2C19 PM [39]. The Clinical Pharmacogenomics Implementation Consortium (CPIC) 2022 guidelines recommend that individuals with or without acute coronary syndrome who are undergoing percutaneous coronary interventions or PCI, getting treatment for peripheral arterial disease or stable coronary artery disease following MI, use a different antiplatelet therapy if they are CYP2C19 PMs. The Dutch Pharmacogenomics Working Group (DPWG) also has recommendations based on CYP2C19 genetic variation [40]. Access to CYP2C19 genetic testing can help personalize the antiplatelet treatment plan for an individual with impaired enzyme activity.

Statins are also among the most common groups of medications for preventing and treating cardiovascular disease. Statin-associated muscle pain or myopathy occurs in about 10% of patients and is related to variations in the PD gene SLCO1B1 [41]. The main role of SLCO1B1 is to transport statins and their metabolites into the liver. Variations in this gene can decrease the transportation of statins into the liver, causing an increase in the systemic concentration of the drug and, therefore, increasing the risk for toxicity [42]. Among the statins, simvastatin is the most researched and known to cause statin-induced toxicity. Although the toxicity can be multifactorial, the discovery of SLCO1B1 gene has provided valuable information [43].

Antiarrhythmic medications metabolized by CYP2D6 include flecainide, procainamide, and propafenone [44].

Elevated propafenone levels and an increased risk of arrhythmias were seen in CYP2D6 PMs using CYP3A4 inhibitors. DPWG has been provided as the only genotype-based guideline for this drug. FDA and DPWG recommend decreasing the dose of flecainide by 50% for CYP2D6 PMs and 25% for CYP2D6 IMs for reasons other than Brugada syndrome. They also recommend decreasing the propafenone dose by 70% in CYP2D6 PMs or utilizing an alternate for CYP2D6 IMs and Ums [45].

7.8 Pharmacogenomics Guidance and Resources

There are many resources available with evidence-based information that a clinician can use when pharmacogenomics is being implemented in clinical practice. Clinical Pharmacogenomics Implementation Consortium (CPIC), Pharmacogenomics Research Network (PGRN), and Pharmacogenomics Knowledgebase (PharmGKB) are a few examples of resources that were started in 2009. Their main goal is to establish practice guidelines to translate the genetics language not only possible but also to provide guidance and recommendations [46].

Pharmacogene Variation Consortium (PharmVar) focuses on maintaining a catalog of allele variations for different genes related to their impact on drug metabolism, response, and a specific nomenclature for the PGx community [47].

Flockhart Table™ is designed as a reference tool for healthcare providers who want to learn more about drug interaction medicated by CYP450 enzymes and know which are inducers and inhibitors [48].

7.9 Pharmacogenomics in Oncology

Precision oncology is a cutting-edge method of treating cancer that tailors therapeutic approaches to the particular genetic and biological features of each patient's disease. Precision oncology acknowledges that every patient's cancer is unique and that treatment plans should be tailored to target the particular genetic mutations and molecular aberrations causing the growth of each patient's malignancies, in contrast to traditional one-size-fits-all treatments. In cancer care, precision medicine has led to an evolution away from the "one size fits all" approach of classifying tumors according to the organ in the body where they originated (e.g., breast, lung, colon) and toward an emphasis on certain discovered mutations that may be the cause of cancer [49].

The European Society for Clinical Oncology defines precision medicine in oncology as "the use of therapies that benefit subsets of patients whose tumors have certain molecular or cellular characteristics, most commonly understood as genomic changes and changes in protein expression" [50]. Oncology is an outstanding example of precision medicine's breakthroughs. On the other hand, "precise" interventions at the individual level are possible. The development of more advanced omics methods has led to the ability to profile complex molecular characteristics of patient tumors, including DNA sequence data, gene expression profiles, protein levels, and more [51]. These methods have the potential to improve our understanding of cancer biology and become an essential element of clinical practice. It holds the potential to fundamentally change our understanding of how to manage and treat cancer patients. Targeted therapeutic approaches have shown significant clinical activity for several tumor types. These include vemurafenib for melanoma with a B-Raf proto-oncogene (*BRAF*) mutation, cetuximab for colorectal cancer with an epidermal growth factor receptor (*EGFR*) mutation, gefitinib and sotorasib for non-small cell lung cancers with EGFR or KRASp.G12C mutations, respectively, and olaparib for ovarian cancers enriched in human growth factor (HRD) [52–55].

Therefore, genomic studies have demonstrated that tumors are extremely diverse and complicated; conventional clinical research and practice models do not generally result in perfect therapy.

Recent data indicates that germline variations should be evaluated in all individuals with solid metastatic cancers. Precision medicine approaches evaluate efficacy and make treatment decisions by analyzing patients' circulating DNA (liquid biopsy), immunological markers, and other biological elements. The most successful biomarkers so far have been genomic biomarkers, although additional biomarkers, such as proteomic and transcriptomics, have been developed and evaluated [56, 57].

In the last decade, there has been a lot of interest in analyzing cancer-related signals utilizing biological fluids (a liquid biopsy). Liquid biopsies can discover a wide range of biomolecular characteristics and may be used to determine disease status. Between 2020 and 2030, the liquid biopsy market is predicted to grow at a 16% annual rate.

Another method for diagnosing cancer is liquid biopsy, which has been developed to find potential molecular abnormalities in cases where tissue biopsy is not practical. Later, liquid biopsy was found to be useful for characterizing the progression of tumors, understanding the mechanisms underlying resistance, and testing associated genomic profiles and multiple metastases [58].

Histological evaluation has historically been used to measure the results of phase II and III cancer clinical trials; however, histological evaluation is not always able to fully capture the effects of immunotherapy or gene-targeted medications. Through the use of high-throughput technology and sequencing, several molecular variations have been found, which has resulted in the FDA's approval of targeted medications [59]. Usually, oncology trials are drug-centered, trying to match patients to a specific medication regimen by identifying common characteristics (such as tumor type or, more recently, a shared genomic aberration) among patients.

Umbrella and basket trials are two new and very different trial strategies for molecular targets in cancer treatment. In an umbrella trial, patients with similar basic cancer types are enrolled, molecular markers for a variety of possible targets are tested, and patients are then assigned to one of the research arms based on whether or not a mutation linked to a potentially beneficial treatment for that marker is present. In a basket trial, patients with the same genetic mutation are enrolled regardless of whether their cancer started in the liver, breast, colon, lung, or any other organ. They are then given the same innovative medication specifically targeting that particular genetic signature [49].

Precision medicine's main objective is a tailored, patient-centered (instead of drug-centered) trial based on the best available biomarkers. Each patient's treatment is considered uniquely in "N-of-1" trials based on molecular, immunological, and other biological factors. Personalized treatment combinations suited to specific patients are used in these trials [60].

In conclusion, precision oncology provides the following achievements: personalized treatment plans, optimizing drug selection, improving treatment response and solving drug resistance problems, and prevention and prediction of cancer at the right time. Tremendous biotechnology advances are revolutionizing cancer treatment. Tumor and cell-free DNA profiling using NGS and proteomic and transcriptomics are improving cancer treatment management. This trend offers new promise for better results and a future in which therapy is personalized to the unique features of each patient's cancer. Tremendous biotechnology advances are revolutionizing cancer treatment. Tumor and cell-free DNA profiling using NGS and proteomic and transcriptomics are improving cancer treatment management.

7.10 Precision Medicine and the Etiology of Cardiovascular Diseases

Cardiovascular diseases (CVDs) have accounted for the majority of all diseases worldwide during the last three decades. An estimated 17.9 million people died from CVDs in 2019, representing 32% of all global deaths. Of these deaths, 85% were due to heart attack and stroke [61, 62]. Cardiovascular diseases (CVD) are conditions affecting the heart and blood vessels, with common types including coronary artery disease (CAD), heart failure, arrhythmias, valve diseases, cardiomyopathies, peripheral artery disease (PAD), deep vein thrombosis and pulmonary embolism (DVT and PE), congenital heart diseases, rheumatic heart disease, and inflammatory heart diseases. These diseases can manifest in various ways and have different underlying causes. Risk factors such as age, family history, smoking, poor diet, lack of exercise, and certain medical conditions can contribute to their development. Early detection, lifestyle modifications, and appropriate medical interventions are crucial in managing and preventing cardiovascular diseases. It is essential to recognize that many individuals may have a combination of these diseases and to address these risk factors. In addition to known risk factors (including smoking, hypertension, diabetes, dyslipidemia, and obesity), a strong genetic basis for cardiovascular diseases has also been established. Extensive analysis of phenotypes and the "omics" data may help cluster patients and reduce inequities while also supporting patient-centered clinical care. Similar to what has happened in the field of oncology, the development of precision medicine and its application to cardiovascular disease holds the promise of changing preventative and treatment strategies while improving health. Precision medicine has the potential to transform the future of cardiovascular disease (CVD) healthcare by utilizing "omics" in cardiology. The investigation of patients' genetic makeup is an essential part of precision medicine in cardiology. Evaluating an individual's genetic profile is an essential component of precision medicine in cardiology. Genetic variants can affect the likelihood of developing cardiovascular issues, drug reactions, and overall heart health. Healthcare experts can detect particular genetic markers associated with higher vulnerability to illnesses such as coronary artery disease, heart failure, and arrhythmias using advanced genomic testing. Various genetic variations have been associated with an increased risk of conditions such as coronary artery disease, heart failure, and arrhythmias. Understanding these genetic factors allows clinicians to determine an individual's susceptibility to various cardiovascular problems, enabling focused therapies and early preventive actions [63]. Precision medicine in cardiovascular disease is based on the development and use of biomarkers, which are biological indications that might indicate the existence of disease or predict therapy responses. Advanced diagnostic tools, such as high-sensitivity troponin assays and molecular imaging techniques, enable more accurate disease identification and risk stratification, allowing physicians to make more informed treatment decisions [63].

The enzymes involved in drug metabolism, particularly those implicated in the cytochrome P450 (CYP) system, show significant genetic variation between individuals. Genetic variations can cause differences in drug metabolism rates, influencing drug concentrations within the blood. In cardiovascular medicine, this can affect the efficacy of medications such as antiplatelet agents, beta-blockers, and statins [36].

The most commonly used pharmacogenomics tests in cardiovascular medicine are those for warfarin and CYP2C9/VKORC1 (vitamin K epOxide reductase complex subunit 1), clopidogrel and CYP2C19, and simvastatin and solute carrier organic anion transporter family member 1B1 (SLCO1B1) genotypes [64].

7.11 Precision Medicine in Endocrinology

Precision medicine can be used to help people with many types of endocrine conditions achieve improved clinical outcomes. Diabetes, a global issue affecting millions, often undiagnosed, causes life-threatening complications involving blood vessel damage and is classified into various types. Precision diabetes medicine aims to understand and leverage the heterogeneity in etiology, clinical presentation, and prognosis, which is a major challenge in diabetes management. The Second international consensus report on precision medicine prevention, diagnosis, treatment, and prognosis2 in monogenic diabetes mellitus (MDM), GDM, T1D, and T2D [65]. This consensus indicates significant advancements in precision diabetes medicine implementation while also revealing numerous knowledge gaps.

7.12 Precision Medicine in Monogenic Diabetes Mellitus (MDM)

7.12.1 Precision Diagnosis, Treatment, and Prognosis in MDM

MDM, a rare genetic disorder, can be diagnosed in the neonatal period or before age 45. It accounts for up to 5% of diabetes and offers precision medicine opportunities. However, many patients are misdiagnosed with T1D or T2D due to overlapping clinical features [65].

The data suggests precision diagnostics testing for neonatal diabetes, metabolic syndrome (MDM), GCK-related hyperglycemia, and GCK-related hyperglycemia in infants, young individuals, and women with GDM without obesity, using ethnic-specific BMI thresholds for overweight and obesity. Testing for neonatal diabetes and metabolic syndrome (MDM) involves targeted next-generation sequencing, genetic panels, a comprehensive panel, multiplex ligation-dependent probe amplification assays, methylation-based assays, and rapid Sanger sequencing of genes like GCK, KCNJ11, ABCC8, and INS. These methods help identify causes of MDM, noncoding mutations, and copy-number variants [65]. In MDM, sulfonylurea use improved diabetes outcomes and was well tolerated. However, evidence for noninsulin therapies was inconsistent later in life. The current evidence is low quality for clinical guidance on non-insulin therapy for 6q24-related diabetes or thiamine-responsive megaloblastic anemia. Sulfonylureas are effective for HNF1A diabetes glycemic control, with the potential for improved control after transitioning from insulin or noninsulin therapies. Alternatives include glinides, GLP1-RAs, and dipeptidyl peptidase four inhibitors. Monogenic disorders of severe insulin resistance include generalized lipodystrophy, partial lipodystrophy, and INSR mutations. The safety and efficacy of recombinant human leptin and thiazolidinediones were analyzed in populations with lipodystrophy syndromes and INSR mutation carriers. Thiazolidinediones lowered HbA1c, triglycerides, and insulin-like growth factor-1 also lowered HbA1c levels—few adverse events were reported [65].

Poor glycemic control in MDM individuals increases the risk of microvascular and macrovascular complications, with studies focusing on younger neonatal diabetes populations showing low rates of severe complications.

7.13 Gestational Diabetes Mellitus (GDM)

GDM is defined as abnormal glucose tolerance that first manifests or occurs during pregnancy. The mother and her offspring may face significant short- and long-term health risks due to GDM, which emphasizes the significance of accessible screening, early diagnosis, and efficient treatment. Based on the available studies, the strength of evidence for GDM risk reduction with lifestyle modification, metformin, and myo-inositol/inositol is considered moderate to very low [65]. New interventions targeting modifiable precision markers like lipid levels and insulin sensitivity during pregnancy should be developed and validated to diagnose GDM precisely. GDM-affected women are more likely to experience T2D and CVD; the degree of pregnancy hyperglycemia and these outcomes are correlated in a dose-dependent manner [65].

7.14 Type 1 Diabetes

T1D, a form of diabetes caused by autoimmune destruction of pancreatic β-cells, is a significant healthcare issue due to its early onset, high insulin costs, and increased risk of complications. Growing understanding of the pathogenesis of T1D prompted the disease to be classified into stages, with stage 0 indicating the presence of one autoantibody among individuals at high genetic risk, stages 1 and 2 indicating the presence of two or more islet autoantibodies in normal (stage 1) or dysglycemia (stage 2) and clinical diabetes (stage 3). Both children and adults demonstrate etiologic heterogeneity. Islet autoantibodies can help define heterogeneity in T1D before stage 3 diagnosis, and incorporating age and genetics into risk scores can help [65].

7.15 Type 2 Diabetes

T2D is believed to affect 500 million individuals worldwide, with the number expected to rise to 1.3 billion by 2050.

Dietary interventions can delay T2D progression, but individual response varies. Precision prevention in T2D requires identifying predictors and benefiting individuals. Precision diagnosis for T2D involves subclassifying individuals based on factors like autoantibodies, BMI, β-cell function, age, lipid profiles, and cardiovascular features to classify and personalize treatment plans accurately. T2D subclassification strategies have been related to significant clinical outcomes. According to the second international consensus report on precision diabetes medicine, a high predictive utility for N-terminal pro-b-type natriuretic peptide, troponin T, and triglyceride-glucose in the precision prognosis of CVD, a moderate utility for coronary computed tomography angiography and single-photon have been introduced emission computed tomography, and low utility for C-reactive protein, coronary artery calcium score, and galectin-3.

7.15.1 Precision Medicine in Osteoporosis

Osteoporosis is a skeletal disease characterized by weakened bones, low bone mass, and deteriorating bone tissue, leading to increased degenerative changes and fracture susceptibility. It causes over 8.9 million fractures worldwide annually, with the most common fractures occurring in the hip, forearm, and vertebral. Current diagnostic approaches, such as bone density testing and fracture algorithms, have not accurately predicted individual fracture risk. Factors contributing to osteoporosis risk include age, gender, nutrition, physical activity, medication usage, and menopausal status. A positive family history is a significant clinical risk factor, emphasizing the importance of genetic factors in predicting disease susceptibility. Identifying genetic markers associated with bone health helps in assessing an individual's predisposition to osteoporosis [66].

7.16 Pharmacogenomics Clinical Trials

With the increased emphasis on personalized medicine (PM), pharmacogenetics and pharmacogenomics (PGx) are becoming increasingly important. In clinical trials, the use of PGx is an additional tool that should be investigated to increase our understanding of new medications' efficacy and safety.

According to an overview report that was conducted on ClinicalTrials.gov, the number of clinical trials carrying PGx information has risen since the 2000s, mostly in Oncology and Mental Health, respectively [67]. A 2007 World Health Organization study explained that PGx would lead to substantial shifts in pharmacological studies at the phase of clinical trials [68].

The most well-known pharmacogenes include CYP2D6, which is investigated in Mental Health and Pain, CYP2C9 that is explored in Hematology, CYP2C19 is studied in Cardiology and Mental Health, and the last ones are ABCB1 and CYP3A5, which is mainly explored in Transplantation and Cardiology [67].

By searching ClinicalTrials.gov from inception through 3 June 2023, The United States, France, Canada, Netherlands, Germany, United Kingdom, Spain, and the Republic of Korea have the biggest number of PGx-related registered clinical trials, respectively [67].

An open-label, multicenter, controlled, cluster-randomized, crossover implementation study of a 12-gene pharmacogenetic panel was conducted in 7 European countries (Austria, Greece, Italy, the Netherlands, Slovenia, Spain, and the UK). Genotype-guided treatment using a 12-gene pharmacogenetic panel significantly decreased the incidence of clinically relevant adverse drug reactions by 30% [69].

A large number of those receiving chemotherapy have side effects, many of which have grade 3 or higher toxicities that require medical interventions. Therefore, in a prospective, randomized Trial known as PhOCus, the implementation of Pharmacogenomic Testing in Oncology Care has been analyzed. Participating patients were genotyped for a panel that comprised UGT1A1 and DPYD variants. The findings of this experiment indicate that integrating germline PGx information could enable individualized chemotherapy dose decisions and develop a new model of care using comprehensive genomic data to improve oncology treatment planning [70].

The PACIFIC-PGx clinical trial was aimed to investigate the feasibility of integrating a pretreatment PGx Screening Program in Australia for cancer patients receiving Fluoropyrimidines or irinotecan-based chemotherapy. The study suggests that PGx testing increases supportive therapy management by minimizing potential drug-gene interactions based on results from the PGx test.

7.17 Precision Medicine Initiatives

Precision medicine can provide new insights into disease mechanisms by improving target identification, drug repurposing, and resource development, as medication with supporting genetic information has better clinical outcomes [71, 72]. Precision medicine also has the potential for precision public health (PPH), in which population health information can be combined with genetics to uncover heterogeneous risk factors for specific treatment and prevention before disease onset [73].

The Australian Genomics Health Alliance established a national framework for transferring -omics discoveries into clinical research and practice. This structure is designed to provide advice based on genetic research and clinical testing/trials. The program "Australian Genomics' work" focuses on 24 priority activities and projects that fulfill strategic needs across the whole value chain of genetics and healthcare system research [74].

The All of Us research program of the National Institutes of Health aims to advance precision medicine by gathering and evaluating a wide range of data from at least one million Americans to accelerate health research and innovations in medicine, allowing personalized prevention, treatment, and care for all Americans [75].

The 100,000 Genome Project in the United Kingdom is an effort that entails sequencing 100,000 genomes and collecting genomic, phenotypic, and clinical information from 70,000 people with rare diseases or cancer [76, 77].

To ensure that all patients have fair access to genomic medicine, France has launched a national strategy called the "2025 France Genomic Medicine Initiative (PFMG2025)," which aims to integrate precision medicine into the healthcare pathway and build a national infrastructure.

To ensure that all patients have fair access to genomic medicine, France has launched a national strategy called the "2025 France Genomic Medicine Initiative (PFMG2025)," which aims to integrate precision medicine into the healthcare pathway and build a national infrastructure [78].

7.18 Ethical Consideration of Personalized Medicine

There are various ethical concerns with personalized medicine, including those related to consent, privacy, ownership of data, equity, and the possibility of unpredictable consequences; here are some of these concerns [79]:

Privacy and Confidentiality The General Data Protection Regulation (GDPR) rule demands "special protection" if personal data is also sensitive data. It distinguishes between "personal data" and "sensitive data." Sensitive data is a collection of unique categories that should be handled with additional caution (Article 6 of the GDPR), with some exceptions indicated [80].

Genetic data are an individual's acquired genetic characteristics that provide unique information about their physiology or health. These characteristics are particularly obtained through the analysis of an individual's biological sample, such as DNA or RNA analysis, or other data that may be inferred from sample analysis. Genetic and health data processing requires additional protections and regulations.

Informed Consent The informed consent adheres to the Helsinki Declaration's Ethical requirements [81]. "Right to human dignity and right to the integrity of individuals." In summary, the three most important goals of informed consent under the Clinical Trials Regulation are as follows: First, offer full details about the study for the suitable participants' information. Second, the future use of data and the dissemination of study findings to patients. Finally, there is the prospect of unforeseen new genetic information and considerations. Patients select whether or not to join a study based on this information. Efforts should be made to ensure that people understand the potential benefits and risks of personalized medicine [79].

Equity and Access The use of PM tools in the health care system aims to investigate health equality as a fair and just opportunity to be as healthy as feasible. Equality implies that each citizen or community obtains the same resources or possibilities. However, Equity understands that each individual has unique needs and, as such, must be provided with the necessary resources and opportunity to achieve an equal outcome [79].

Psychological Consequences The disclosure of some genetic information may have psychological and emotional consequences for individuals and their families. Ethical considerations include giving proper counseling and support services to individuals to assist them in dealing with the potential psychological implications of personalized medicine.

Discrimination and Stigmatization Genetic Stigma: Individuals who are predisposed to a specific disease may fear genetic prejudice. Concerns about potential stigmatization or discrimination in sectors such as work, insurance, and social connections should be addressed in ethical guidelines.

Data Ownership and Research Consent Establishing clear criteria for data ownership and gaining explicit authorization for the use of data in research are both ethical considerations.

Regulatory Control As personalized medicine advances, stringent regulatory frameworks and standards are required to assure the responsible and ethical development, marketing, and use of tailored medical interventions.

During a Global Precision Medicine Council in 2019, the World Economic Forum developed a summary of the key policy and governance gaps for PM implementation, as well as potential ways to overcome them: Interoperability and data sharing, ethical application of technology, trust and engagement in society, easy access and reasonable pricing reliable regulatory systems [79].

7.19 Potential Future of Personalized Medicine

Personalized medicine considers individuals' genetic characteristics to predict susceptibility to particular diseases and customize the proper preventive and/or therapeutic strategy for the right person at the right time. Pharmacogenetics is an essential component of personalized medicine since it utilizes reliable, precise, and evidence-based data to prescribe the right drug at the right dosage for the right patient. This approach allows medicine to transform from reactive to proactive, resulting in better therapeutic outcomes. Pharmacogenetics and other "omics" such as transcriptomics, proteomics, and metabolomics will significantly impact clinical decision-making and modern medicine.

7.20 Conclusion

Personalized medicine is at the cutting edge of transformational developments in healthcare, holding tremendous relevance and exhibiting an urgent need for its wider, more comprehensive implementation in clinical practice. As our awareness of the complex interaction between genetics, environmental factors, and individual lifestyles expands, the paradigm one-size-fits-all approach to personalized and precision medicine becomes more and more shift from a one-size-fits-all approach to personalized and precision medicine becomes increasingly crucial.

The importance of personalized medicine derives from its promise to transform patient care by personalizing interventions to each individual's unique traits. Clinicians can go beyond conventional therapeutic approaches by utilizing the potential of genomic data, multi-omics integration, and modern technologies to offer targeted therapies with increased efficacy and lower adverse reactions. Personalized medicine addresses the limitations of conventional treatments by providing a deep understanding of genetic variations, enabling healthcare providers to make informed decisions that align with each patient's unique needs and characteristics. Personalized medicine in clinical practice holds the potential of improved therapeutic outcomes, earlier disease detection, and preventive strategies customized to individual risks. This strategy not only improves therapeutic efficacy but also provides the broad goal of better patient outcomes and quality of life. Precision medicine has emerged as a new paradigm in healthcare, with significant implications across multiple fields of medicine. Its impact is especially prominent in oncology, psychiatry, cardiology, and endocrinology. A key component of precision medicine is the pharmacogenomics approach, which focuses on understanding how an individual's genetic makeup impacts their reaction to medications. By identifying genetic variants that impact drug metabolism and efficacy, physicians can prescribe drugs more accurately, thereby reducing adverse reactions and maximizing therapeutic outcomes. As personalized medicine improves, collaboration among clinicians, researchers, policymakers, and patients becomes increasingly important. Prioritizing ethical concerns about privacy, informed consent, and fair access is necessary to ensure sustainable implementation. Furthermore, continued research and technological advancements will deepen and broaden the scope of personalized medicine, making it an essential resource in the rapidly changing environment of modern healthcare. Implementing personalized medicine is a commitment to providing patient-centered care that has the potential to transform the way we prevent, diagnose, and treat diseases, eventually improving the well-being of individuals and communities globally.

References

1. Hasanzad M. Precision medicine in clinical practice. Berlin: Springer; 2022.
2. Naithani N, Sinha S, Misra P, Vasudevan B, Sahu R. Precision medicine: concept and tools. Med J Armed Forces India. 2021;77(3):249–57.
3. Hood L, Friend SH. Predictive, personalized, preventive, participatory (P4) cancer medicine. Nat Rev Clin Oncol. 2011;8(3):184–7.
4. Hasanzad M, Sarhangi N, Ehsani Chimeh S, Ayati N, Afzali M, Khatami F, et al. Precision medicine journey through omics approach. J Diabetes Metab Disord. 2022;21(1):881–8.
5. Pirmohamed M. Pharmacogenomics: current status and future perspectives. Nat Rev Genet. 2023;24(6):350–62.
6. Hasanzad M, Sarhangi N, Hashemian L, Sarrami B. Principles of pharmacogenomics and pharmacogenetics. In: Precision medicine in clinical practice. Berlin: Springer; 2022. p. 13–32.
7. Mirnezami R, Nicholson J, Darzi A. Preparing for precision medicine. N Engl J Med. 2012;366(6):489–91.
8. Nkhoma ET, Poole C, Vannappagari V, Hall SA, Beutler E. The global prevalence of glucose-6-phosphate dehydrogenase deficiency: a systematic review and meta-analysis. Blood Cells Mol Dis. 2009;42(3):267–78.
9. Johnson JA, Weitzel KW. Advancing pharmacogenomics as a component of precision medicine: how, where, and who? Clin Pharmacol Ther. 2016;99(2):154–6.
10. Hippman C, Nislow C. Pharmacogenomic testing: clinical evidence and implementation challenges. J Pers Med. 2019;9(3):40.
11. Gray KA, Yates B, Seal RL, Wright MW, Bruford EA. Genenames.org: the HGNC resources in

2015. Nucleic Acids Res. 2015;43(Database issue):D1079–85.
12. Umamaheswaran G, Kumar DK, Adithan C. Distribution of genetic polymorphisms of genes encoding drug metabolizing enzymes & drug transporters—a review with Indian perspective. Indian J Med Res. 2014;139(1):27–65.
13. Derendorf H, Lesko LJ, Chaikin P, Colburn WA, Lee P, Miller R, et al. Pharmacokinetic/pharmacodynamic modeling in drug research and development. J Clin Pharmacol. 2000;40(12 Pt 2):1399–418.
14. Giri J, Moyer AM, Bielinski SJ, Caraballo PJ. Concepts driving pharmacogenomics implementation into everyday healthcare. Pharmgenomics Pers Med. 2019;12:305–18.
15. Sachidanandam R, Weissman D, Schmidt SC, Kakol JM, Stein LD, Marth G, et al. A map of human genome sequence variation containing 1.42 million single nucleotide polymorphisms. Nature. 2001;409(6822):928–33.
16. Nelson MR, Marnellos G, Kammerer S, Hoyal CR, Shi MM, Cantor CR, Braun A. Large-scale validation of single nucleotide polymorphisms in gene regions. Genome Res. 2004;14(8):1664–8.
17. Ma MK, Woo MH, McLeod HL. Genetic basis of drug metabolism. Am J Health Syst Pharm. 2002;59(21):2061–9.
18. Li Y, Meng Q, Yang M, Liu D, Hou X, Tang L, et al. Current trends in drug metabolism and pharmacokinetics. Acta Pharm Sin B. 2019;9(6):1113–44.
19. Roots I, Laschinski G, Arjomand-Nahad F, Kirchheiner J, Schwarz D, Brockmöller J, et al. Genotype and phenotype relationship in drug metabolism. Ernst Schering Res Found Workshop. 2007;59:81–100.
20. Gaedigk A, Sangkuhl K, Whirl-Carrillo M, Klein T, Leeder JS. Prediction of CYP2D6 phenotype from genotype across world populations. Genet Med. 2017;19(1):69–76.
21. Gong P, Liu J, Blue PR, Li S, Zhou X. Serotonin receptor gene (HTR2A) T102C polymorphism modulates individuals' perspective taking ability and autistic-like traits. Front Hum Neurosci. 2015;9:575.
22. McMahon FJ, Buervenich S, Charney D, Lipsky R, Rush AJ, Wilson AF, et al. Variation in the gene encoding the serotonin 2A receptor is associated with outcome of antidepressant treatment. Am J Hum Genet. 2006;78(5):804–14.
23. Kaye AD, Garcia AJ, Hall OM, Jeha GM, Cramer KD, Granier AL, et al. Update on the pharmacogenomics of pain management. Pharmgenomics Pers Med. 2019;12:125–43.
24. Saba R, Kaye AD, Urman RD. Pharmacogenomics in pain management. Anesthesiol Clin. 2017;35(2):295–304.
25. Peiró AM, Planelles B, Juhasz G, Bagdy G, Libert F, Eschalier A, et al. Pharmacogenomics in pain treatment. Drug Metab Pers Ther. 2016;31(3):131–42.
26. Bateman DN, Eddleston M, Sandilands E. Codeine and breastfeeding. Lancet. 2008;372(9639):625; author reply 6.
27. Sheikh S, Smotherman C, Patel M, Langaee T, Wang D, Swaray E, et al. Characterizing OPRM1 DNA methylation in prescription opioid users with chronic musculoskeletal pain. Pain Rep. 2022;7(6):e1046.
28. Crist RC, Berrettini WH. Pharmacogenetics of OPRM1. Pharmacol Biochem Behav. 2014;123:25–33.
29. Corponi F, Fabbri C, Serretti A. Pharmacogenetics in psychiatry. Adv Pharmacol. 2018;83:297–331.
30. Pelgrim TAD, Philipsen A, Young AH, Juruena M, Jimenez E, Vieta E, et al. A new intervention for implementation of pharmacogenetics in psychiatry: a description of the PSY-PGx clinical study. Pharmaceuticals. 2024;17(2):151.
31. Amare AT, Schubert KO, Baune BT. Pharmacogenomics in the treatment of mood disorders: strategies and opportunities for personalized psychiatry. EPMA J. 2017;8(3):211–27.
32. Sinyor M, Schaffer A, Levitt A. The sequenced treatment alternatives to relieve depression (STAR*D) trial: a review. Can J Psychiatry. 2010;55(3):126–35.
33. Müller DJ, Rizhanovsky Z. From the origins of pharmacogenetics to first applications in psychiatry. Pharmacopsychiatry. 2020;53(4):155–61.
34. Boland JR, Duffy B, Myer NM. Clinical utility of pharmacogenetics-guided treatment of depression and anxiety. Pers Med Psychiatry. 2018;7–8:7–13.
35. Kumar Y, Kung S, Shinozaki G. CYP2C19 variation, not citalopram dose nor serum level, is associated with QTc prolongation. J Psychopharmacol. 2014;28(12):1143–8.
36. McDonough CW. Pharmacogenomics in cardiovascular diseases. Curr Protoc. 2021;1(7):e189.
37. Banharak S, Metprommarat A, Mahikul W, Jeamjitvibool T, Karaket A. Effectiveness of acute myocardial infarction interventions on selected outcomes among community dwelling-older adults: a systematic review and meta-analysis. Sci Rep. 2023;13(1):18538.
38. Roden DM, Van Driest SL, Wells QS, Mosley JD, Denny JC, Peterson JF. Opportunities and challenges in cardiovascular pharmacogenomics: from discovery to implementation. Circ Res. 2018;122(9):1176–90.
39. Pereira NL, Rihal CS, So DYF, Rosenberg Y, Lennon RJ, Mathew V, et al. Clopidogrel pharmacogenetics. Circ Cardiovasc Interv. 2019;12(4):e007811.
40. Dean L, Kane M. Clopidogrel therapy and CYP2C19 genotype. In: Pratt VM, Scott SA, Pirmohamed M, Esquivel B, Kattman BL, Malheiro AJ, editors. Medical genetics summaries. Bethesda, MD: National Center for Biotechnology Information (US); 2012.
41. Kitzmiller JP, Mikulik EB, Dauki AM, Murkherjee C, Luzum JA. Pharmacogenomics of statins: understanding susceptibility to adverse effects. Pharmgenomics Pers Med. 2016;9:97–106.
42. Weeke P, Roden DM. Pharmacogenomics and cardiovascular disease. Curr Cardiol Rep. 2013;15(7):376.

43. Kee PS, Chin PKL, Kennedy MA, Maggo SDS. Pharmacogenetics of statin-induced myotoxicity. Front Genet. 2020;11:575678.
44. Lim KS, Jang I-J, Kim B-H, Kim J, Jeon J-Y, Tae Y-M, et al. Changes in the QTc interval after administration of flecainide acetate, with and without coadministered paroxetine, in relation to cytochrome P450 2D6 genotype: data from an open-label, two-period, single-sequence crossover study in healthy Korean male subjects. Clin Ther. 2010;32(4):659–66.
45. Group TDPW. The Dutch Pharmacogenomic Working Group. 2022. Phamacogenomic recommendations, farmacogenetica-update. www.knmpnl/.
46. Caudle KE, Klein TE, Hoffman JM, Muller DJ, Whirl-Carrillo M, Gong L, et al. Incorporation of pharmacogenomics into routine clinical practice: the Clinical Pharmacogenetics Implementation Consortium (CPIC) guideline development process. Curr Drug Metab. 2014;15(2):209–17.
47. Pharmacogene Variation Consortium. www.pharmvar.org/.
48. Clinical Pharmacology. School of Medicine. Drug Interactions Flockhart Table ™. https://drug--interactions.medicine.iu.edu/MainTable.aspx.
49. West H. Novel precision medicine trial designs: umbrellas and baskets. JAMA Oncol. 2017;3(3):423.
50. Yates LR, Seoane J, Le Tourneau C, Siu LL, Marais R, Michiels S, et al. The European Society for Medical Oncology (ESMO) precision medicine glossary. Ann Oncol. 2018;29(1):30–5.
51. D'Adamo GL, Widdop JT, Giles EM. The future is now? Clinical and translational aspects of "omics" technologies. Immunol Cell Biol. 2021;99(2):168–76.
52. Chapman PB, Hauschild A, Robert C, Haanen JB, Ascierto P, Larkin J, et al. Improved survival with vemurafenib in melanoma with BRAF V600E mutation. N Engl J Med. 2011;364(26):2507–16.
53. Cunningham D, Humblet Y, Siena S, Khayat D, Bleiberg H, Santoro A, et al. Cetuximab monotherapy and cetuximab plus irinotecan in irinotecan-refractory metastatic colorectal cancer. N Engl J Med. 2004;351(4):337–45.
54. Skoulidis F, Li BT, Dy GK, Price TJ, Falchook GS, Wolf J, et al. Sotorasib for lung cancers with KRAS p.G12C mutation. N Engl J Med. 2021;384(25):2371–81.
55. Ledermann J, Harter P, Gourley C, Friedlander M, Vergote I, Rustin G, et al. Olaparib maintenance therapy in platinum-sensitive relapsed ovarian cancer. N Engl J Med. 2012;366(15):1382–92.
56. Schwaederle M, Zhao M, Lee JJ, Lazar V, Leyland-Jones B, Schilsky RL, et al. Association of biomarker-based treatment strategies with response rates and progression-free survival in refractory malignant neoplasms: a meta-analysis. JAMA Oncol. 2016;2(11):1452–9.
57. Duan M, Wang Y, Qiao Y, Wang Y, Pan X, Hu Z, et al. Pan-cancer identification of the relationship of metabolism-related differentially expressed transcription regulation with non-differentially expressed target genes via a gated recurrent unit network. Comput Biol Med. 2022;148:105883.
58. Cobain EF, Wu YM, Vats P, Chugh R, Worden F, Smith DC, et al. Assessment of clinical benefit of integrative genomic profiling in advanced solid tumors. JAMA Oncol. 2021;7(4):525–33.
59. Long GV, Stroyakovskiy D, Gogas H, Levchenko E, de Braud F, Larkin J, et al. Combined BRAF and MEK inhibition versus BRAF inhibition alone in melanoma. N Engl J Med. 2014;371(20):1877–88.
60. Sicklick JK, Kato S, Okamura R, Schwaederle M, Hahn ME, Williams CB, et al. Molecular profiling of cancer patients enables personalized combination therapy: the I-PREDICT study. Nat Med. 2019;25(5):744–50.
61. Roth GA, Mensah GA, Johnson CO, Addolorato G, Ammirati E, Baddour LM, et al. Global burden of cardiovascular diseases and risk factors, 1990–2019: update from the GBD 2019 study. J Am Coll Cardiol. 2020;76(25):2982–3021.
62. World Health Organization (WHO). Cardiovascular diseases (CVDs). https://www.who.int/news-room/fact-sheets/detail/cardiovascular-diseases-(cvds).
63. Leopold JA, Loscalzo J. Emerging role of precision medicine in cardiovascular disease. Circ Res. 2018;122(9):1302–15.
64. Zhou Z. Pharmacogenomics in cardiovascular precision medicine. J Lab Precis Med. 2020;5:30.
65. Tobias DK, Merino J, Ahmad A, Aiken C, Benham JL, Bodhini D, et al. Second international consensus report on gaps and opportunities for the clinical translation of precision diabetes medicine. Nat Med. 2023;29(10):2438–57.
66. Aghaei Meybodi HR, Hasanzad M, Sarhangi N, Larijani B. Precision medicine in endocrinology practice. In: Precision medicine in clinical practice. Berlin: Springer; 2022. p. 67–91.
67. Nogueiras-Álvarez R. Pharmacogenomics in clinical trials: an overview. Front Pharmacol. 2023;14:1247088.
68. Boulyjenkov V, Schapper C. The ethical, legal and social implications of pharmacogenomics in developing countries: report of an international group of experts. Geneva: World Health Organization; 2007.
69. Swen JJ, van der Wouden CH, Manson LE, Abdullah-Koolmees H, Blagec K, Blagus T, et al. A 12-gene pharmacogenetic panel to prevent adverse drug reactions: an open-label, multicentre, controlled, cluster-randomised crossover implementation study. Lancet. 2023;401(10374):347–56.
70. Reizine N, Vokes EE, Liu P, Truong TM, Nanda R, Fleming GF, et al. Implementation of pharmacogenomic testing in oncology care (PhOCus): study protocol of a pragmatic, randomized clinical trial. Ther Adv Med Oncol. 2020;12:1758835920974118.
71. Cully M. Target validation: genetic information adds supporting weight. Nat Rev Drug Discov. 2015;14(8):525.
72. Ochoa D, Karim M, Ghoussaini M, Hulcoop DG, McDonagh EM, Dunham I. Human genetics evidence

supports two-thirds of the 2021 FDA-approved drugs. Nat Rev Drug Discov. 2022;21(8):551.
73. Olstad DL, McIntyre L. Reconceptualising precision public health. BMJ Open. 2019;9(9):e030279.
74. Australian Genomics. https://www.australiangenomics.org.au/.
75. National Institutes of Health (NIH). The future of health begins with you. https://allofus.nih.gov.
76. Smedley D, Smith KR, Martin A, Thomas EA, McDonagh EM, Cipriani V, et al. 100,000 genomes pilot on rare-disease diagnosis in health care—preliminary report. N Engl J Med. 2021;385(20):1868–80.
77. Turnbull C, Scott RH, Thomas E, Jones L, Murugaesu N, Pretty FB, et al. The 100000 genomes project: bringing whole genome sequencing to the NHS. BMJ. 2018;361:k1687.
78. Lévy Y. Genomic medicine 2025: France in the race for precision medicine. Lancet. 2016;388(10062):2872.
79. Alvarez MJR, Griessler E, Starkbaum J. Ethical, legal and social aspects of precision medicine. In: Precision medicine in clinical practice. Berlin: Springer; 2022. p. 179–96.
80. General Data Protection Regulation (GDPR). GDPR personal data—what information does this cover? https://www.gdpreu.org/the-regulation/key-concepts/personal-data/.
81. World Medical Association. World Medical Association Declaration of Helsinki. Ethical principles for medical research involving human subjects. Bull World Health Organ. 2001; 79(4):373.

Digital Health

8

Marius Geanta, Adriana Boata, Bianca Cucos, and Cosmina Cioroboiu

What Will You Learn in This Chapter?
This chapter outlines the integration and advancements of digital technologies in healthcare, from the transformative impact of electronic health records and telemedicine in modernizing patient care and healthcare delivery to point-of-care tools for disease prevention and best practices for integrating LLMs in healthcare. Moreover, it highlights successful digital health strategies from global frontrunners, discussing examples of best practice which can be adapted and implemented in different countries. Through this chapter, readers will gain insights into how digital health technologies are not just augmenting existing healthcare practices but are also paving new pathways for personalized medicine, patient care, and the broader implications of these technologies on society and ethical considerations.

Rational and Importance
The importance of this chapter lies in its ability to provide readers with a holistic view of the current digital health landscape, highlighting the significant benefits and challenges associated with the integration of advanced technologies in healthcare. It underscores the necessity of navigating the ethical and regulatory considerations to leverage digital health innovations effectively. The chapter emphasizes the critical role of digital health in improving health outcomes, making healthcare more accessible and personalized, and the potential of these technologies to address pressing healthcare challenges. By showcasing examples of successful digital health implementations and the potential of emerging technologies, the chapter serves as a valuable resource for understanding the future direction of healthcare and the importance of digital health in achieving better health outcomes for individuals and populations globally.

M. Geanta (✉)
Centre for Innovation in Medicine, Bucharest, Romania

KOL Medical Media, Bucharest, Romania

United Nations University-Maastricht Economic and Social Research Institute on Innovation and Technology, Maastricht, The Netherlands
e-mail: marius.geanta@ino-med.ro

A. Boata
Centre for Innovation in Medicine, Bucharest, Romania

Hesy Biotech Consulting, Bucharest, Romania

B. Cucos
Centre for Innovation in Medicine, Bucharest, Romania

Genomed Consulting, Bucharest, Romania

C. Cioroboiu
Centre for Innovation in Medicine, Bucharest, Romania

KOL Medical Media, Bucharest, Romania

M. Hasanzad (ed.), *A Glimpse at Medicine in the Future*,
https://doi.org/10.1007/978-981-97-4974-4_8

8.1 The Evolution of Digital Health

8.1.1 Historical Perspective and Recent Advances

The evolution of digital health is underpinned by a series of technological milestones that have collectively enabled its development. It begins with the invention of the computer, a cornerstone that transformed data processing and management. The early computers, though bulky and limited in function, introduced the concept of digitizing information, setting the stage for further technological advancements. The next significant milestone was the development of the Internet and networking technologies. This breakthrough revolutionized information sharing and connectivity, laying the groundwork for remote access to healthcare data and telemedicine. The Internet also facilitated the widespread dissemination of medical knowledge, democratizing access to health information for professionals and patients alike [1].

The advent of personal computing devices, particularly smartphones and tablets, marked another critical development. These devices brought computing power to the hands of individuals, making health information and tools more accessible than ever. This era also saw the rise of health-specific applications and platforms, further integrating digital technology into everyday health management.

Simultaneously, advancements in data storage and cloud computing played a crucial role. These technologies allowed for the secure storage and management of vast amounts of health data, enabling real-time data analysis and supporting the transition to Electronic Health Records (EHRs). EHRs not only improved the efficiency of data handling but also laid the groundwork for more advanced digital health applications [2].

Further, the integration of artificial intelligence (AI) and machine learning has been pivotal. AI has transformed various aspects of healthcare, from predictive analytics in public health to personalized treatment plans in clinical settings. AI's ability to analyze complex datasets has been instrumental in advancing diagnostic tools, drug development, and even operational efficiencies within healthcare systems [3].

In recent years, wearable technology and the Internet of Medical Things (IoMT) have emerged as key contributors to digital health. These innovations provide continuous health monitoring and generate a wealth of health data, offering insights into patient health trends and enabling proactive healthcare management [4].

Each of these technological milestones has played a fundamental role in shaping the landscape of digital health. Together, they have transformed healthcare from a traditionally reactive, one-size-fits-all system into a more personalized, efficient, and accessible domain, marking a new era in the history of medicine. These technologies extended the capabilities of digital health from mere data storage and management to advanced analytics and predictive modeling. Healthcare providers could now leverage these tools for more than administrative tasks; they could predict disease outbreaks, understand patient behavior and trends, and tailor treatments to individual needs.

The last decade has been particularly transformative for digital health, marked by rapid advancements and widespread adoption. These innovations have handed individuals unprecedented control over citizen health management, providing real-time access to health metrics and fostering a culture of proactive health engagement. During critical times, such as the COVID-19 pandemic, digital health tools like telemedicine emerged not just as conveniences but as necessities, enabling care delivery while adhering to social distancing norms [5].

Moreover, the integration of AI in healthcare has been nothing short of revolutionary. AI's role in diagnostic procedures, drug development, and personalized treatment plans has dramatically enhanced the precision and efficiency of healthcare services. AI algorithms are now capable of analyzing complex medical data sets, providing diagnostic insights, and suggesting treatment options with increasing accuracy.

8.1.2 Integration of Digital Technologies in Healthcare

The integration of digital technologies into healthcare signifies a transformative shift, touching every facet of the healthcare ecosystem. This integration is multidimensional, impacting patient records management, diagnostic procedures, treatment plans, patient engagement, and health education.

Personalized medicine, powered by digital technologies, is one of the most significant outcomes of this integration. The convergence of genomics, biotechnology, and advanced data analytics has enabled the tailoring of medical treatments to individual genetic profiles, marking a shift from a one-size-fits-all approach to a more personalized healthcare paradigm. This personalization extends beyond treatment to preventive care, where technologies like wearable devices and health apps offer individualized health monitoring and lifestyle recommendations [6].

Accessibility in healthcare has been greatly enhanced by digital technology. Telemedicine has broken down geographical barriers, providing quality medical consultations to patients in remote or underserved areas. Mobile health applications have brought health education to people worldwide, enhancing health literacy and promoting self-care [7].

Additionally, the efficiency of healthcare systems has seen significant improvement. EHRs and health information systems have streamlined healthcare operations, reducing paperwork, and minimizing error rates. AI-driven models contribute to efficient resource allocation, patient flow management, and effectively handling public health crises and epidemic responses [8].

The evolution of digital health is a narrative of continuous innovation and adaptation. From early digitization efforts to the latest AI-driven personalized medicine approaches, digital health is redefining the landscape of healthcare delivery. It promises a future where healthcare is not only more accessible and efficient but also more attuned to the individual needs and nuances of each patient. The journey has witnessed the transformation of digital tools from being mere adjuncts to becoming central components in the delivery of healthcare.

The story of digital health is also one of collaboration and interdisciplinary fusion. The fields of information technology, biotechnology, medical science, and public health have converged, leading to innovations like genomic sequencing, telehealth, and AI in diagnostics, which were unimaginable a few decades ago. This collaborative approach has paved the way for breakthroughs in disease management and preventive healthcare [9].

The future trajectory of digital health is poised to be as dynamic as its past. Emerging technologies like blockchain, augmented reality, and the Internet of Medical Things (IoMT) are set to further revolutionize healthcare. Blockchain promises enhanced security and transparency in health data management, while augmented reality and IoMT hold potential for advanced patient education, remote surgeries, and real-time health monitoring [10].

In conclusion, the evolution of digital health has become a journey that has not only transformed healthcare systems but also fundamentally altered how individuals engage with and experience health and wellness. As digital health continues to evolve, it holds the promise of redefining the boundaries of what is possible in healthcare delivery and management.

8.2 Enhancing Healthcare Efficiency: The Synergy of Electronic Health Records, Telemedicine, and Connectivity

The contemporary healthcare landscape is witnessing a significant transformation, driven by the synergistic integration of Electronic Health Records (EHRs), telemedicine, and enhanced connectivity with patients. This chapter delves into how these components collectively contribute to a more efficient and responsive healthcare system.

8.2.1 Electronic Health Records (EHRs): Foundation of Modern Healthcare Efficiency

The development and implementation of EHR systems have been a cornerstone in enhancing healthcare efficiency. EHRs have revolutionized data management in healthcare, transitioning from traditional paper-based records to digital formats. This digital transformation has led to improved accuracy in patient information, streamlined workflows, and better coordination among healthcare providers [11]. EHRs serve as a comprehensive database, offering clinicians a holistic view of a patient's medical history, current treatments, and potential health risks. This integrated approach significantly reduces medical errors, enhances diagnostic accuracy, and facilitates faster and more effective treatment decisions.

One notable example of EHRs enhancing healthcare is in the management of chronic diseases. EHRs enable healthcare providers to track patients' health metrics over time, adjust medications and treatments as needed, and monitor the effectiveness of management plans. This ongoing monitoring and adjustment improve patient outcomes and reduces hospital readmissions. Additionally, EHRs facilitate better coordination among different healthcare providers, especially for patients consulting multiple specialists, by providing a shared platform for information exchange [12].

EHRs also play a significant role in health promotion. They generate reminders for both patients and healthcare providers about preventive screenings and routine check-ups, aiding in early disease detection and encouraging proactive health management. Moreover, the aggregated data from EHRs can be utilized for public health research, identifying trends and developing targeted health promotion campaigns [13].

However, EHRs are not without their challenges, especially in supporting behavioral health interventions. Initiatives are being undertaken to improve their capacity in this area, such as integrating tools for screening, counseling, and referral processes directly into the EHR workflow. This integration would enable healthcare providers to more effectively address behavioral health issues, a vital aspect of preventative healthcare. Additionally, many EHR systems include patient portals that empower patients by allowing them to access their medical records, view test results, and communicate with healthcare providers, thereby encouraging them to take an active role in their healthcare [14].

EHRs are dynamic tools that go beyond being mere digital repositories of patient data. They enhance healthcare performance by improving information accuracy, streamlining workflows, and facilitating proactive healthcare management. Their role is crucial in the ongoing efforts to improve healthcare outcomes and promote public health, making them an indispensable component of modern healthcare systems [15].

8.2.2 Telemedicine: Expanding Access and Streamlining Care

Telemedicine has emerged as a critical component in modernizing healthcare delivery. It encompasses a range of services, from virtual consultations to remote monitoring, fundamentally changing the interaction between healthcare providers and patients [16]. Telemedicine extends the reach of healthcare services, making them accessible even in remote or underserved areas. It also offers convenience and flexibility for both patients and healthcare providers, reducing the need for physical visits and thereby decreasing the burden on healthcare facilities. Through telemedicine, continuous monitoring and management of chronic conditions have become more feasible, improving patient outcomes and overall health management [17].

Preventive medicine, increasingly prioritized over curative approaches, has demonstrated its effectiveness in reducing healthcare costs and improving patient outcomes, particularly evident during the COVID-19 pandemic. Telemedicine plays a crucial role in this shift, offering innovative solutions for early disease detection, accessible treatment, and promoting healthier lifestyles. Despite its potential, telemedicine's implementation faces challenges such as the need for extensive training, limited internet access, and engagement barriers among specific populations. Addressing these challenges through community engagement, training, and multi-sector collaboration is essential to enhance telemedicine's role in preventive care, aligning with the broader goal of advancing public health and individual well-being [18].

Telemedicine is significantly enhancing cardiovascular health promotion through innovative approaches such as remote monitoring and patient education. For example, patients with cardiovascular diseases can use wearable devices that track vital signs like heart rate, blood pressure, and oxygen saturation levels. This data is transmitted in real-time to healthcare providers via telemedicine platforms. Physicians can monitor these vital signs remotely, allowing for timely intervention if any abnormal patterns are detected.

Additionally, telemedicine enables patients to participate in virtual cardiac rehabilitation programs from the comfort of their homes. These programs typically include exercise sessions, dietary counseling, and education about heart health, all facilitated through video conferencing and digital communication tools. This approach not only increases accessibility for patients who might have difficulty traveling to a physical location but also ensures continuous support and monitoring of their health.

Telehealth interventions can be an effective tool in reducing the incidence of CVD, with studies showing multiple evidence on its potential in reducing mortality and hospitalization for health failure, for example [19].

In the context of primary CVD prevention, THI can facilitate remote monitoring of vital signs and heart health, allowing healthcare providers to track patients' blood pressure, cholesterol levels, and heart rhythms regularly. This continuous monitoring enables early identification and management of risk factors like hypertension and arrhythmias, which are crucial in preventing CVD. Additionally, THI can support lifestyle modification programs, such as smoking cessation and weight management, which are integral to reducing CVD risk. Patients can receive counseling and coaching through telehealth platforms, providing them with the guidance and support needed to make healthier lifestyle choices [20].

Regarding cancer risk factor prevention, THI can play a key role in promoting regular screenings and early detection strategies. Telehealth platforms can be used to educate patients about the importance of screenings for cancers like breast, colorectal, and cervical cancer and facilitate the scheduling of these screenings. Furthermore, THI can support patients in adopting behaviors that reduce cancer risk, such as dietary changes, regular exercise, and avoidance of tobacco use. By offering remote counseling sessions and digital health tools for tracking and managing health behaviors, THI empowers patients to take proactive steps in cancer prevention [21].

8.2.3 Connectivity with Patients: Personalizing and Democratizing Healthcare

The enhanced connectivity with patients, facilitated by digital health technologies, is reshaping the patient-provider relationship. Digital platforms, mobile health applications, and patient portals have opened new channels of communication, making healthcare more patient-centric. These tools empower patients to actively participate in their health management, access their health records, and communicate more effectively with their healthcare providers. This level of connectivity has also led to an increase in health awareness and self-management among patients, contributing to better health outcomes.

8.3 Citizen-Generated Data in Healthcare

The landscape of healthcare is rapidly evolving with the integration of citizen-generated data, marking a significant shift towards more personalized and preventive care.

8.3.1 Wearable Devices in Healthcare

Wearable devices have become ubiquitous in modern healthcare, providing continuous health monitoring and generating vast amounts of data. These devices range from fitness trackers and smartwatches to more specialized medical wearables that monitor specific health parameters like heart rate, blood glucose levels, and sleep patterns. They offer real-time insights into an individual's health status, enabling both patients and healthcare providers to track and manage health conditions more effectively. Wearable devices also play a crucial role in preventive healthcare by identifying potential health issues before they become serious, thereby facilitating early intervention and reducing the need for more invasive treatments [22].

Fitness trackers and smartwatches, commonly used for monitoring basic health metrics like heart rate and steps, are now sophisticated enough to detect heart arrhythmias with built-in ECG functions. Continuous Glucose Monitors (CGMs), offer real-time blood sugar tracking for diabetes patients, eliminating frequent finger-prick tests. Wearable ECG monitors and blood pressure devices provide extended heart monitoring and blood pressure measurements, aiding in the management of chronic conditions [23].

The advent of wireless ultrasound monitoring for subjects in motion represents a significant technological breakthrough in medical diagnostics and patient care. This innovative approach leverages cutting-edge wireless technologies and advanced ultrasound techniques to monitor internal body structures and functions in real-time, even as the subjects engage in physical activities. Traditionally, ultrasound monitoring has been confined to stationary settings due to the bulky nature of the equipment. However, the development of compact, wireless ultrasound devices has overcome this limitation, allowing for dynamic and continuous monitoring of patients while they are moving [24].

For example, engineers at the University of California San Diego have developed a groundbreaking wearable ultrasound system, the first of its kind, designed for deep-tissue monitoring. This system, integrating an ultrasonic system-on-patch, allows for continuous tracking of deep physiological signals such as central blood pressure, heart rate, and cardiac output for up to 12 h [25]. It's a significant advancement in cardiovascular monitoring, particularly useful for detecting heart failure and personalizing exercise plans based on cardiovascular response. This innovation is also a key development in the Internet of Medical Things (IoMT), allowing for wireless transmission of physiological data to the cloud for analysis and diagnosis [26].

Additionally, emerging technologies like smart clothing with integrated sensors can monitor vital signs. These wearable technologies signify a shift towards more proactive health management, offering continuous data for both patients and healthcare providers to enable more personalized care and timely interventions.

The landscape of wearable technologies is rapidly evolving to overcome the limitations of conventional wearables, which often suffer from a lack of flexibility and bulkiness, hindering their effectiveness in continuous health monitoring. This evolution is markedly seen in the rise of e-textiles, where the ingenious integration of miniature electronic devices into fabric has given birth to smart clothing systems. These systems are designed for remote health monitoring, representing a significant leap from traditional wearables. The market for smart clothing is on an impressive trajectory, projected to reach a value of USD 8.8 billion by 2024. This surge in demand for smart garments spans various industries, including retail, military, sports, and healthcare. The growing interest in these sectors reflects the broad applicability and potential of smart clothing in enhancing health monitoring and personal well-being [27]. Hexoskin is a prime example of how e-textiles are revolutionizing the field of wearable health monitoring. This innovative smart clothing system is designed for precise and continuous tracking of vital health metrics.

8.3.2 Understanding Behaviors in Social Media Using AI Models

The rise of social media has provided an unprecedented opportunity to understand public health trends and individual behaviors. AI models are increasingly being employed to analyze social media data for insights into mental health, disease outbreaks, and public responses to health campaigns. These models can identify patterns and trends in the vast amounts of data generated on social media platforms, offering valuable insights into public health concerns. For instance, AI-driven analysis of social media posts can help in early detection of mental health issues, enabling timely intervention. Additionally, social media data, when analyzed effectively, can aid in understanding the spread of misinformation regarding health, guiding public health officials in crafting targeted informational campaigns [28].

Social media data plays a crucial role in shaping digital health strategies, offering valuable insights across various aspects of healthcare. It serves as a vital tool for public health monitoring and disease surveillance, enabling health officials to detect emerging health issues and disease outbreaks by analyzing trends in social media discussions. This data is also instrumental in understanding patient sentiments, as analyzing online discussions and feedback can help healthcare organizations tailor their services to better meet patient needs and improve satisfaction. Furthermore, social media is a rich source for health behavior research, providing large-scale insights into public health behaviors and aiding in the development of targeted health promotion campaigns. An important application of social media data lies in tracking and addressing health misinformation, ensuring the dissemination of accurate health information to the public. It also supports mental health initiatives by helping in the early identification of mental health struggles through user posts. In the realm of pharmaceuticals, social media offers valuable real-world data on drug efficacy and side effects, contributing to pharmacovigilance and guiding drug development. Additionally, healthcare organizations leverage social media analytics for targeted marketing and patient engagement, building trust and strengthening patient-provider relationships [29].

8.3.3 Point of Care

Point-of-care technologies represent a significant advancement in providing immediate and location-flexible healthcare services. These technologies include portable diagnostic devices, mobile health applications, and telehealth services, which enable healthcare delivery outside traditional clinical settings. Point-of-care tools are particularly beneficial in remote or underserved areas, where access to healthcare facilities is limited. They allow for quick diagnostics and decision-making at the patient's location, improving the speed and efficiency of healthcare delivery. Moreover, these technologies support the management of chronic conditions by enabling regular monitoring and immediate feedback to patients, thus enhancing patient engagement and adherence to treatment plans [15].

A notable example of a point-of-care tool for disease prevention is a mobile health (mHealth) application used for diabetes risk assessment and management. This app is designed to be used directly by patients, providing them with a convenient and efficient way to monitor their health status and take preventive measures against diabetes. The app works by allowing users to input key health metrics such as their weight, age, diet, physical activity levels, and family history of diabetes. Utilizing a built-in algorithm, the app then assesses the user's risk of developing Type 2 diabetes. For those identified as high-risk, the app provides personalized lifestyle modification recommendations, such as diet plans, exercise routines, and weight management strategies, to help reduce their risk. Additionally, the app includes features like blood glucose tracking, where users can log their blood sugar levels, and the app can alert them if the readings indicate a pre-diabetic state. It also provides educational resources about diabetes, including tips for prevention and management, as well as the importance of regular check-ups [30]. Furthermore, the app can facilitate remote consultations with healthcare providers. Users can share their health data through the app with their doctors, who can then provide more tailored advice or intervene as necessary.

In summary, citizen-generated data through wearable devices, AI-powered analysis of social media, and point-of-care technologies are transforming the healthcare landscape. They are shifting the focus from reactive to proactive care, emphasizing prevention, early detection, and personalized treatment. This shift not only improves patient outcomes but also contributes to the overall efficiency of the healthcare system. As these technologies continue to evolve, they promise to further empower patients, enhance healthcare delivery, and shape the future of public health.

8.4 Data Analytics and AI

8.4.1 Big Data in Healthcare: Opportunities and Challenges

In the realm of healthcare, the integration of big data is opening a world of opportunities while presenting distinct challenges. Big data in healthcare encompasses the vast accumulation of information from varied sources such as electronic health records, medical imaging, genomic sequencing, and patient-generated data through wearables. The application of big data presents significant opportunities. It paves the way for precision medicine, where treatments and therapies are tailored to individual patients based on their unique data profiles. Big data also enhances predictive analytics, allowing healthcare providers to anticipate health issues and intervene preventively. On a larger scale, it offers invaluable insights for public health management, helping in formulating effective disease control and prevention strategies. Additionally, big data contributes to cost efficiency in healthcare by optimizing treatment approaches and reducing unnecessary healthcare spending [31].

However, the use of big data in healthcare is not without its challenges. Data privacy and security emerge as primary concerns, as the handling of sensitive health data necessitates stringent protective measures. Integrating and standardizing data from diverse sources also presents significant technical challenges. There is a growing demand for skilled personnel proficient in data analytics and capable of interpreting complex health data. Moreover, ethical concerns arise regarding patient consent, data ownership, and the potential biases in the interpretation of this data.

8.4.2 Artificial Intelligence in Diagnostics and Treatment Planning

Artificial intelligence is significantly reshaping diagnostics and treatment planning in healthcare. In diagnostics, AI's impact is profoundly seen in the realm of medical imaging. AI algorithms are being increasingly employed to interpret complex images such as X-rays, CT scans, and MRIs, often with greater speed and accuracy than traditional methods. AI also plays a crucial role in disease prediction, analyzing patient data to identify potential health issues before they manifest, thereby enabling early intervention. In pathology, AI enhances the accuracy of tissue sample analysis, particularly in the detection and treatment of diseases like cancer [3, 32].

In treatment planning, AI's influence is marked by the advent of customized treatment regimens. Leveraging AI algorithms, healthcare providers can now devise personalized treatment plans based on the unique health profiles of individual patients, enhancing the efficacy of treatments. AI is also accelerating drug development processes, predicting the success rates and potential side effects of new drugs, and is integral to the burgeoning field of personalized medicine. Beyond diagnostics and treatment, AI improves operational efficiency in healthcare settings, streamlining processes such as patient scheduling, resource allocation, and hospital management.

The advent of big data and AI in healthcare is a landmark development, bringing transformative changes with immense potential benefits. While the opportunities they present—from personalized medicine to enhanced operational efficiency—are vast, the challenges, particularly in data privacy, technical integration, and ethical concerns, are significant hurdles that need to be addressed. Navigating these challenges effectively is key to fully leveraging the potential of big data and AI in advancing healthcare [3, 15].

8.5 Large Language Models (LLMs) in Healthcare

The advent of Large Language Models (LLMs) like GPT (Generative Pre-trained Transformer) has introduced a new dimension to healthcare, offering unique opportunities and challenges in patient interaction, health literacy, clinical practice, and establishing best practices.

8.5.1 Chat GPT and Emerging AI Tools for Interaction (Health Impact)

Chat GPT and similar AI tools are redefining patient-provider interactions. These LLMs, capable of understanding and generating human-like text, are being used for patient engagement, health education, and support. They can handle initial patient inquiries, provide information on health conditions, and guide patients through the healthcare system [33]. For example, AI chatbots can answer routine health queries, schedule appointments, and even provide preliminary advice, which can be especially beneficial in areas with limited access to healthcare providers. However, this brings challenges in ensuring the accuracy of information, maintaining patient confidentiality, and establishing clear boundaries for AI's role in healthcare.

The successful integration of LLMs in healthcare is likely for conditions where interventions are backed by comprehensive data from clinical trials and observational studies. Initial focus on routine medical encounters, supported by ample data, can address bias issues, with larger data sets helping to identify gaps in underlying data. Currently, LLMs in healthcare are mainly used as digital assistants for patient data collection and structured information presentation to clinicians, without offering direct diagnostic or treatment recommendations to avoid FDA regulation. However, these tools are poised to enhance medical care, overseeing routine aspects of health care and broadening patient engagement [34].

Data on the social determinants of health can be extracted from electronic patient records using an LLM model. A recent study published in Digital Medicine highlights the potential of deep learning LLM (large language models) systems to improve the identification of social determinants of health (SDoH) in electronic health records.

This innovative model demonstrated remarkable accuracy, identifying 93.8% of patients with an SDoH risk profile. Considering that using traditional diagnostic codes only captures about 2% of SDoH data, the LLM model would enable the identification of patients who could benefit from additional health services and social care, while also highlighting the often inadequately documented impact of SDoH on patient outcomes.

For the development of this AI tool, the Mass General Brigham team analyzed data from 770 cancer patients, identifying 6 major categories of SDoH: employment, housing, transportation, parental status, relationships, and social support. They trained language models using this dataset to identify SDoH references in clinician-reported data.

The team also paid particular attention to mitigating algorithmic bias, a known issue in some generative AI models in healthcare. Comparing their tool with OpenAI's GPT-4, they found that their model is less likely to alter the identification of an SDoH based on the patient's race/ethnicity or gender. The authors emphasize the importance of monitoring algorithmic bias in the development of LLMs to avoid exacerbating existing health inequalities [33, 35].

8.5.2 Health Literacy

LLMs have the potential to significantly enhance health literacy. By providing easily accessible information tailored to an individual's comprehension level, these models can help bridge the gap between complex medical information and patient understanding. LLMs can generate patient education materials, answer FAQs, and even translate medical jargon into layman's terms. However, there is a need to monitor the quality of information and prevent the dissemination of misinformation.

An example of using LLMs in prevention and health promotion is in the creation of personalized health education content. For instance, an LLM could be trained on vast amounts of medical literature and patient data to generate tailored educational material for individuals with specific health risks. If a patient has a family history of heart disease, the LLM can create a personalized guide on heart-healthy habits, dietary recommendations, and warning signs to watch for. This approach not only educates the patient in a way that's directly relevant to their health situation but also promotes proactive measures to prevent the onset or worsening of conditions [36].

8.5.3 Embedding in Clinical Practice: Diagnostics

In diagnostics, LLMs are beginning to find a place as supportive tools. They can assist in processing patient history and presenting potential diagnoses based on symptoms and medical records. This can aid clinicians in decision-making by providing quick access to a vast database of medical knowledge. However, integrating LLMs into clinical practice requires careful consideration of their limitations and should not replace the expert judgment of healthcare professionals.

In a recent study focusing on optimizing large language models (LLMs) for differential diagnosis (DDx), a newly developed LLM integrated with an interactive interface demonstrated significant improvements in generating accurate DDx lists. Utilizing challenging cases from the New England Journal of Medicine (NEJM), the study compared the diagnostic capabilities of 20 board-certified clinicians using the LLM against traditional information retrieval methods. Results indicated that the optimized LLM outperformed the previous GPT-4 model in both accuracy and quality of DDx generation. Clinicians reported improved diagnostic capacity and diversity in DDx lists when using the LLM tool. The study highlights the potential of LLMs in enhancing the DDx process, suggesting a more comprehensive and appropriate generation of DDx lists compared to those produced by physicians alone [37].

8.5.4 Best Practices

Establishing best practices for the use of LLMs in healthcare is critical. This includes setting guidelines for their use, ensuring the reliability and accuracy of the information provided, and safeguarding patient privacy. Training healthcare professionals on how to effectively utilize these AI tools in conjunction with traditional medical practices is also essential. Furthermore, ongoing evaluation and updating of these AI systems are necessary to keep pace with the rapidly evolving medical field.

One notable best practice example of using Large Language Models (LLMs) in healthcare is the implementation of AI-driven chatbots for patient triage and initial consultation in hospitals and clinics. These chatbots, powered by advanced LLMs, are being used to interact with patients upon their initial contact with healthcare services, either through hospital websites, apps, or virtual health platforms.

In this implementation, patients can describe their symptoms and concerns in natural language, and the LLM-powered chatbot interprets this information to provide initial guidance. The chatbot can ask follow-up questions to gather more detailed information, much like a human medical professional would. Based on the responses, the chatbot assesses the urgency of the patient's condition and guides them to the appropriate level of care—be it scheduling an in-person appointment, advising a visit to the emergency room, or providing home care recommendations.

For example, the AI chatbot developed by Babylon Health uses LLMs to evaluate patients' symptoms and suggests potential diagnoses and treatment options. This system not only improves efficiency by reducing wait times for patients but also helps to triage cases effectively, ensuring that urgent cases receive prompt attention [38].

In conclusion, LLMs like Chat GPT offer exciting possibilities in healthcare, from enhancing patient interactions and health literacy to assisting in diagnostics. However, their integration into healthcare must be navigated carefully, with a focus on maintaining accuracy, privacy, and complementing rather than replacing human medical expertise. Establishing best practices and continuous evaluation will be key in leveraging the full potential of LLMs in healthcare.

8.6 Digital Health Policy and Ethics

The final chapter of this exploration into digital health delves into the crucial aspects of policy and ethics, focusing on data governance, the regulatory landscape, and ethical considerations surrounding digital health innovations.

8.6.1 Data Governance and Innovative Models of Health Data Sharing

Data governance in the context of digital health is pivotal, encompassing the stewardship of health data to ensure its quality, security, and privacy. Effective data governance models are crucial for facilitating health data sharing while protecting sensitive patient information. This involves establishing clear policies and standards for data collection, storage, and dissemination. Innovative models for health data sharing are also emerging, such as health information exchanges (HIEs) that allow seamless sharing of patient data across different healthcare systems, improving care coordination and patient outcomes. Blockchain technology is another innovative approach being explored for secure and transparent health data sharing. These models, however, must navigate complex issues such as patient consent, data ownership, and interoperability challenges [39].

8.6.2 Regulatory Landscape for Digital Health Solutions

The regulatory landscape for digital health solutions is continuously evolving. This encompasses the frameworks and guidelines set by governmental and regulatory bodies to ensure the safety, efficacy, and ethical deployment of digital health technologies. As digital health innovations proliferate, regulators face the challenge of keeping pace with rapid technological advancements while ensuring public safety. Regulations must address a range of issues, from the approval of AI-driven diagnostic tools to the use of wearable devices and telehealth services. Ensuring compliance with laws such as the Health Insurance Portability and Accountability Act (HIPAA) in the United States and the General Data Protection Regulation (GDPR) in Europe is crucial for maintaining patient privacy and data security [40].

The regulatory landscape of digital health in Europe is rapidly evolving, with several initiatives and frameworks being established to guide the development and implementation of digital health technologies. A key component of this landscape is the European Health Data Space (EHDS), which is a significant effort by the European Union to promote health data sharing and interoperability across member states.

The EHDS aims to facilitate the exchange and access to different types of health data (electronic health records, genomics data, etc.) to support healthcare delivery, research, and policymaking. It focuses on ensuring data privacy and security, fostering a trustworthy digital health environment. The EHDS is part of the broader European strategy for data, which seeks to create a single market for data within the EU, allowing it to flow freely across borders, yet under strict control and in full compliance with EU data protection laws [41].

In addition to the EHDS, Europe has other regulatory frameworks and initiatives that impact digital health. The General Data Protection Regulation (GDPR) plays a crucial role in protecting personal data, including health data. The Medical Device Regulation (MDR) and the In Vitro Diagnostic Regulation (IVDR) are also important, as they set stringent standards for the safety and performance of medical devices, including software and AI-driven tools used in healthcare [42].

Germany has been successful in its digital health strategies, particularly with the introduction of the Digital Healthcare Act (DVG). This act facilitates the integration of digital health applications into standard care, allowing doctors to prescribe digital health applications (DiGA) that are reimbursed by health insurance. Germany has also made significant strides in the adoption of electronic health records and telemedicine services [43].

Estonia is often hailed as a leader in digital health, with a highly advanced e-health system. The country has a nationwide Electronic Health Record (EHR) system, where patients own their data and can give selective access to healthcare providers. Estonia's e-Prescription service, telemedicine, and digital health data exchange under the X-road platform are other notable components of their successful digital health strategy [44].

8.6.3 Ethical Considerations in Digital Health Innovations

Ethical considerations are at the heart of digital health innovations. As technology reshapes healthcare delivery, ethical dilemmas arise around patient autonomy, equity of access, and the potential for bias in AI algorithms. Ensuring that digital health solutions are accessible to all segments of the population, including underserved communities, is a significant ethical concern. Furthermore, there is a need to address potential biases in AI-driven healthcare solutions, which could lead to disparities in patient care. Patient consent and transparency in the use of health data are also critical ethical issues that need to be addressed, ensuring patients are informed and have control over their health information.

In conclusion, while digital health presents a frontier of opportunities for enhancing healthcare delivery, it also brings forth complex challenges in data governance, regulation, and ethics. Navigating these challenges is crucial for realizing the full potential of digital health innovations while ensuring patient safety, privacy, and equitable access to healthcare. As the field continues to evolve, robust policies and ethical frameworks will be essential in guiding the responsible and effective integration of digital technologies into healthcare [15, 45, 46].

References

1. Ambinder EP. A history of the shift toward full computerization of medicine. J Oncol Pract. 2005;1(2):54–6.
2. Surantha N, Atmaja P, David, Wicaksono M. A review of wearable internet-of-things device for healthcare. Procedia Comput Sci. 2021;179:936–43.
3. Takahashi EA, Schwamm LH, Adeoye OM, Alabi O, Jahangir E, Misra S, et al. An overview of telehealth in the management of cardiovascular disease: a scientific statement from the American Heart Association. Circulation. 2022;146(25):e558–68.
4. Pastorino R, De Vito C, Migliara G, Glocker K, Binenbaum I, Ricciardi W, et al. Benefits and challenges of Big Data in healthcare: an overview of the European initiatives. Eur J Pub Health. 2019;29(Suppl 3):23–7.
5. Blockchain technology in healthcare: a systematic review—PMC. [cited 2024 Jan 22]. https://www.ncbi.nlm.nih.gov/pmc/articles/PMC9000089/.
6. ChatGPT. [cited 2024 Jan 22]. https://chat.openai.com/.
7. Shaffer KM, Turner KL, Siwik C, Gonzalez BD, Upasani R, Glazer JV, et al. Digital health and telehealth in cancer care: a scoping review of reviews. Lancet Digit Health. 2023;5(5):e316–27.
8. Dittrich F, Mielitz A, Pustozerov E, Lawin D, von Jan U, Albrecht UV. Digital health applications from

a government-regulated directory of reimbursable health apps in Germany—a systematic review for evidence and bias. Mhealth. 2023;9:35.
9. Digital health system—Estonia—human rights and biomedicine. [cited 2024 Jan 23]. https://www.coe.int/en/web/bioethics/-/estonia-digital-health-system.
10. Marques I, Ferreira JJ. Digital transformation in the area of health: systematic review of 45 years of evolution digital transformation in the area of health: systematic review of 45 years of evolution. Heal Technol. 2020;10:575.
11. Gu X, Zhu Y, Zhang Y, Sun L, Bao ZY, Shen JH, et al. Effect of telehealth interventions on major cardiovascular outcomes: a meta-analysis of randomized controlled trials. J Geriatr Cardiol. 2017;14(8):501–8.
12. Babatunde AO, Shobanke HA, Akinade AA, Michael AJ, Osadare M, Akanbi OK, et al. Enhancing preventive medicine over curative medicine: role of telemedicine. Public Health Pract. 2021;2:100130.
13. European Health Data Space—European Commission. 2024 [cited 2024 Jan 23]. https://health.ec.europa.eu/ehealth-digital-health-and-care/european-health-data-space_en.
14. Reddy S. Evaluating large language models for use in healthcare: a framework for translational value assessment. Inform Med Unlocked. 2023;41:101304.
15. Frontiers | A comparison of artificial intelligence and human doctors for the purpose of triage and diagnosis. [cited 2024 Jan 23]. https://www.frontiersin.org/articles/10.3389/frai.2020.543405/full.
16. Frontiers | Identifying developments over a decade in the digital health and telemedicine landscape in the UK using quantitative text mining. [cited 2024 Jan 22]. https://www.frontiersin.org/journals/digital-health/articles/10.3389/fdgth.2023.1092008/full.
17. Frontiers | The sociotechnical ethics of digital health: a critique and extension of approaches from bioethics. [cited 2024 Jan 23]. https://www.frontiersin.org/journals/digital-health/articles/10.3389/fdgth.2021.725088/full.
18. HIPAA & GDPR—what are the differences? [cited 2024 Jan 23]. https://blog.medicai.io/en/hipaa-gdpr-what-are-the-differences.
19. Bradway M, Morris RL, Giordanengo A, Årsand E. How mHealth can facilitate collaboration in diabetes care: qualitative analysis of co-design workshops. BMC Health Serv Res. 2020;20(1):1104.
20. Gottlieb S, Silvis L. How to safely integrate large language models into health care. JAMA Health Forum. 2023;4(9):e233909.
21. Krause-Jüttler G, Weitz J, Bork U. Interdisciplinary collaborations in digital health research: mixed methods case study. JMIR Hum Factors. 2022;9(2):e36579.
22. Singhal K, Azizi S, Tu T, Mahdavi SS, Wei J, Chung HW, et al. Large language models encode clinical knowledge. Nature. 2023;620(7972):172–80.
23. Guevara M, Chen S, Thomas S, Chaunzwa TL, Franco I, Kann BH, et al. Large language models to identify social determinants of health in electronic health records. npj Digit Med. 2024;7(1):1–14.
24. Boyles O. Learn the history of EHR electronic health records. ICANotes. 2019 [cited 2024 Jan 22]. https://www.icanotes.com/2019/04/16/a-history-of-ehr-through-the-years/.
25. National Institute of Biomedical Imaging and Bioengineering. [cited 2024 Jan 22]. Wireless wonder: wearable ultrasound patch goes completely cable-free. https://www.nibib.nih.gov/news-events/newsroom/wireless-wonder-wearable-ultrasound-patch-goes-completely-cable-free.
26. Fu J, Li C, Zhou C, Li W, Lai J, Deng S, et al. Methods for analyzing the contents of social media for health care: scoping review. J Med Internet Res. 2023;25:e43349.
27. Amato MS, El-Toukhy S, Abroms LC, Goodfellow H, Ramsey AT, Brown T, et al. Mining electronic health records to promote the reach of digital interventions for cancer prevention through proactive electronic outreach: protocol for the mixed methods OptiMine study. JMIR Res Protoc. 2020;9(12):e23669.
28. Myskja BK, Steinsbekk KS. Personalized medicine, digital technology and trust: a Kantian account. Med Health Care Philos. 2020;23(4):577–87.
29. Hohman KH, Martinez AK, Klompas M, Kraus EM, Li W, Carton TW, et al. Practice full report: leveraging electronic health record data for timely chronic disease surveillance: the multi-state EHR-based network for disease surveillance. J Public Health Manag Pract. 2023;29(2):162.
30. Giansanti D. Precision medicine 2.0: how digital health and AI are changing the game. J Pers Med. 2023;13(7):1057.
31. Regulation (EU) 2017/746 of the European Parliament and of the Council of 5 April 2017 on in vitro diagnostic medical devices and repealing Directive 98/79/EC and Commission Decision 2010/227/EU (text with EEA relevance). OJ L Apr 5, 2017. http://data.europa.eu/eli/reg/2017/746/oj/eng.
32. Self-care behaviors and technology used during COVID-19: systematic review—PubMed. [cited 2024 Jan 22]. https://pubmed.ncbi.nlm.nih.gov/35442904/.
33. Signals | Free full-text | Smart clothing framework for health monitoring applications. [cited 2024 Jan 22]. https://www.mdpi.com/2624-6120/3/1/9.
34. Ma Y, Zhao C, Zhao Y, Lu J, Jiang H, Cao Y, et al. Telemedicine application in patients with chronic disease: a systematic review and meta-analysis. BMC Med Inform Decis Mak. 2022;22(1):105.
35. Haleem A, Javaid M, Singh RP, Suman R. Telemedicine for healthcare: capabilities, features, barriers, and applications. Sens Int. 2021;2:100117.
36. The benefits of AI in healthcare—IBM Blog. [cited 2024 Jan 23]. https://www.ibm.com/blog/the-benefits-of-ai-in-healthcare/.
37. Smuck M, Odonkor CA, Wilt JK, Schmidt N, Swiernik MA. The emerging clinical role of wearables: factors for successful implementation in healthcare. npj Digit Med. 2021;4(1):1–8.

38. Maeckelberghe E, Zdunek K, Marceglia S, Farsides B, Rigby M. The ethical challenges of personalized digital health. Front Med (Lausanne). 2023;10:1123863.
39. Clusmann J, Kolbinger FR, Muti HS, Carrero ZI, Eckardt JN, Laleh NG, et al. The future landscape of large language models in medicine. Commun Med. 2023;3(1):1–8.
40. Huffstetler AN, Epling J, Krist AH. The need for electronic health records to support delivery of behavioral health preventive services. JAMA. 2022;328(8):707–8.
41. Davenport T, Kalakota R. The potential for artificial intelligence in healthcare. Future Healthc J. 2019;6(2):94–8.
42. Ghahramani A, de Courten M, Prokofieva M. The potential of social media in health promotion beyond creating awareness: an integrative review. BMC Public Health. 2022;22(1):2402.
43. Kamga P, Mostafa R, Zafar S. The use of wearable ECG devices in the clinical setting: a review. Curr Emerg Hosp Med Rep. 2022;10(3):67–72.
44. McDuff D, Schaekermann M, Tu T, Palepu A, Wang A, Garrison J, et al. Towards accurate differential diagnosis with large language models. arXiv; 2023 [cited 2024 Jan 23].http://arxiv.org/abs/2312.00164.
45. Gedikci Ondogan A, Sargin M, Canoz K. Use of electronic medical records in the digital healthcare system and its role in communication and medical information sharing among healthcare professionals. Inform Med Unlocked. 2023;42:101373.
46. Kenny JES. Wearable ultrasound for continuous deep-tissue monitoring. Nat Biotechnol. 2024;16:1–2.

Gene Therapy and Gene Editing

9

Niloofar Bazazzadegan, Mandana Hasanzad, and Seyedeh Sedigheh Abedini

What Will We Learn in This Chapter?
This chapter will delve into gene therapy and editing, two prominent treatments that have recently demonstrated efficacy in addressing a spectrum of diseases. These include cancer, metabolic diseases, viral infections, ocular disorders, hematological disorders, Duchenne muscular dystrophy, and primary immunodeficiencies. The chapter will commence with an introduction to gene therapy, offering insights into various gene delivery methods and approaches for treating infectious diseases, as well as complex and monogenic disorders. The discussion will extend to gene editing, incorporating various editing tools for both therapeutic applications and disease modeling. Lastly, the chapter will explore several issues related to therapeutic targeting and ethical considerations.

N. Bazazzadegan
Genetics Research Center, University of Social Welfare and Rehabilitation Sciences, Tehran, Iran

M. Hasanzad
Personalized Medicine Research Center, Endocrinology and Metabolism Clinical Sciences Institute, Tehran University of Medical Sciences, Tehran, Iran

S. S. Abedini (✉)
School of Biotechnology and Biomolecular Sciences, University of New South Wales, Sydney, NSW, Australia
e-mail: m.abedini@unsw.edu.au

Rationale and Importance
Genome editing, a revolutionary technique in genetic engineering, enables the precise manipulation of DNA. This method is instrumental in modifying diverse model organisms and establishing animal models to enhance our understanding of various human diseases' development. The application of genome editing tools in biological therapy has become increasingly common, aiming to correct genetic errors underlying specific phenotypes. This clinical use seeks to address these errors, enhancing the efficacy of biological therapy. To unlock the potential of genome editing technologies in disease treatment and the evolution of genome editing medications, a solid grasp of current biological therapy, including genome editing tools, is essential. Furthermore, understanding the diverse methods of DNA alteration and the challenges associated with employing genome editing in humans is crucial.

9.1 Section 1: Introduction to Gene Therapy and Gene Editing

Gene therapy, also known as genetic therapy, holds great promise as a therapeutic strategy for treating various diseases by altering or modifying defective genes. With the introduction of foreign genetic materials (e.g., DNA or RNA) to cells,

M. Hasanzad (ed.), *A Glimpse at Medicine in the Future*,
https://doi.org/10.1007/978-981-97-4974-4_9

tissues, or organs, promising therapeutic outcomes are being achieved. Successful gene therapy requires selecting a secure and efficient procedure to transfer manipulated DNA into targeted cells or tissues [1, 2]. This transfer can occur directly or through vectors. The schematic procedure is illustrated in Fig. 9.1. A delivery system is necessary because naked DNA molecules face challenges in penetrating cells due to their large size and hydrophilic compositions, making them prone to fragmentation by nucleases [3]. This technology allows for precise and efficient modifications to the genome, enabling researchers to study the effects of specific genetic alterations on disease development and progression [4, 5].

With the evolving landscape of gene therapy, it becomes increasingly evident that these approaches have the potential to revolutionize the treatment of a wide range of diseases, particularly those once deemed incurable, thereby offering newfound hope for both patients and clinicians.

Some examples include the use of gene editing in cancer research to understand the underlying genetic mechanisms and develop targeted therapies. In addition to cancer research, gene therapy and gene editing have shown promising results in treating other diseases as well. For example, gene therapy has been explored as a potential treatment for cardiovascular abnormalities, neurological disorders, genetic disorders, and infectious diseases [4, 5].

Moreover, various editing techniques, including homologous recombination, base editing, and knockout, have advanced gene therapy. An overview of the progress of gene therapy techniques and concepts has been shown in Fig. 9.2. These technologies enable precise epigenetic modifications, allowing for the exact regulation of gene expression through operations like activation, repression, and demethylation [1, 6].

Two different types of gene therapy are pivotal in the realm of molecular medicine and clinical treatment: Germinal Gene Therapy (GGT) and Somatic Gene Therapy (SGT). In GGT, the focus

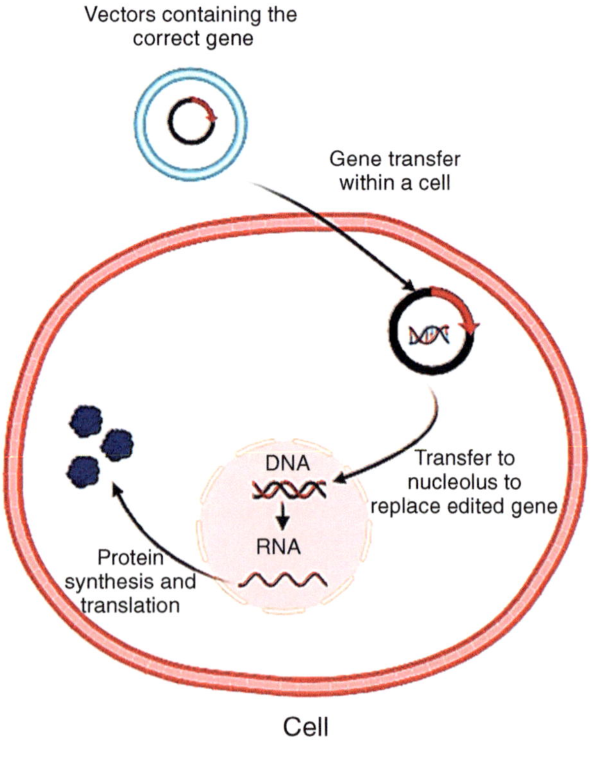

Fig. 9.1 The picture shows gene therapy, replacing a diseased gene with a healthy one to restore normal protein and essential functions

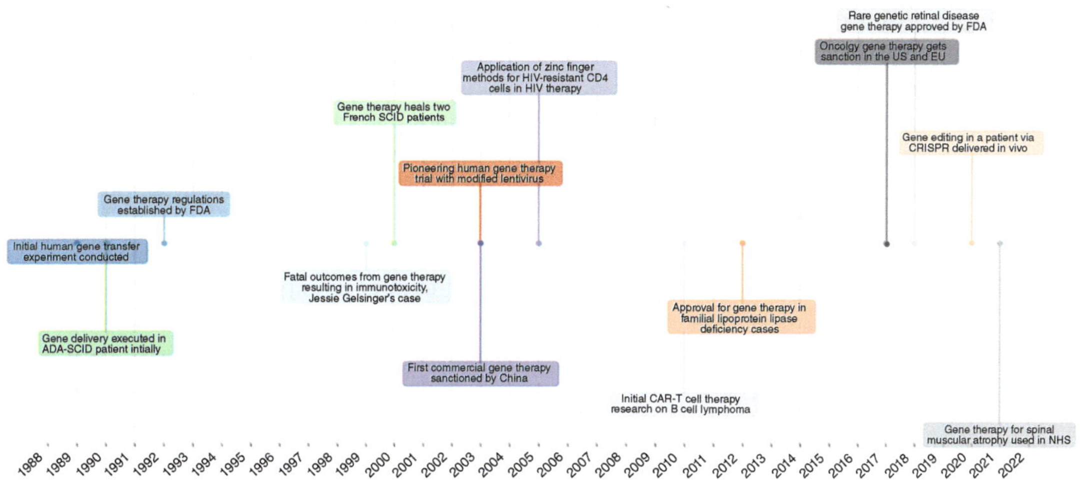

Fig. 9.2 Gene therapy and gene-editing milestones timeline

is on reproductive cells, aiming to correct mutations early in embryonic development and thereby prevent the transmission of genetic diseases across generations. This strategy would theoretically be extremely efficient in combating genetic and inherited disorders [7]. However, GGT has its own set of challenges for clinical use. SGT, which is extensively applied in clinical settings, addresses genetic abnormalities without introducing changes that can be inherited by future generations [8]. In both cases, therapeutic genes are delivered using advanced tools, such as naked oligonucleotides, viral vectors, and nonviral vectors [2, 8].

9.1.1 Different Types of Gene Delivery in Gene Therapy

Gene delivery in gene therapy can take various forms, with in vivo and ex vivo methods being the two primary approaches. In in vivo gene therapy, gene products are directly administered into the body through systemic or in situ delivery methods. On the other hand, ex vivo gene therapy involves modifying a patient's stem cells externally before reintroducing them into the body. Both in vivo and ex vivo gene therapy have their advantages and limitations [3, 9, 10].

In vivo gene therapy is often used when target cells are accessible throughout the body, allowing for a systemic delivery of genetic materials [3, 11]. Ex vivo gene therapy, on the other hand, is performed on specific types of patient-derived cells, such as immune cells, adult stem cells, or germ cells. This approach allows for a more localized and targeted delivery of gene therapies, minimizing the risk of complications associated with systemic administration [10].

The choice between in vivo and ex vivo gene therapy depends on factors such as the accessibility of target cells and the nature of the target disease. Additionally, the development of efficient delivery vehicles, such as viral or non-viral vectors, plays a critical role in successful gene transfer during both in vivo and ex vivo gene therapy [5, 11].

As a vehicle for gene delivery, the ideal genetic vector should accurately carry non-toxic genetic material to specified cells. It should also be able to achieve optimum and sustainable transgenic expression for deficiency correction, accommodate significant foreign genes, and be safe and non-immunogenic [12].

9.1.1.1 Viral-Based Gene Delivery Methods

Various viral vectors with single-stranded or double-stranded RNA and DNA genomes are used for long- and short-term gene expression. A perfect gene therapy vector substitutes viral features with a therapeutic gene cassette, guaranteeing long-term expression in target cells that is consistent and efficient. Viral vectors are at the forefront of gene delivery due to their biological ability to infiltrate cells efficiently [13, 14]. They are classified as integrating (e.g., retroviral, lentiviral, and adeno-associated viral vectors) or non-integrating (e.g., adenoviral vectors). Integrating vectors can become part of the human genome, but non-integrating vectors stay in the nucleus without integration, resulting in transitory expression of the foreign gene due to potential loss during cell division [3, 14].

- **Retroviral** vectors are a popular choice due to their ability to easily integrate with the host's genome. These diploid viruses contain 2 RNA molecules per virion that range in size from 7 to 10 kb. They contain the gag, pol, and env genes surrounded by long terminal repeats (LTRs). These LTRs play an important role in the retrovirus lifecycle, including viral DNA incorporation into the host genome. While retroviruses are typically non-toxic, there is a significant concern about insertional mutagenesis, which might potentially lead to oncogenesis [14, 15].
- **Lentivirus**, a subtype of retrovirus, can transduce both dividing and non-dividing cells, supporting long-term gene expression with a lower risk of genotoxicity. This makes lentiviruses an attractive option for stable gene expression. However, similar to retroviruses, the risk of insertional mutagenesis exists. The use of lentivirus vectors extends to the treatment of single-gene and long-lasting diseases, such as those affecting the nervous system and metabolism [14, 16, 17].
- **Adenoviruses** have a large packaging capacity and high gene expression, but they may cause strong immune responses and do not integrate into the host's genome, making them effective only for short-term therapies. These non-enveloped DNA viruses have genomes ranging from 34 to 43 kb, and there are 57 recognized human adenovirus serotypes. The development of gutless adenoviral vectors addresses concerns about anti-adenoviral vector antibodies, and secure administration methods with established dosage protocols enhance safety considerations. These vectors are mainly applied in infectious disease vaccination and cancer treatment [16, 18].
- **Adeno-associated virus (AAVs) vectors** are emerging as the most preferred viral vectors due to their lower pathogenicity, ability to infect both dividing and non-dividing cells, long-term gene expression, and lower immunogenicity. The limitation with AAVs is their relatively small packaging size. This non-pathogenic parvovirus has a 4.7 kb DNA genome encased in a nonenveloped icosahedral capsid. They have a wide range of clinical applications, primarily focusing on treating different monogenic diseases such as neurological, metabolic, musculoskeletal, and hematological disorders [19, 20].
- The **herpes simplex virus (HSV)** is an enveloped virus with a genome exceeding 150 kb, providing ample space for foreign DNA in modified vectors. HSV vectors demonstrate features like immune avoidance, the capacity to transport large DNA payloads and multiple genes, as well as natural or modified cell-specific lytic properties. HSV vectors are also an option for long-term gene delivery to neurons, appropriate for treating neurodegenerative diseases and cancer [21, 22].

While viral vectors are proven to be effective, recent studies have highlighted several limitations associated with their use. A significant complicating factor is the presence of viral genetic material in the plasmid, which can trigger an immediate immune response and lead to oncogenic transformations [14].

9.1.1.2 Non-viral Gene Delivery Methods

Non-viral methods are safer, provide long-term gene expression potential, and are easier to produce on a large scale. These delivery methods include electroporation, microinjection, and other non-viral vectors. Despite their lower efficiency compared to viral vectors, non-viral vectors offer potential advantages such as increased safety and reduced immune responses. This has led to a growing demand for non-viral delivery systems with high effectivity and biocompatibility. Various non-viral vectors, including lipid, polymers, and inorganic particles, have been extensively utilized in recent years, either alone or in combination [12, 23].

- **Lipid-based vectors** use lipids to encapsulate and deliver DNA or RNA sequences. They often leverage different types of lipids, such as neutral lipids and ionizable lipids. Liposomes are artificial vesicles transporting lipophilic and hydrophilic agents into cells, offering protection for encapsulated DNA/RNA from degradation until reaching target cells. Composed of phospholipid bilayers, they allow easy production and modification for enhanced efficiency and stability. Targeting molecules can be attached for specific cell targeting. However, challenges such as instability, uncontrolled release, and potential toxicity hinder their effectiveness, along with possible immune system responses [24–26].
- Lipid Nanoparticles (LNPs) are potent non-viral gene delivery systems with high gene loading capacity and targeted delivery potential. Despite complex manufacturing and storage issues, they excel at encapsulating diverse RNA formats. LNPs, a lipid nanoparticle subclass, are crucial for mRNA vaccine delivery but face concerns about unintended tissue accumulation and potential side effects [24, 25, 27].
- **Polymer-based gene delivery vectors** offer advantages such as easy production and modification, stability, and low immunogenicity. However, the transfection efficiency of polymeric systems is usually lower than that of viral systems [5]. The two categories are non-biodegradable and biodegradable polymers. Non-biodegradable polymers like Polyethylenimine are known for high transfection efficiency but can lead to cytotoxicity over time. Biodegradable polymers such as polylactide and chitosan offer advantages in terms of biocompatibility, low toxicity, and potential for molecular modification. They can be easily broken down and eliminated by the body, reducing the risk of long-term side effects [5, 12, 28]. These vectors include natural polymers like chitosan and alginate, as well as synthetic ones like polyethyleneimine and poly(lactic-*co*-glycolic acid). They provide protection for nucleic acids with the ability to be modified for enhanced performance while minimizing toxicity [23]. Polyethyleneimine (PEI) is one of the most extensively studied cationic polymers for gene delivery due to its strong DNA condensing capacity and its ability to facilitate endosomal escape. However, it exhibits significant cytotoxicity [23, 29].
- Poly(amidoamine) Dendrimers, with high control over size and shape, have been utilized as an effective non-viral gene vector. These are complex, tree-like polymers that can compact DNA in supernanomolar sizes. The generation number of a dendrimer determines its transfection efficiency—a partially degraded dendrimer always shows higher transfection efficiency. However, there are still difficulties in terms of complex preparation, inability for large-scale production, and potential high toxicity that restrict the clinical translation of dendrimer [30, 31].
- **Inorganic nanoparticles**, such as gold nanoparticles, quantum dots, and magnetic nanoparticles, have unique properties that make them promising for various biomedical applications. These properties, including easy functionalization, excellent biocompatibility, and the potential to improve drug delivery, gene therapy, biosensing, and bioimaging, make them valuable tools. Their attributes can be harnessed for specific purposes, like targeted delivery or imaging processes. Importantly, these nanoparticles can be customized in terms of size, shape, and surface features, enhancing their interaction with cells and biological molecules. However, a more thorough evaluation of their safety profile and degradation characteristics within the body is still necessary [32, 33].
- Gold nanoparticles are widely researched for gene therapy due to their biocompatibility and ease of synthesis. However, they may have

lower nucleic acid delivery efficiency than alternative vectors. DNA can be complexed with these particles, enabling transportation within cells using the photothermal effect. Nevertheless, their degradation inside the cell is challenging and may lead to accumulation and potential toxicity [34].

- Magnetic nanoparticles, another subset of inorganic nanoparticles, play a crucial role in magnetic-targeted gene delivery, enhancing efficacy through an external magnetic field. Nevertheless, substantial challenges exist, including potential toxicity concerns and difficulties in achieving target-specific delivery without causing harm to surrounding tissues [5, 32].
- **Natural material-based vectors** offer potential for gene therapy due to their high biocompatibility and biodegradability. They face challenges such as batch-to-batch variability but provide promising avenues for non-viral gene delivery strategies [35]. Cell-derived vesicles, tiny particles released from cells, including exosomes and microvesicles, transport biological materials like proteins, lipids, and nucleic acids. Exosomes, nanometer-scale vesicles from endocytic sources, are easily taken up by cells and non-immunogenic, making them appealing for gene therapy. Functionalizing exosomes with ligands improves targeting ability, but challenges include high development costs and limited characterization [36, 37]. In gene therapy, cell-derived vesicles can be robust vectors with inherent cell-penetrating ability and cargo transport capacity. However, challenges involve isolating vesicles from cell cultures and maintaining precise control over their contents, impacting gene delivery efficiency [35].
- Virus-Like Particles (VLPs), lacking the viral genome, are non-infectious entities composed solely of viral structural proteins. They effectively mimic the infection process, enabling efficient gene delivery by encapsulating genetic material like DNA, RNA, siRNA, or antisense oligonucleotides. Despite their safety for gene therapy, challenges include complex assembly, optimizing therapeutic gene loading, and managing potential immune responses that could affect their efficacy [38].
- **Hybrid vectors** seek to combine the best properties of different materials—such as lipids, polymers, and inorganic nanoparticles—to optimize gene transfer [39]. By integrating materials, hybrid systems can potentially exhibit improved stability, enhanced targeting abilities, or higher transfection efficiencies. For instance, lipid-polymer hybrid particles combine the advantages of both materials, allowing for improved stability and sustained release [40]. Another example involves combining a cationic amphipathic peptide and RNA through a hyaluronic acid nanoparticle system, resulting in enhanced effectiveness at a lower dosage compared to other formulations [12, 29]. These hybrid systems show potential in overcoming individual material limitations and creating customized solutions for specific therapeutic applications. Despite their effectiveness, further studies are needed to understand the assembly, stability, and interaction of these hybrid systems with the biological milieu [39].

9.1.2 Clinical Gene Therapy

Clinical gene therapy involves the delivery of therapeutic genes into cells to treat various diseases, including monogenic disorders, cancers, and infectious diseases. Advances in this area have led to the exploration of innovative strategies such as chemically modified antisense deoxy oligonucleotides and inhibitory mutant proteins derived from regulatory proteins observed to inhibit replication and gene expression. Therefore, gene therapy involves not just interventions at the DNA level but also alterations at the mRNA level, given its capacity to convey genetic information [4, 41]. Nucleic acid therapeutics have attracted scientists worldwide due to their potential in mitigating dreaded inherited and acquired genetic diseases. In the field of clinical gene therapy, non-viral vectors such as lipid-based nanoparticles and polymer-based carriers have become appealing choices for gene delivery because of their safety, scalability, and ease of production in comparison to viral vectors [23].

A notable application of this approach in addressing emerging diseases is observed in the use of nanotechnology-based mRNA vaccines during the COVID-19 pandemic. These mRNA-based vaccines, developed using lipid nanoparticles, have showcased remarkable efficacy and have been authorized for emergency use by regulatory agencies such as the US-FDA (Food and Drug Administration) and the European Medicines Agency [25].

Noteworthy demonstrations have highlighted the potential applications of gene therapy in various conditions such as neurodegenerative disorders, hemophilia B, pancreatic cancer, muscular dystrophy type 2D (alpha-sarcoglycan deficiency), leukemia (acute myeloid and chronic myeloid), and β-thalassemia major [1, 2, 4, 39]. For optimal efficacy outcome, it is crucial to consider certain disease attributes while selecting a specific gene therapy approach. Depending on the method and extent of the manipulation of genetic information, three approaches are used to treat various genetic disorders. The following is a description of these methods.

9.1.2.1 Gene Replacement Therapy

Gene replacement therapy is a new strategy that is now being employed in the treatment of genetic diseases. This approach involves substituting a harmful gene with a healthy one or replacing a natural gene with an engineered one that offers enhanced benefits, commonly done by transferring exogenous cDNA using viral vectors [41, 42]. Hence, it is particularly appropriate for monogenic disorders, when the phenotype is modified by a single gene mutation or deletion. Gene replacement therapy faces theoretical challenges in missense mutations, where the abnormal gene product persists and disrupts the cell even with the presence of the correct product. A suggested two-step approach involves initially removing the mutant gene and subsequently introducing the normal gene [42, 43].

Vectors such as lentiviruses or AAVs are used in gene replacement therapy to efficiently deliver the therapeutic gene. These viruses have been genetically modified to contain the therapeutic gene while losing their disease-causing capabilities [44, 45]. Certain characteristics of the disease, such as the targeted tissue to be transduced, should be considered. For instance, viral vectors such as AAVs are preferred for long-term transgenic expression in non-dividing cells due to their lack of incorporation into the DNA of the host cell. The structure of the transgenic can affect the duration of the modified gene's activity. In cases where genes need prolonged activity, a robust, continuous engine or promoter is essential for uninterrupted transcription [46]. Conversely, for life-threatening conditions demanding swift and elevated gene expression during early development, opting for an alternative promoter or a combination may be more suitable. In addition, the choice of viral vector impacts potential immunogenicity, which could compromise safety and efficacy outcomes of the therapy [44–46].

Another approach to gene replacement is the use of RNA oligonucleotides, which are tiny RNA molecules intended to bind with certain pre-mRNA sites. It is through these interactions that either troublesome exons can be the target of removal and skip or exons can be kept in mRNA, increasing the likelihood of creating functional proteins [47]. RNA-level functional replacement has become prominent in gene therapy, particularly after advancements in COVID-19 vaccine research. Therapeutic mRNA, delivered through lipid nanoparticles, enables direct translation into functional proteins previously absent in patients. Synthetic mRNAs generated in vitro are modified with a cap analogue at their 5′ terminus to facilitate cellular recognition. In order to deliver mRNA effectively, delivery vehicles that encapsulate it are needed, such as polymeric and lipid nanoparticles as well as cationic polypeptides [48, 49]. Additionally, modified mRNAs can express tools for genetic editing, like clustered regularly interspaced palindromic repeats (CRISPR)/CRISPR-associated protein 9 (Cas9). Using mRNA for therapeutic tools is considered safer than direct genome manipulation due to the transient nature of nucleic acids, offering increased flexibility for in vivo applications [47].

This medical strategy is majorly suitable for treating monogenic diseases such as sickle cell disease, cystic fibrosis, and spinal muscular atrophy, where the disease condition is driven by a mutation or deletion in a single gene [42].

9.1.2.2 Epigenetic Modification Therapy

Epigenetic modification therapy is an emerging field with enormous promise for treating a variety of disorders. Epigenetic modifications can be broadly classified into three groups: DNA and RNA methylations, histone modifications, and non-coding RNAs. Exploring the dynamic field of epigenetics reveals its essential role in mitosis, meiosis, and various biological processes. The biological processes that are regulated by epigenetic changes, including gene expression regulation, suppression of transposable components, genetic imprinting, genomic stability, X-chromosome inactivation, aging, and some diseases, including cancer [50, 51]. These mechanisms play important roles in disease progression. Epigenetic modification treatment tries to change gene expression patterns and restore normal cellular function by targeting these modifications. This involves addressing gene expression imbalances, deactivating a gene involved in a disease, affecting abnormal chemical changes, or triggering antagonist genes [50–53].

The complex interaction of epigenetic events is dependent on three groups of epigenetic proteins: writers, readers, and erasers. Readers recognize and interpret the chemical changes in DNA and histones that writers make, while erasers remove the markings. Surprisingly, epigenetic medicines can target all of these epigenetic regulators [54]. A variety of substances that interact with DNA are used to manipulate gene expression, including probiotic bacteria, drugs, enzymes, and other compounds. For instance, a probiotic or gut microbiota can affect how epigenetically silenced genes in cancer cells are activated, thereby altering the epigenome of the host [55].

With advancements in gene therapy technologies, such as CRISPR/Cas, precise epigenetic modifications have become more achievable compared to regulatory noncoding RNAs like small interfering RNA (siRNA) and short hairpin RNA (shRNA). Consequently, standard operating procedures are being developed for epigenetic regulation [50]. This method is now being studied in the treatment of malignant tumors and a number of other disorders, including lupus, diabetes, and a range of neurological problems [51].

Epigenetic modification therapies show promise in cancer treatment, primarily due to the reversible nature of these modifications. Despite epigenetic medication therapy's success in treating myelodysplastic syndromes and leukemias, it is not effective against solid malignancies due to its resistance [50, 56]. A new generation of epigenetic medicines is being developed with improved half-life, safety characteristics, and bioavailability. Treatment strategies involve the co-administration of different epigenetic drugs or the conjunction of epigenetic drugs with chemotherapy to sensitize cancer cells and maximize treatment efficacy. Additionally, strategies also focus on targeting the metabolic causes of epigenetic changes, known as oncometabolites [6, 57]. The advent of CRISPR/Cas9 technology has revolutionized epigenetic editing, allowing precise changes to particular chromatin marks to enhance targeted epigenetic treatment options. However, there remain obstacles, notably in establishing high specificity, non-immunogenicity, and effective administration [53, 57].

9.1.2.3 Genome Editing Therapy

One option for reducing the expression of the disease-causing gene is to modify the local DNA sequence to counteract its effects. In the 1970s, pioneers in genome engineering first pushed for genome-wide editing, and substantial progress has been made since then [58]. In this procedure, healthy or modified sequences are inserted into the genome, and disease-causing sequences are modified, eliminated, or replaced, as well as targeted mutations or tiny indels are added [41].

The first clinical study incorporating gene-editing therapy for therapeutic purposes was approved in 1990 for adenosine deaminase-severe combined immunodeficiency (ADA-

SCID). Retroviruses were used to deliver wild-type adenosine deaminase genes to autologous T lymphocytes of patients with ADA-SCID. Although early experiments yielded disappointing results, the practicality of using viral-mediated gene editing led to further gene therapy trials [59].

In a study, researchers demonstrated that integrating homologous DNA fragments into a cell facilitates homologous recombination (HR), which allows for targeted changes in the genome. A number of issues and limitations were introduced when using HR alone for genetic alteration, such as poor integration of exogenous DNA and random inclusion in undesirable genomic regions. The creation of a double-stranded break (DSB) in eukaryotic cells proved crucial for overcoming limitations such as ineffective integration and random incorporation [60]. In vivo applications are generally considered dangerous due to the unexpected DSB, indel, and deletion consequences caused by Cas9. Further investigation is therefore usually not possible until these issues and ethical concerns are resolved. The majority of gene editing is currently applied to ex vivo product development in order to prevent potentially harmful side effects [61]. Additionally, the base editing method may be considered to be safer since it doesn't cause DNA strand breaks. For instance, lipid nanoparticle-encapsulated CRISPR base editors were used to efficiently decrease cholesterol levels in the livers of cynomolgus monkeys and treat atherosclerotic cardiovascular disease with no off-target consequences. These developments demonstrate potential breakthroughs in the field of gene-editing products [62].

Recently, inserting an artificial DNA restriction enzyme into cells allows for targeted cleavage of DNA at recognition sites, boosting both HR and NHEJ (non-homologous end joining) repair. Various gene-editing strategies use site-directed nucleases to build precise endonuclease-based mechanisms to create targeted breaks. The process includes nucleases cleaving site-specific DNA, which activates the cell's repair processes, NHEJ and HR. This technique is limited by the fact that it requires the amount of activation depends on the cell type and target gene location and activation in growing cells [63].

In conclusion, genome editing revolutionizes exact alterations for specified qualities in cells or organisms, bringing potential applications in human health and agriculture by inserting constructs into the target entities.

9.2 Section 2: Gene-Editing Technologies

In vivo gene therapy has great potential for treating a wide range of disorders by directly introducing genetic material into the human body. The modification of genes is a critical step in every gene therapy treatment, with numerous methods available for altering gene structure and function. In recent decades, gene editing has largely relied on exploiting cellular processes, specifically natural DSBs. However, to enhance specificity and efficiency, molecular tools are required to introduce DSBs at specific sites. There are now two basic methods: "targeted gene editing," which allows for restricted modifications in existing gene sequences without creating mutations, and another way that involves more comprehensive but subtle changes in native genome species. The employment of programmable nucleases, which cause DSBs at specified genomic locations, is critical for precise genome editing [64].

There are four basic platforms capable of creating site-specific DSBs. Gene-editing approaches typically use these platforms since they are generally applicable to all cells that use endogenous DNA repair mechanisms. Each platform, including Meganucleases (MegNs), Zinc Finger Nucleases (ZNFs), Transcription Activator-Like Effector Nucleases (TALENs), and CRISPR-Cas9, has its own distinct characteristics and applications. When considering which platform to use for genome editing, it's important to weigh factors such as recognition site lengths, specificity, targeting constraints, ease of engineering, immunogenicity, and delivery methods (Table 9.1) [58, 64].

Table 9.1 Programmable nucleases analysis on different platforms

	CRISPR-Cas9	ZNF	TALEN	Meganucleases
Site of recognition	22 base pairs, extending up to 44 bp for double nicking	18–36 bp per pair, 9–18 bp per monomer	14–20 bp per monomer, adding up to 28–40 bp per pair	From 14 to 40 bp
Action mechanism	wtCas9 introduces single-strand nicks or Cas9 nickase introduces DSBs into target DNA or	DSBs are introduced into target DNA	DSBs are introduced into target DNA	DSBs are introduced into target DNA
Immunogenicity	Uncertain, originates from diverse bacteria	Immunogenicity possibly low for human protein; but bacterial-derived protein could be immunogenic	Uncertain, with protein sourced from *Xanthamonas* sp.	Uncertain; originates from a variety of species, such as eukaryotes
Specificity	Tolerates positional and several consecutive mismatches	A few positional mismatches are acceptable	A few positional mismatches are acceptable	A few positional mismatches are acceptable
Cytotoxicity	Low	Variable to high	Low	Low
Engineering simplicity	Re-targeting simplified with oligo synthesis and standard cloning	Complex, often requiring significant protein modification	Moderate complexity in engineering due to intricate cloning methods	Complex, often requiring significant protein modification
Targeting restrictions	Sequence to be targeted must be followed by a PAM	Targeting non-G-rich sequences is challenging	Each TALEN monomer requires a T at the targeted 5′ base	Efficiency drops when targeting new sequences
Multiplexing simplicity	High	Low	Low	Low
In vivo delivery simplicity	In vivo challenges due to viral vectors' poor packing of the large Cas9; however, smaller variants are available	Small size of ZFN expression cassettes enables diverse viral vector compatibility, easing in vivo delivery	TALEN's size and repetitive DNA encoding complicate packaging into lentiviral vectors for in vivo delivery	Small size facilitates use in diverse viral vectors for in vivo delivery
Ex vivo delivery simplicity	Viral methods and electroporation facilitate easy ex vivo delivery	Viral methods and electroporation facilitate easy ex vivo delivery	Viral methods and electroporation facilitate easy ex vivo delivery	Viral methods and electroporation facilitate easy ex vivo delivery
Effectiveness of cleavage	Highly effectiveness	Effectiveness	Effectiveness	Effectiveness
Rate of success	High	Low	High	High
Configurable	Easy	Very difficult	Difficult	Difficult

9.2.1 MegNs

Meganucleases are endodeoxyribonucleases found in microbes and specific eukaryotic cell compartments. They can cleave specific sites (between 14 and 40 bp) on double-stranded DNA, known as "homing," allowing genetic elements to move laterally within an organism. MegNs' precise DNA specificity makes them powerful tools for genome manipulation [4, 58].

MegNs are categorized into five major families: LAGLIDADG, GIYYIG, HNH, His-Cys box, and PD-(D/E) XK. These classifications are based on sequence and structure motifs. The ability of MegNs to recognize DNA varies among families, requiring a balance between identifying a suitably long target to minimize host toxicity and allowing some sequence drift. For example, I-Cre I, a member of the LAGLIDADG family (the largest family of homing endonucleases), possesses the ability to recognize and cleave a pseudo-palindromic target sequence of 22 bases. The substrate sequences in the central region play a critical role in influencing the binding and cleavage efficiency of these enzymes, underscoring the important role of indirect readout in modifying protein-DNA interactions [65, 66].

Meganucleases show great promise in various applications, particularly in the development of therapies for a broad spectrum of inherited diseases caused by frameshift or nonsense mutations, where their specificity surpasses that of other genetic editing tools [67]. Some MegNs, particularly those from the LAGLIDADG family, have been modified to target novel sequences, enabling their use in genome editing. To overcome traditional constraints, chimeric proteins were generated by combining MegN capabilities with additional editing techniques that utilize the DNA-binding capacities of ZNFs and TALENs along with the cleavage specificity of MegNs [68].

Furthermore, the compact size of MegNs makes them well-suited for gene delivery, including incorporation into single viral vectors to induce multiple DNA breaks. This approach opens up possibilities for homology-directed repair, enabling precise genetic changes. However, the customization of MegNs for each new genome engineering target poses an initial hurdle, limiting their routine application and presenting technical challenges [58, 69]. A key obstacle is the intricate task of differentiating between the DNA-binding and DNA-cleavage domains, making the engineering of MegNs a complex and highly inefficient process. Moreover, the requirement to insert known cleavage sites into specific regions of interest, along with the costly and time-consuming production of sequence-specific enzymes, has impeded the widespread adoption of MegNs in various applications [67, 70].

Despite these obstacles, MegNs, with their high specificity, enhanced DNA recombinant production through the creation of a 3′ overhang after cleavage, and easy delivery due to their small size, hold promise as valuable tools in genetic manipulation [67, 69, 70].

9.2.2 TALENs

Transcription activator-like effector nucleases (TALENs) have become a prominent tool in genome editing, owing to their distinctive properties and capabilities. These engineered chimeric proteins are constructed by linking a nonspecific *FokI* restriction endonuclease domain to a DNA-binding domain. This domain is composed of highly conserved repeat sequences, spanning 33–35 amino acids, derived from transcription activator-like effectors, originally produced by the plant pathogen *Xanthomonas campestris*. These repeat domains contact DNA through two amino acid residues at positions 12 and 13, termed the repeat variable di-residues (RVDs) [70–72].

TALE repeats can be assembled into arrays with tailored DNA-binding abilities, refining their performance. Various engineering methods are available, including standard cloning from plasmid archives, advanced techniques like Golden Gate cloning, and high-throughput approaches such as ligation-independent cloning or solid-phase assembly [67].

A pivotal feature in the mechanism of action of TALENs is the recognition of a single nucleotide within the DNA-binding domain, established by their RVDs. This differs from the method utilized in ZNFs, in which the first zinc-finger domain identifies three base pairs. This functional difference simplifies TALEN design and lowers possible complications in the interactions between these proteins and their target nucleotides [73].

Moreover, TALENs are highly valuable for precisely inducing double-strand breaks (DSBs) in various cells, encompassing somatic and pluripotent stem cells. These DSBs are essential for successful gene editing as they facilitate the 'knock-in' or 'knock-out' process at specific genome locations [74]. Their effectiveness in executing this crucial step in gene editing, coupled with a documented lower occurrence of off-target events, expands their applicability. For instance, TALENs are less toxic to human cell lines, minimizing the potential for off-target effects. Furthermore, TALEN mRNA is also more successful at genome editing in livestock embryos than ZFN induction when it is injected cytoplasmically. This extends to the creation of genetically engineered organisms and potential therapeutic applications [70, 74].

Recent advancements in assembling TALENs have aimed to boost their performance. This involves improving specificity through the identification of nonconventional RVDs and utilizing directed evolution to refine TALEN specificity [70].

Although TALENs have many advantages, they are also not without challenges. Due to the larger size and repetitive structure of these entities, they do not typically deliver well in vivo, especially when delivered by viral vehicles, which are inherently limited in their packaging capabilities. However, there are also a few innovative approaches being developed to overcome these issues; for instance, the use of mRNA and proteins to transfer TALENs to cells is an innovative approach [67, 69].

Although the design of the TALEN transgene is, in comparison to that of the ZFN transgene, relatively simpler, it still poses some complexities for those who are not familiar with molecular biology experiments. Nevertheless, evolving strategies have been developed to enhance TALEN specificity, including phage-assisted continuous evolution, a laboratory evolutionary process that is used to fine-tune TALEN specificity [75]. Therefore, TALENs are dynamic and promising for a wide range of therapeutic and research applications as well.

9.2.3 ZNFs

Zinc-finger nucleases possess unique structures and functions that make them important tools for genome editing. These proteins are fusions of Cys2-His2-zinc finger proteins found in transcription factors and FokI restriction endonuclease cleavage domains. Combining these domains yields precision "genomic scissors," allowing well-designed ZFN pairings to target any gene in any organism [76]. A group of artificially engineered nucleases, renowned for their unique binding properties to DNA, are capable of causing DSBs in the genome at certain locations, contributing to gene modification through endogenous repair systems such as NHEJ and HR repair systems [77].

ZFNs possess a modular structure, allowing optimization in both their DNA-binding and catalytic domains. Typically, a ZFN comprises three to four zinc-finger domains arranged in a ββα motif, with each domain consisting of 30 amino acids interacting with specific parts of the DNA sequence. A ZFN monomer recognizes a half-site sequence spanning 9–18 base pairs (bps) of genome, enabling ZFN dimers to specify large DNA sections (~18 to 36 bp) at intervention sites through the FokI cleavage domain. Specificity is determined by the amino acid sequence of each finger, the number of fingers, and interactions within the nuclease domain. The dimeric FokI domain, which is responsible for DNA cleavage, requires two ZFN molecules to bind in the right orientation for efficient cleavage [67, 70, 78].

Expanding on this design, incorporating multiple zinc-finger domains and the FokI endonuclease domain has resulted in potent tools for custom site-specific genome editing. A current concept for improving ZFN targeting accuracy incorporates architectural diversity using "selection-based methods." Adding a new linker boosted the number of ZFN combinations available for targeting breakage at any particular DNA base by 64-fold [78].

Despite their promise, ZFNs pose a risk of off-target mutations, prompting research into enhancing specificity. This includes designing heterodimeric ZFN architectures that prevent undesired homodimerization of the FokI cleavage domain through charge repulsion [78, 79]. Other strategies involve delivering ZFNs into cells as proteins to reduce off-target effects or creating modified ZFNs with improved intracellular penetrating activity. Gene editing can be performed more effectively with purified ZFN proteins applied directly to cells than when they are expressed from nucleic acids within cells [80, 81].

Even though progress has been made in the development of ZFNs, there are still some obstacles to overcome, especially in the development of zinc-finger domains that are capable of detecting all DNA triplets, including those associated with 5′-TNN-3′ and 5′-CNN-3′ variants. There's another limitation of ZFNs: their context-dependent specificity can impact cleaving efficiency depending on adjacent zinc fingers and DNA context, which can lead to genome instability. Additionally, they are less efficient for knockout editing, as they target only one location at a time as well as exhibit overt toxicity due to off-target cleavage, increasing the risk of incorrect DNA sequences [67, 78, 80].

9.2.4 CRISPR/Cas9

CRISPR-Cas9, developed in 2013, is a groundbreaking third-generation gene-editing technology. Its simplicity and unique attributes have made it an indispensable tool in biological research and a key strategy against diseases like COVID-19 [4].

CRISPR-Cas9 is derived from the natural defensive systems of archaea and bacteria, constituting approximately 90% and 40% of them, respectively. As part of their adaptive immune system, these organisms incorporated short sequences of foreign DNA, such as viruses, into their CRISPR loci. A CRISPR RNA (crRNA) was transcribed from integrated segments, working with trans-activating crRNAs (tracrRNAs) alongside Cas proteins, identifying the specificity of DNA cleavage. To transform this defensive mechanism into a genomic editing tool, researchers optimized the system with the Cas9 nuclease and a specially created single-guide RNA (sgRNA). The sgRNA combines both tracrRNA and crRNA, enabling the CRISPR/Cas9 system to perform precise, site-specific gene editing [82, 83].

The *Streptococcus pyogenes* Cas9 protein, commonly used in gene editing, requires a Protospacer Adjacent Motif (PAM) sequence of 5′-NGG-3′ (two guanine bases come after any nucleotide base) downstream of the target sequence for effective binding and cleavage. The PAM sequence is crucial for recruiting Cas9 and enabling its conformational change for interaction with the target DNA [82, 84]. CRISPR/Cas9 are classified into 2 classes, 6 types, and up to 29 subtypes based on characteristic Cas genes and locus arrangement. Class 1 (types I, III, and IV) consists of multiprotein complexes, while Class 2 (types II, V, and VI) features single protein complexes. This diversity opens opportunities for emerging gene-editing technologies beyond Cas9-based systems [85].

Three main strategies are employed in CRISPR/Cas9 genome editing: (1) delivering sgRNA and Cas9 mRNA intracellularly, with transient expression and mRNA instability; (2) plasmid-based, in which a plasmid encodes both sgRNA and Cas9 protein, thus providing prolonged expression but facing challenges in nuclear introduction; and (3) delivering Cas9 protein directly into cells, which offers stability, fast action, and reduced antigenicity. It's important to note that not all CRISPR bacterial defense systems function similarly [86].

The superiority of the CRISPR/Cas9 system over traditional genome editing tools lies in its reliance on RNA-DNA interactions instead of protein-DNA recognition. This RNA-centric approach not only simplifies design but also facilitates the predictability of off-target sites and the simultaneous modification of multiple genomic loci, a phenomenon known as multiplexing. Strategies such as modulating Cas9 protein or sgRNA dosage, employing paired Cas9 nickases, and utilizing truncated gRNAs sensitive to mismatches enhance specificity, addressing off-target effects [83, 87]. In several species, using techniques like transient transfection and nucleofection of a pre-assembled Cas9 ribonucleoprotein complex has shown significantly decreased off-target effects, enabling DNA-free gene editing [67, 70]. The activity of Cas9 ribonucleoprotein is enhanced further by chemical nodification that protects RNA and stabilizes secondary structures. Additionally, the hybrid approach, fusing the FokI nuclease domain with an inactivated Cas9 variant, theoretically enhances specificity by requiring protein dimerization for DNA cleavage [88].

Despite these advances, challenges persist, notably the off-target effect, distinguishing CRISPR/Cas9 from its predecessors TALEN and ZFN, which exhibit less frequent off-target effects due to accurate pair binding. Strategies to mitigate off-target effects include using a mutant Cas9 nickase and employing "base editing" technologies, demonstrating promise in gene therapy [89]. Base editing employs Cas9 D10A nickases combined with designed base deaminase enzymes, enabling precise single-base modifications without producing DSBs or requiring an external repair template. The Cas9 nickase directs deaminase activity to the desired location, allowing for precise editing without the need for complicated repair mechanisms. The CRISPR-SKIP variant, as new methods of base editing tools, leveraging cytidine deaminase single-base editors, manifests effective exon skipping, holding substantial promise in the domain of gene therapy [89, 90].

Ongoing research aims to improve specificity and efficiency while addressing broader biological implications. Despite challenges, CRISPR/Cas9 has revolutionized genomic editing, opening up unprecedented possibilities across various sectors, including healthcare, agriculture, and environmental conservation. As comprehension and advancements in CRISPR/Cas9 technology progress, its potential applications become virtually limitless.

9.3 Section 3: Genome-Editing Tools for Disease Modelling and Gene Therapy

Various genome editing tools, including CRISPR/Cas9, have played a crucial role in disease modelling and gene therapy research. These tools enable precise modifications to the genome, facilitating the creation of disease models for studying underlying mechanisms. This has resulted in substantial progress in comprehending the pathogenesis of diverse diseases, such as cancer, cardiovascular diseases, and neurodegenerative disorders. The utilization of CRISPR/Cas9 technology has notably advanced cancer research.

9.3.1 An Extensive Study of Various Cancers

Genome editing, particularly CRISPR/Cas9, is a critical area of interest in cancer research. It enables researchers to precisely change genetic elements, focusing on tumor suppressors and oncogenes. This approach enables targeted breakage events, which is important in a variety of research, from evaluating the efficacy of gene editing in eukaryotes to designing T cells for tumor targeting and hematopoietic stem cells (HSCs) [91, 92].

ZFN-mediated gene targeting enabled precise disruption of the BCR-ABL oncogene in a mouse cell line [93]. Custom-designed ZFNs were then used to generate site-specific DSBs in a gene associated with mixed lineage leukemia (MLL), facilitating the study of chromosomal rearrangements in leukemogenes [94]. In addition, ZFNs have had a major impact on the improvement of T cell-mediated anticancer treatment. In a mouse model of adult T cell leukemia (ATL), this therapeutic method targeting the long terminal repeat promoter of HTLV-1 (human T cell leukemia virus type 1) selectively removing HTLV-1-infected cells [95]. Furthermore, a specialized yeast-one-hybrid fourfinger ZFN was utilized to induce replacement of mutant p53 with wild-type p53 in various cancer cell lines, such as breast cancer and glioblastoma, through ZFN-induced HR [96].

Concurrently, TALENs have shown efficacy in human leukemia cells by altering endogenous genes, *NTF3* and *CCR5*, by targeted coding sequence modifications. Subsequent applications of TALEN technology led to precise disruptions in the T cell receptor α constant gene and the *CD52* gene in allogeneic T cells [97]. In addition, a site-specific TALEN was used to disrupt a single allele of the *FLT3* (Fms-related tyrosine kinase 3) gene, resulting in isogenic leukemia cell clones. Inducing *FLT3* haploinsufficiency significantly impaired colony formation and cell proliferation in vitro, maintaining these suppressive effects in vivo [98].

In contrast to traditional methods, CRISPR/Cas9, a revolutionary gene-editing tool, has transformed cancer research by enabling precise genomic modifications, prompting DNA responses, facilitating repairs, and effectively reshaping genomic structures crucial for unraveling the genetic basis of cancer. This technology exhibits substantial potential in the targeted therapy of diverse cancers. This ground-breaking method has proven beneficial in cancer cell dependence research, exposing key genes such as *topA* and *CDK,* which are involved in DNA mismatch repair system, as well as synthetic linkages [99].

The impact of CRISPR extends to the discovery of cancer cell dependencies and critical genes in distinct human Acute Myeloid Leukemia (AML) cell lines. It gives information on important pathways mediated by transcription factors *JUNB*, *FRA1*, and *ERK2*, providing insights into *BRAF* inhibitor-dependent tumor cell proliferation [100]. Furthermore, the adaptability of CRISPR is illustrated by the ability of the *GOT1* deletion to sensitize human Jurkat leukemia T cells to the mitochondrial complex I inhibitor phenformin. In addition, CRISPR-edited human stem cell-derived organoids help in the tracking of cancer mutational signals to DNA repair gene deletions, therefore improving our understanding of cancer biology [101, 102]. Detailed studies have shown the significance of CRISPR/Cas9 in a variety of cancer types, including breast cancer, oral cancer, leukemia, and ovarian cancer.

In breast cancer, critical genes such as *CDKN2A* and the *HER2* gene have been identified as major candidates for gene editing. CRISPR/Cas9 is used to address mutations in these particular genes, inhibiting the proliferation of HER2-expressing cells and therefore reducing tumorigenicity [103]. The potential of CRISPR technology also extends to tackling drug resistance, a significant roadblock in cancer therapy. A specific example is the BRCA1-delta11q alternative splice isoform in breast cancer [104, 105]. Using CRISPR, researchers confirmed that this isoform serves as a primary factor in breast cancer resistance to treatment. Furthermore, CRISPR's therapeutic potential extends beyond gene replacement or inhibition. In breast cancer, mutated BRCA1 or BRCA2 genes can be replaced with their wild-type counterparts, restoring functional DNA repair proteins and acting as tumor suppressors, offering a promising strategy against the disease [106, 107].

In the context of Acute Lymphoblastic Leukemia, a unique approach using the CRISPR-Cas9 system has been employed. Researchers introduced the cDNA that encodes *RUNX1* exons into the native ETV6 locus of iPSC (induced pluripotent stem cells) using homologous recombination knock-ins. This methodology signifies a crucial step towards incorporating gene therapy in leukemia treatment, setting the basis for the application of CRISPR/Cas9 in this domain [108]. In addition, CRISPR/Cas9 has the potential to treat certain types of leukemia caused by overexpressed oncogenes, which contribute to cancer development. In T-cell acute lymphoblastic leukemia, utilizing CRISPR/Cas9 to modify and target oncogenes like *TAL1* and *LMO2* could inhibit cancer cell development, suggesting a possible technique to prevent or reverse disease progression [109, 110].

Genome editing is possible to modify the target genetic material with or without DBSs using CAR T-cell (chimeric antigen receptor T lymphocyte cell) either with CRISPR-Cas9 or novel editing techniques. CAR T-cell therapy, in which T lymphocytes are modified with chimeric protein receptor genes to improve their capacity to target and destroy cancer cells, has shown remarkable success. This innovative approach aims to enhance T lymphocytes' ability to recognize and destroy cancer cells effectively. A variety of T-cell modifications, such as removing HLA-I antigens, inactivating T-cell receptor (TCR) genes, and destroying the CD52 gene, allow CAR-T cells to be made universally applicable to a wide variety of cancers, especially solid tumors [111].

CAR T-cell advancement is notable in TALEN-driven research, leading to the development of "universal" CAR T cells (dKO-CART19) via a retroviral vector engineered TALEN. This TALEN expressed a chimeric antigen receptor that targeted CD19+ leukemic B cells [112].

In conclusion, gene therapy stands at the forefront of cancer treatment with its versatile potential to modify various genes in different types of cancers. The promise of a comprehensive and innovative approach to cancer treatment is within reach, as techniques like ZFN, TALEN, and CRISPR/Cas9 reveal immense possibilities. While their potential is vast, further exploration and refinement are needed for a transformative impact on cancer therapy.

9.3.2 Metabolic Disorders

Metabolic disorders are caused by a protein, fat, and carbohydrate imbalance within the body. A group of conditions known as metabolic diseases may result from both genetic and environmental factors [113–115]. Since the invention of gene editing, several uses for this technique have been possible, including functional gene screening, gene therapy, and the creation of models for metabolic diseases like obesity, diabetes, and hyperlipidemia.

The study of lipid metabolism has also made use of gene-editing technology [116]. A transcription factor called CREB3L3 (CAMP responsive element-binding protein-like 3) is expressed in the liver and regulates how the body uses energy when it is fasting. For the first time, a CREB3L3-floxed mouse model was created, and mice lacking CREB3L3 in the small intestine and liver were then produced using a one-step CRISPR/Cas9 method. The proof that was previously offered makes CREB3L3's function in plasma triglyceride metabolism and its involvement in liver and intestinal cholesterol metabolism even more clear [116].

Familial hypercholesterolemia is an autosomal dominant disorder brought on by a low-density lipoprotein receptor (LDLR) gene deficiency. Additionally, LDLR is being targeted by TALEN technology in a study to create familial hypercholesterolemia models in porcine fetal fibroblasts, which may hold promise for the treatment of lipid metabolic syndrome.

The gene's fat mass and obesity-related gene (FTO) allele has been associated with obesity and suppresses mitochondrial thermogenesis in adipose precursor cells. The FTO gene mutation hinders the process of conversion of white fat into brown fat. The FTO gene-regulated thermogenic pathway is influenced by a number of factors, including ARID5B, rs1421085, IRX3, and IRX5. The CRISPR/Cas9 platform allows for the editing of the rs1421085 gene to fix the pattern structure of ARID5B. This helps to decrease the expression of IRX3 and IRX5, resulting in reduced weight loss [117]. CRISPR/Cas9 technology is a promising tool in studying the molecular mechanisms of diabetes by targeting diabetes-related genes in human iPSCs. This technology is used to edit mutations in the HNF4A, GCK, PDX-1, and INS genes of iPSCs obtained from MODY patients with single-gene diabetes. The edited iPSCs can be differentiated into pancreatic progenitor cells, which can be transplanted into patients [102].

The CRISPR/Cas9 technology has been employed to create a rat model lacking the cytochrome P450 (CYP) 2E1 gene. This model aims to investigate the involvement of the CYP2E1 gene in various aspects such as biochemical metabolism, toxicology, and diseases like diabetes and alcoholic cirrhosis [118].

The animal model of diabetes is a valuable tool for conducting various experiments related to pathology, preclinical research, and drug testing. In a recent study utilizing CRISPR/Cas9 technology, researchers injected pX330 (a combination of gRNA and Cas9 sequences along with the donor DNA plasmid) into the oocyte to create new Cre tool mice. This allowed them to genetically manipulate pancreatic β cells [119]. The Ins1 promoter and stop codon sequences were used as targets for the insertion of the recombinase Cre. The resulting F1 mice were labeled as Cre-loxP recombination, which was observed in all islets that expressed insulin-positive cells and not observed in other tissues. These genetically edited mice showed no significant difference in glucose tolerance compared to wild-type mice.

9.3.3 Viral Diseases

Gene-editing platforms have recently emerged as potential antiviral treatments for infectious diseases. These platforms work by either modifying the genes of the host that the virus relies on or targeting the viral genes needed for replication [120]. Genome-based HIV therapy to far has involved modifying infection-related genes to create HIV-resistant CD4+ T cells and then re-infusing the modified cells into patients. Ebina and Hu et al.'s laboratory findings have shown that CRISPR/Cas9 could not only effectively remove latent HIV infection but also suppress new HIV infection [121, 122]. Similar to this, Hendel et al. recently showed that chemically modified CCR5 sgRNA delivered along with Cas9 mRNA/protein increased the efficacy of genome editing in human primary CD4+ T cells and CD34+ HSPCs without causing any damage related to DNA delivery. Cervical cancer is closely linked to the persistent expression of the high-risk human papillomavirus (HPV) oncogenes E6 and E7, which are implicated in malignant transformation [123]. Cervical cancer in situ may be reverted by the targeted mutagenesis of those high-risk HPV genes using gene editing techniques. A ZFN that could precisely detect and cleave HPV16/18 E7 DNA was created by Ding et al. In their research, HPV16/18-positive cervical cancer cells underwent type-specific apoptosis and effective growth inhibition as a result of ZFN-mediated HPV16/18 E7 DNA breakage [124]. The therapeutic effects of ZFNs were further confirmed when various plasmid-encoded zinc finger modules were introduced in vivo. These modules successfully inhibited tumor growth in mice that had cervical cancer cells. The apoptotic rate in HPV16/18-positive cervical cancer cells was significantly increased when ZFNs were combined with two chemotherapeutic drugs, cisplatin and trichostatin A. This combination resulted in a twofold increase in the apoptotic rate compared to using ZFNs alone. TALEN technology was used to specifically target the HPV oncogenes E6 and E7 in the DNA of the host. This led to the reactivation of the host tumor suppressors p53 and RB1. As a result, the ability of HPV-positive cell lines to form tumors was reduced, and the development of cervical cancer related to HPV was improved in transgenic mouse models [125]. In addition, the combination of CRISPR-Cas9 and HPV16 E6/E7 has been found to increase the sensitivity of cervical cancer cells to cisplatin. This suggests that it could be a promising approach for treating cervical cancer [126]. The most significant liver disease-causing pathogen is the hepatitis B virus (HBV) [127]. In order to deliver the sgRNA/Cas9 system to HepG2 cells that had been infected with HBV, Rananan et al. created and tested an effective gRNA targeting the HBV genomic locus. Finally, HBV gene expression and replication were blocked, and the quantity of covalently closed circular DNA (cccDNA) rapidly decreased, reducing by 92% on the 36th day [128]. In one study, latent EBV infections were treated by employing CRISPR/cas9 technology to delete genes relevant to the Epstein-Barr virus (EBV) [129]. The development of therapeutic strategies against viral infections, which rank among the most difficult diseases to treat, has enormous potential thanks to genomic editing technology, which enables us to better understand the mechanisms underlying the variety of diseases caused by viral infection. As a result, although being a powerful gene-editing technique, CRISPR/Cas9 still has certain issues with HIV use.

9.3.4 Hereditary Eye Disorders

A number of hereditary eye diseases, including congenital cataract, congenital glaucoma, retinitis pigmentosa (RP), congenital corneal dystrophy, Leber congenital amaurosis (LCA), retinoblastoma (RB), and Usher syndrome, can now be genetically diagnosed more clearly thanks to advances in gene sequencing technology [130]. CRISPR/Cas9 has already been utilized to create animal models of RP. One particular protein called Receptor Expression Enhancer Protein 6 (REEP6), which belongs to the REEP/Yop1 family of proteins, plays a role in shaping the structure of the endoplasmic reticulum [131]. Arno et al. discovered that autosomal-recessive retinitis pigmentosa is caused by biallelic mutations in REEP6 [132]. They found variations in REEP6 in RP patients from different families. Additionally, they used CRISPR/Cas9 to create a mouse model with the Reep6 p.Leu135Pro mutation. The Reep6L135P/L135P homozygous knock-in mice successfully reproduced the clinical characteristics of RP, including the degeneration and dysfunction of rod photoreceptors. This finding suggests that these mice can serve as a more effective animal model for future RP studies. Since the RP phenotype that causes blindness is yet unknown, the rodless (rd1) mouse, the most widely used preclinical model of RP, has been the subject of heated dispute. The Pde6b locus on chromosome five contains two homozygous variations in the rd1 mouse: a nonsense mutation (Y347X) and an insertion of the murine leukemia virus (Xmv-28) in intron 1 with the reverse orientation [133, 134]. By using CRISPR/Cas9 to correct the nonsense point mutation, Wu et al. showed that the Y347X mutation in rd1 mice causes disease and that it can be repaired [135]. The use of AAVs and iPSCs as well as other techniques in conjunction with CRISPR/Cas9 technology opens up new therapy options for eye diseases that are related to them. RPGF point mutations, which cause X-linked RP, were successfully repaired by CRISPR/Cas9 in patient-specific iPSCs, according to Bassuk et al. This suggests that combining gene editing with autologous iPSCs could be an individualized iPSC transplantation strategy for treating various retinal degenerations [136].

The most typical developing retinal tumor in children is RB [137]. By using the CRISPR/Cas9 method to cause the deletion of rb1 and retinoblastoma-like 1 (rbl1) in Xenopus tropicalis, Naert et al. established a highly penetrant and quick RB preclinical model [138]. The animal model demonstrated quick RB development, making it a useful tool for early drug discovery and quick identification of therapeutic targets. Jian Tu and colleagues successfully created a pluripotent H1 human embryonic stem cell line by using CRISPR/Cas9 nickase to knockout one copy of the RB1 gene. This achievement is significant as it offers a valuable cell resource for researching hereditary retinoblastoma [139]. Elevated intraocular pressure (IOP) is a defining feature of glaucoma, which is the second most common cause of blindness globally [140]. Primary open-angle glaucoma (POAG) has been caused by gain-of-function mutations in myocilin (MYOC) [141–143]. When mutant myocilin builds up inside of cells, the unfolded protein response (UPR) cascade and endoplasmic reticulum stress are activated in the trabecular meshwork (TM) [143, 144]. TM cells are prone to chronic ER stress, which eventually leads to their demise. This, in turn, causes an elevation in IOP and the development of glaucoma [145, 146]. By using CRISPR/Cas9 to reduce the expression of mutant MYOC in a mouse model of POAG, Jain et al. were able to reduce ER stress, lower IOP, and the capacity of the mouse eyes to avoid further glaucomatous damage [147]. Importantly, scientists also proved that using CRISPR/Cas9 in glaucoma-affected human eyes was possible [148]. A humanized MECE mouse model's corneal stroma after receiving combined Cas9/sgRNA injections revealed frame-shifting deletions of the mutant KRT12 allele. This study is the first to show how a new PAM made by a heterozygous disease-causing SNP was edited utilizing CRISPR/Cas9 in vivo gene editing [149].

9.3.5 Hematological Disorders

Nearly 50% of hemophilia A patients Factor VIII (F VIII) cases are caused by the incorrect expression of this protein as a result of chromosomal inversion [150]. In one study, hemophilia A patients' somatic cells were used to create iPSCs, which were then used to modify the F VIII gene using the CRISPR/Cas9 system [151]. The altered iPSCs were subsequently transplanted into hemophilia mice lacking factor VIII after being stimulated to develop into mature endothelial cells capable of expressing factor VIII. The outcomes demonstrated that the transplanted mice started producing factor VIII, which successfully reduced the symptoms of bleeding. A lack of factor IX (F IX) leads to hemophilia B. By raising plasma FIX levels, coagulation activity can be recovered. In a family of hemophilia B patients, Guan et al. revealed that the F9 gene has a novel mutation, Y371D, which results in a more severe hemophilia B phenotype than the previously identified Y371S variant. They transferred Cas9 to adult F9 Y371D mutant mice using adenoviral vectors and naked DNA constructs [152]. After therapy, it was discovered that even though the mutant gene was highly effectively repaired when adenovirus was utilized as a vector to transport cas9, the hepatotoxicity was significant. However, more than 0.56% of the F9 alleles in the hepatocytes of hemophilia B mice were effectively repaired by Cas9 with a naked DNA structure, which was sufficient to regain hemostasis. Sickle cell anemia is the first genetic disease whose pathogenesis is well understood. It is caused by a mutation in a single nucleotide, where the nucleotide A is changed to T in the first exon of the human β-globin gene. This mutation leads to a specific abnormality [153]. In 2016, a team from Stanford University published a report on their use of CRISPR/Cas9 technology to fix mutations in the β-globin gene (HBB) in patient-derived HSCs in vitro [154]. After the modified induced pluripotent stem cells (iPSCs) transformed into red blood cells, they were able to detect normal HBB mRNA. In order to enhance the efficiency of gene editing and minimize unintended effects, modified versions of Cas9 enzymes have been extensively used to edit genes in HSPCs. For instance, Park used a highly accurate Cas9 enzyme called SpyCas9 to correct the b-globin gene, which resulted in fewer unintended effects [155]. Similarly, Anzalone employed a tool called prime editing to edit genes without causing double-strand breaks [156]. This preclinical experiment offers theoretical backing for the use of gene-editing technology in treating sickle cell anemia. By utilizing CRISPR/Cas9 technology, it became feasible to accurately guide the calibrated DNA sequence to the specific HBB mutation site. This allowed for the successful correction of two distinct β-thalassemia mutations found in the HBB gene of patient iPSCs through HR. Similar to SCD, b-thalassemia, another uncommon but serious monogenetic disease, is aberrant b-globin expression. Reexpressing HbF is therefore a treatment option for β-thalassemia as well. Additionally, the 200 bp genomic area of the BCL11A enhancer was knocked out in order to stimulate the formation of γ-globin, one of the HbF precursors [157], and 13 kb of the β-globin locus was identified in HSPCs by using CRISPR systems designed for both upstream and downstream cleavage locations [158]. This provided experimental support for the use of CRISPR/Cas9 in the treatment of β-thalassemia.

9.3.6 Duchenne Muscular Dystrophy

In 2016, Long and his colleagues used CRISPR/Cas9 along with AAVs vectors to investigate effective therapeutic strategies for DMD. They aimed to skip the mutated dystrophin exons in vivo, resulting in the production of dystrophin protein [159]. Later in 2018, Ryu used base editing tools to accurately fix the dystrophin gene, offering a more secure approach to DMD treatment [160]. CRISPR/Cas9 technology offers a huge opportunity to treat other severe X-linked disorders in addition to DMD. For instance, Kuo reported that CRISPR/Cas9 might be used to treat X-Linked Hyper-IgM Syndrome (XHIM) by knocking in the cDNA of the CD40 ligand [161]. Additionally, one likely therapeutic approach for X-linked chronic granulomatous disease (X-CGD) is the repair of the CYBB gene in HSPCs [162]. Pavel-Dinu's team recently showed that long-term hematopoietic stem cells (LT-HSCs) can be used to integrate a matching cDNA for the treatment of X-linked severe combined immunodeficiency (SCID-X1) using CRISPR/Cas9 [163]. In a similar vein, Goodwin treated the immune dysregulation polyendocrinopathy enteropathy X-linked (IPEX) syndrome by editing the forkhead box protein 3 (FOXP3) gene using CRISPR/Cas9. This research suggested gene correction using CRISPR/Cas9 is one of the most promising therapeutic modalities for treating genetic autoimmune diseases [164]. CRISPR/Cas9 was also used to target the monkey dystrophin gene and produce DMD-causing mutations. The detection of the relative targeting rate revealed that up to 87% of the dystrophin alleles in monkey muscle could develop mosaic alterations as a result of CRISPR/Cas9 [165, 166]. Recently, three groups of researchers have made significant progress in using CRISPR/Cas9 to eliminate mutations in the DMD gene [167–169]. The researchers utilized the CRISPR/Cas9 system to remove the defective part of DMD in the mdx mouse model. As a result, they were able to produce a shorter form of the dystrophin protein in the muscle fibers, leading to a partial restoration of muscle function. This approach shows great potential for treating genetic mutations that cause muscle diseases in patients.

9.3.7 Primary Immunodeficiencies

Allogeneic HSC transplantation can be used to treat patients with primary immunodeficiencies who are immune system-deficient or immune system-dysfunctional [170]. The interleukin 2 receptor gamma (IL2RG) gene mutation that leads to severe combined immunodeficiency (SCID), the most severe immunodeficiency, causes the developmental arrest of T cells as well as additional primary or secondary defects in B cells. The HDR at the IL2RG gene has been effectively induced in a variety of human cell types, including HSCs and embryonic stem cells (ESCs), using the ZFN and TALEN approaches [171–173]. Previous studies have used endonucleases to create various types of animal models with weakened immune systems. These models were difficult to develop before because there was no efficient way to modify their genes [173–176]. Other animal disease models, including those for Rett syndrome [177], hereditary deafness [178], Wilson disease [179], Laron syndrome [180], Niemann-Pick disease [181], Netherton syndrome [182], and others, have been created as a result of designed nuclease-mediated editing of genomic alterations. Animal models will increasingly be used in research on disease mechanisms and the development of new treatments as a result of advancements in genome editing technologies.

HIV infection may be prevented with the help of CRISPR/Cas9 approaches that have been discovered and developed. Long terminal repeat (LTR) promotor, which is regarded as one of the therapeutic targets, is crucial to the expression of the HIV gene during the course of HIV infection. As a result, Ebina first modified the LTR promotor using CRISPR/Cas9 in 2013, which resulted in a marked reduction in LTR-driven expression [121]. The CRISPR system, which has been recently developed, has also been used in the treatment of HIV. One specific system called CRISPR-Cas12a, known for its high activity and specificity, was employed by Gao to target and cut the LTR sequence without introducing any additional DNA [183]. In addition to LTR, the pol gene and tat/rev gene have also been recognized as potential targets for therapy. By using CRISPR/Cas9, these genes can be deactivated, leading to a reduction in the expression of HIV genes to varying degrees [184]. Notably, in 2017, Yin's lab substantially reduced HIV RNA expression using CRISPR/Cas9-based editing of the LTR and Gag/Pol genes in vivo by AAVs delivery, which greatly advances the development of Cas9-based HIV therapy.

9.4 Section 4: Exploring Gene Editing in Clinical Trials

The use of genome editing as a therapeutic strategy has the potential to prevent or reverse the mutations that cause cancer and other genetic illnesses. Ex vivo genome editing, or genetically modifying cells in vitro before re-engrafting them into patients, has been the method that has received the most attention to date. A number of gene-editing clinical trials have recently been carried out by teams from China and the United States, including the development of more potent CAR T cells for the treatment of cancer and the knockout of the BCL11A gene that regulates the erythroid-specific enhancer to increase gamma globulin in autologous erythroid HSCs as a potential treatment for sickle cell disease and beta-thalassemia.

9.4.1 Anticancer Clinical Trials

Only two studies have been reported that used TALENs in CAR T cells. One of these studies (NCT02808442) focused on creating a group of allogeneic, universal CAR T cells called UCART19. These cells were designed to target relapsed or refractory CD19-positive B acute lymphoblastic leukemia. This study aimed to remove alloreactivity and alemtuzumab sensitivity by disrupting the loci that encode TRAC and CD52. A similar approach was used to create allogeneic TALEN-edited CAR T cells that specifically target CD123 (known as UCART123) in acute myeloid leukemias (AMLs) and blastic plasmacytoid dendritic cell neoplasms. The clinical trial identifier for this study is NCT03190278. The CRISPR/Cas9 system is a valuable tool in cancer therapy because it has a straightforward design process and allows for multiple gene edits simultaneously. So far, 11 clinical trials have been conducted to evaluate the effectiveness of the CRISPR system in cancer therapy. Out of these trials, seven focus on immunotherapies that aim to target the expression of the PD-1 protein. In 2016, the first patient was enrolled in a groundbreaking clinical trial at West China Hospital, Sichuan University. This trial aimed to utilize the innovative CRISPR/Cas9 technique for treating cancer [185]. This content describes a phase I study (NCT02793856) that evaluated the safety of ex vivo engineered PD-1 knockout T cells in the treatment of metastatic non-small cell lung cancer. The patients in this study had already undergone all standard treatments, but their cancer had continued to progress. In this trial, the PD-1 expression in the patients' peripheral lymphocytes was disabled using CRISPR/Cas9 technology. The patients received a second infusion of the expanded and edited lymphocytes. The similar idea of PD-1 knockout autologous T cells is currently being tested in human trials to treat different cancer types, such as prostate cancer (NCT02867345), esophageal cancer (NCT03081715), and renal cell carcinoma (NCT02867332). The in vitro CRISPR/Cas9 gene deletion technology is being used in these trials as the first proof-of-concept investigations for cancer therapy. There is currently research combining PD-1 knockout with other forms of targeted editing in the development of therapies, which may increase effectiveness for use in clinical settings. One illustration is the phase I/II clinical trial (NCT03044743) adding PD-1 deletion to autologous T cells specific for the Epstein-Barr virus (EBV) for the treatment of EBV-positive malignancies. A focus of study in clinical trials is the creation of CAR T cells using the CRISPR/Cas9 system to combat cancerous cells. Several doses of universal CD19-specific CAR T cells (UCART 019) were administered to patients with recurrent or CD19+ refractory leukemia and lymphoma in a clinical phase I/II trial (NCT 03166878) to assess their safety and tolerability. In this study, lentiviral delivery of CAR receptors and CRISPR RNA electroporation to simultaneously damage endogenous TCR and B2M genes were combined to create UCART019 cells. These cells, which come from one or more unrelated, healthy donors, may help patients experience anti-leukemic effects while reducing the risk of graft-versus-host disease (GVHD) and host-mediated immunity. Unfortunately, a few individuals experienced relapses because tumor cells did not express CD19. As a result, another clinical trial (NCT03398967) that was designed to be more suitable for a wider spectrum of patients focused on allogenic CRISPR-edited bispecific CD19+ CD20+ or CD19+ CD22+ CAR T cells, which could identify and eliminate CD19-negative malignant cells through the identification of CD20 or CD22. Another investigation has proposed a new clinical trial (NCT03057912) to assess the safety and effectiveness of combining genome editing with TALENs and CRISPR/Cas9 by targeting HPV16 and HPV18 E6/E7 DNA in the treatment of HPV-associated cervical intraepithelial neoplasia. The target gene expression of the CAR T cells was disrupted in this experiment by using both editing methods, which were given twice a week for 4 weeks. This was done in an effort to minimize side effects. The neurofibromatosis type 1 (NF1) gene has one of the greatest mutation rates in the human genome, which increases the likelihood that it will result in a variety of benign or malignant tumors [186]. CRISPR/Cas9 technology was created for one trial (NCT03332030) to screen and find NF1-specific medications. First, a human iPSC library was created from NF1 patients with good phenotypic traits. Next, CRISPR/Cas9 was used to create the NF1+/+, NF1+/, and NF1/ cell lines. Then, by looking at the reversal or remission characteristics following certain drug use, possible therapeutic drugs could be found. Although genome editing clinical trials have shown encouraging results, further research is required to ensure the safety and efficacy of this technology in treating human tumors.

9.4.2 Antiviral Clinical Trials

In the early stages of HIV infection, CCR5 plays a significant coreceptor role, whereas CXCR4 contributes significantly as an auxiliary receptor when stable infections are established [187]. The main method for treating HIV with genome editing is to create immune cells that are resistant to HIV infection or replication. The most typical procedure consists of two steps: in vitro modification of the cells (CD4+ T cells and CD34+ hematopoietic stem/progenitor cells) and subsequent reinfusion of the changed cells into the patients [188, 189]. In the context of HIV infection, CD4+ T cell modulation has already been studied in a number of clinical trials. In order to assess the security and anti-HIV benefits of modified autologous CD4+ T cells in HIV-1-infected patients, the first approved genome editing trial involving the treatment of HIV with ZFNs (NCT00842634) started in 2009. Each participant received a single infusion of 5–10 billion ZFN-modified CD4+ T cells, which were administered ex vivo to autologous CD4+ T cells via adenoviral vectors for CCR5 gene deletion. Chinese researchers developed a CRISPR/Cas9-modified CCR5 gene-editing system for adult HPSCs to achieve long-term and stable hematopoietic system reconstruction following infusion of modified CD34+ cells into patients with HIV-1 infection and ALL (NCT03164135), according to a recently published article [190]. This study helped forward the development of gene-editing technology for use in clinical settings by providing preliminary evidence of the viability and safety of adult HPSC gene-editing transplantation in the human body. There is significant interest in altering patient-specific iPSCs and reprograming them to HSPCs because HSPC-based gene therapy is frequently constrained by ex vivo culture methods and challenges in HPSC proliferation [191].

9.4.3 Clinical Trials of Hematological Disorders

Five clinical gene-therapy trials for hematological illnesses, including hemophilia B, -thalassemia, and sickle cell disease, have used ZFN with CRISPR/Cas9 so far [192]. By using SB-FIX, an AAVs-delivered ZFN, which is intended to be intravenously administered to the subject's own hepatocytes to insert a corrective FIX transgene into the albumin locus, Sangamo Therapeutics Inc. began a phase I clinical trial (NCT02695160) with the anticipated 12 participants in November 2016. Their goal is to permanently induce FIX clotting factor production in the liver of severe hemophilia B patients. The goal of this ascending dose phase I trial, which is anticipated to be finished in January 2021, is to evaluate the safety and tolerability of SB-FIX when used to treat hemophilia B patients. In a 12-subject early phase I trial that began in January 2019, Allife Medical Science and Technology Co., Ltd. used CRISPR/Cas9 to edit the *HBB* gene in vitro in patient-specific induced hematopoietic stem cells (iHSCs). The edited cells were then intravenously transfused back into the HBB-mutated -thalassemia subjects. The clinical severity of -thalassemia and sickle cell disease is linked to BCL11A, a crucial modulator in hemoglobin diseases characterized by suppressing fetal hemoglobin (HbF) [193]. Therefore, gene therapy that targets BCL11A has been tried in trials to treat the two disorders mentioned above. The BCL11A gene has been suppressed in autologous CD34+ hematopoietic stem/progenitor cells in vitro in three trials so far, with the goal of intravenously reintroducing the altered cells to the participants. The initial trial, NCT03432364, a single dosage phase I/II study with six transfusion-dependent -thalassemia (TDT) patients, was directed by Sangamo Therapeutics Inc. The gene-edited therapeutic cell ST-400 was created using ZFN; its safety, tolerability, and effects on HbF are now being studied, as well as its transfusion needs. Vertex Pharmaceuticals Inc. started a single-dose phase I/II study trial (NCT03655678) with up to 45 participants that focuses on transfusion-dependent -thalassemia (TDT). They used the CRISPR/Cas9-modified cell CTX001 to examine the safety and effectiveness of the technique. Vertex Pharmaceuticals Inc. has previously investigated CTX001 in severe sickle cell disease (NCT03745287) with identical trial designs and start and completion durations. The Food and Drug Administration (FDA) has given its approval for two gene therapies, CASGEVY™ (exagamglogene autotemcel) and LYFGENIA™ (lovotibeglogene autotemcel), to be used in the treatment of sickle cell disease. These therapies are specifically for patients aged 12 and above who experience recurrent vaso-occlusive crises (VOCs). The Children's Hospital of Philadelphia (CHOP) is a Qualified Treatment Center that provides LYFGENIA, developed by bluebird bio, and also plans to offer CASGEVY, which is manufactured by Vertex Pharmaceuticals. Sickle cell disease can now be treated with a newly approved therapy that modifies the patient's own DNA. This therapy allows the patient's red blood cells to produce a different type of hemoglobin. Clinical trials conducted at CHOP and other locations have demonstrated that effective gene-editing techniques can prevent cells from adopting the characteristic crescent shape associated with sickle cell disease. Furthermore, this approach has successfully eliminated pain episodes in nearly all patients. It is worth noting that CASGEVY, which utilizes CRISPR technology, is the pioneering therapy to receive FDA approval. LYFGENIA is a gene therapy that aims to treat sickle cell disease by addressing its root cause. It does this by introducing a functional gene that allows the production of adult hemoglobin, which does not take on the crescent shape typically seen in this disease. LYFGENIA is a gene therapy that aims to treat sickle cell disease by introducing a functional gene that allows the production of adult hemoglobin, which does not take on the crescent shape typically seen in this disease (https://www.chop.edu/news/fda-approves-two-gene-therapies-sickle-cell-disease).

9.4.4 Clinical Trials of Hereditary Eye Diseases

A significant number of clinical trials of gene augmentation for LCA, choroideremia, achromatopsia, X-linked retinoschisis, and RP are now underway. Gene augmentation is successfully used to treat inherited retinal illnesses [194]. There has only been one clinical trial employing gene editing in LCA10 up to this point. Editas and Allergan recently started a clinical research (NCT03872479) to test the safety, tolerability, and effectiveness of single-dose AGN-151587 (EDIT-101), an AAVs vector with 3 components: a *S. aureus* Cas9 and 2 gRNAs, gRNA-323 and gRNA-64. To treat LCA10, AGN-151587 could reverse the c.2991 + 1655A > G mutation in intron 26 of the CEP290 gene. Despite the fact that ophthalmic disease gene-editing clinical trials have just started, the special characteristics of eyes, such as their accessibility and relative immune privilege, make CRISPR-Cas a viable and practical approach for the treatment of ocular diseases in the near future [194, 195].

9.4.5 Current FDA-Approved Cell and Gene Therapies

The following table includes all licensed products that are categorized as "Approved Cellular and Gene Therapy Products" on fda.gov. This information was last updated on 12/26/2023 (https://www.mirusbio.com/fda-approved-gene-cell-therapies/) (Table 9.2).

Table 9.2 Approved cellular and gene therapy products

Name	Type	Manufacturer	Description
ABECMA (2021)	Cell (autologous, modified)	Celgene Corporation, a Bristol-Myers Squibb Company	CAR T cells recognize the B-cell maturation antigen (BCMA) and kill BCMA-expressing tumor cells. A lentiviral vector, developed by bluebird bio, is used to engineer the CAR T cells
ADSTILADRIN (2022)	Adenovirus	Ferring Pharmaceuticals A/S	A recombinant adenovirus (rAd-IFNa/Syn3) delivers human interferon *alfa-2b* cDNA into the bladder epithelium. Treats patients with certain types of bladder cancer
ALLOCORD (2013)	Cell (allogeneic)	SSM Cardinal Glennon Children's Medical Center	Blood cell transplant derived from umbilical cord and placenta
BREYANZI (2021)	Cell (autologous, modified)	Juno Therapeutics, Inc., a Bristol-Myers Squibb Company	CAR T cells engineered with lentivirus to attack CD19-expressing tumor cells. Treats patients with certain kinds of large B-cell lymphoma
CARVYKTI (2022)	Cell (autologous, modified)	Janssen Biotech, Inc.	CAR T cells engineered with lentivirus to attack BCMA-expressing tumor cells. Treats patients with certain kinds of relapsed or refractory multiple myeloma
CASGEVY (2023)	Cell (autologous, modified)	Vertex Pharmaceuticals, Inc.	CD34+ hematopoietic stem cells engineered with electroporation of CRISPR/Cas9 RNP complexes to decrease expression of *BCL11A* to increase fetal hemoglobin production. Treats patients with certain kinds of sickle cell disease
CLEVECORD (2016)	Cell (allogeneic)	Cleveland Cord Blood Center	Transplant of hematopoietic progenitor cells (HPCs) from human cord blood
DUCORD (2012)	Cell (allogeneic)	Duke University School of Medicine	Transplant of HPCs from human cord blood
ELEVIDYS (2023)	AAVrh74	Sarepta Therapeutics, Inc.	Recombinant AAVrh74 delivers a micro-dystrophin gene to treat patients with certain types of Duchenne muscular dystrophy
GINTUIT (2012)	Cell (allogeneic)	Organogenesis Inc.	Cellularized scaffold for topical application to a surgically created wound bed to treat certain mucogingival conditions
HEMACORD (2011)	Cell (allogeneic)	New York Blood Center	Transplant of HPCs from human cord blood
HEMGENIX (2022)	AAV5	CSL Behring LLC, UniQure	Recombinant AAV5 delivers *F9* to treat patients with certain kinds of hemophilia B
HPC, Cord Blood (2012)	Cell (allogeneic)	Clinimmune Labs, University of Colorado Cord Blood Bank	Transplant of HPCs from human cord blood
HPC, Cord Blood, MD Anderson Cord Blood Bank (2018)	Cell (allogeneic)	MD Anderson Cord Blood Bank	Transplant of HPCs from human cord blood
HPC, Cord Blood, LifeSouth (2016)	Cell (allogeneic)	LifeSouth Community Blood Centers, Inc.	Transplant of HPCs from human cord blood
HPC, Cord Blood, Bloodworks (2016)	Cell (allogeneic)	Bloodworks	Transplant of HPCs from human cord blood

(continued)

Table 9.2 (continued)

Name	Type	Manufacturer	Description
IMLYGIC (2015)	HSV-1	BioVex, Inc., a subsidiary of Amgen Inc.	Recombinant HSV-1 delivers *GM-CSF* to treat patients with certain unresectable melanoma lesions
KYMRIAH (2017)	Cell (autologous, modified)	Novartis Pharmaceuticals Corporation	CAR T cells engineered with lentivirus to attack CD19-expressing tumor cells. Treats patients with certain kinds of relapsed or refractory follicular lymphoma
LANTIDRA (2023)	Cell (allogeneic)	CellTrans Inc.	Pancreatic islet cells from a deceased donor are transplanted to treat patients with certain types of type I diabetes
LAVIV (2011)	Cell (autologous, expanded)	Fibrocell Technologies	Fibroblasts to improve the appearance of moderate to severe nasolabial fold wrinkles
LUXTURNA (2017)	AAV2	Spark Therapeutics, Inc.	Recombinant AAV2 delivers *RPE65* to treat patients with RPE65-mutation-associated retinal dystrophy
LYFGENIA (2023)	Cell (autologous, modified)	bluebird bio, Inc.	Hematopoietic stem cells are engineered with a lentivirus, BB305, expressing a modified *β^A-globin* gene. Treats patients with certain kinds of sickle cell disease
MACI (2016)	Cell (autologous, expanded)	Vericel Corp.	Cellularized scaffold to repair cartilage defects of the knee
OMISIRGE (2023)	Cell (allogeneic, modified)	Gamida Cell Ltd.	Transplant of nicotinamide-modified HPCs from human cord blood
PROVENGE (2010)	Cell (autologous, activated/ expanded)	Dendreon Corp.	Antigen-presenting cells are activated and used to treat certain types of prostate cancer
RETHYMIC (2021)	Cell (allogeneic)	Enzyvant Therapeutics GmbH	Thymus tissue implant to improve immune function in patients with certain types of congenital athymia
ROCTAVIAN (2023)	AAV5	BioMarin Pharmaceutical Inc.	Recombinant AAV5 delivers a B-domain deleted SQ form of *F8* to treat patients with certain types of hemophilia A
SKYSONA (2022)	Cell (autologous, modified)	bluebird bio, Inc.	Hematopoietic stem cells are engineered with Lenti-D vectors that express the *ABCD1* gene. Treats patients with a kind of cerebral adrenoleukodystrophy
STRATAGRAFT (2021)	Cell (allogeneic)	Stratatech Corporation	Cellularized scaffold of keratinocytes and dermal fibroblasts in murine collagen to treat patients with some types of thermal burns
TECARTUS (2020)	Cell (autologous, modified)	Kite Pharma, Inc.	CAR T cells engineered with gammaretrovirus to attack CD19-expressing tumor cells. Treats patients with certain kinds of relapsed or refractory mantle cell lymphoma. Differs from YESCARTA in that the engineered T cells are enriched for activity against certain B cells
VYJUVEK (2023)	HSV-1	Krystal Biotech, Inc.	Recombinant HSV-1 delivers the collagen type VII gene (*COL7*) to treat certain wounds of patients with dystrophic epidermolysis bullosa with mutations in the *COL7A1* gene

(continued)

Table 9.2 (continued)

Name	Type	Manufacturer	Description
YESCARTA (2017)	Cell (autologous, modified)	Kite Pharma, Inc.	CAR T cells engineered with gammaretrovirus to attack CD19-expressing tumor cells. Treats patients with certain kinds of relapsed or refractory large B-cell lymphoma
ZYNTEGLO (2022)	Cell (autologous, modified)	bluebird bio, Inc.	Hematopoietic stem cells are engineered with a lentivirus, LentiGlobin BB305, to express *HBB*. Treats patients with certain types of ß-thalassemia
ZOLGENSMA (2019)	AAV9	Novartis Gene Therapies, Inc.	Recombinant AAV9 delivers *SMN1* to treat patients with certain types of spinal muscular atrophy

9.5 Section 5: Future Prospects

The research evidence accumulated to date has shown significant contributions of genome editing systems to exploit therapeutic strategies for different types of human diseases, among which the CRISPR/Cas9 system has been particularly effective by directly affecting target gene loci or resulting in multifunctional tools. Future research will be able to detect synthetic fatal interactions in the genome and make it easier to find new therapeutic targets by combining pooled CRISPR screening with knowledge of the genetic and epigenetic properties of cancer cell lines. Additionally, the CRISPR/Cas9 technology offers a fresh method for modifying non-coding areas of the cancer genome, accelerating the functional investigation of up-to-now understudied features. Genome editing has moved from being a theoretical idea to being used in clinical settings thanks to the enormous improvements in the creation of synthetic nucleases, particularly ZFNs, TALENs, and CRISPR/Cas9.

9.5.1 Genome Editing in Cancer Immunotherapy

Cancer immunotherapy, which uses the patient's own immune system to fight tumor cells, has recently sparked a lot of interest [196]. Utilizing genetically modified T cells, also known as chimeric antigen receptor (CAR) T cells, which allow the targeting of tumor-associated antigens and may improve the therapy response, is one promising field of immunotherapy [197, 198]. There are several crucial phases involved in creating functional CAR T cells. CARs are artificial receptors that typically include a T cell signaling domain, a costimulatory molecule area that can connect to receptors like CD28, 4-1BB, or CD278;220, and an antibody-derived targeting ectodomain that detects tumor antigens [199]. The CAR can send signals and activate modified T cells after attaching to a specific antigen. Genetically modified CAR T cells have a key antitumor advantage thanks to the independence of CAR recognition, which gets over the restriction normally imposed by the major histocompatibility complex (MHC) [200]. The usefulness of this revolutionary product is severely constrained by the difficulty of the manufacturing process, the small number of target antigens available, and the poor antitumor responses to solid tumors. Flexible gene-editing methods have developed over the last few years into important engineering tools to address these constraints and enhance CAR T designs. The creation of allogeneic CAR T cell treatment would streamline and address several difficulties in the creation of autologous CAR T cells [201]. For the identification of major and minor histocompatibility antigens, the endogenous T cell receptor (TCR) is responsible. Allogeneic T cells can have various components of the TCR complex and/or the HLA class I loci genetically disrupted in order to produce a universal cellular therapy product that confers a wider range of application capability with a low risk of associated side effects, such as graftversus-host disease (GVHD). In order to prevent the expression of or chains in endogenous TCRs and the subsequent loss of TCR function in CD19 CAR T-cells, Torikai et al. employed modified ZFNs in 2012 [202]. The development of universal allogeneic tumor-associated antigen-specific CAR T cells resulted from these altered T cells' inability to react to TCR-specific stimuli while retaining the capacity to identify and target CD19. The development of universal allogeneic tumor-associated antigen-specific CAR T cells resulted from these modified T cells' inability to react to TCR-specific stimuli while retaining the capacity to identify and target CD19. The selective deletion of HLA expression in CD19-specific T cells and embryonic stem cells was accomplished using the same method, which broadened the utility of this approach by preventing the infusion of HLA-diverse immune cells [203]. The advancement of the CRISPR/Cas9 system has successfully accelerated and simplified multiple gene editing in CAR T cells. Liu et al. successfully produced CAR T cells with two (TRAC and B2M) or three (TRAC, B2M, and PD-1) disrupted genes using this technology, and then they assessed their antitumor activity in vitro and in vivo [204]. More significantly, the CRISPR/Cas9-mediated multiplex gene-edited CAR T cells kept their CD19-specific antitumor activity in a lymphoma xenograft mouse model, indicating that they are potential cancer therapeutic tools. Another intriguing study [205] used a one-shot CRISPR technique to manufacture allogeneic universal CAR T cells by including several gRNAs into a CAR lentiviral vector. This led to the efficient double knockout of endogenous TCR and HLA class I molecules. Adenosine 2a receptor (A2aR) (230), T cell membrane protein-3 (TIM-3) (229), and lymphocyte-activation protein 3 (LAG-3) (231) are examples of immune receptors that inhibit the immune response. Recently, using CRISPR/Cas9 to create CAR T cells, which are designed to target and disable these immune receptors, has shown promising results in xenograft mouse models, with a higher percentage of complete remission. IFN-g, GM-CSF, and MIP-1b are only a few of the tumor-fighting cytokines that are secreted more frequently by CAR T cells made by CRISPR/Cas9. Since checkpoint inhibitors and CAR T cells may increase the antitumor efficacy of acute myeloid leukemia (AML) and other hematological malignancies, these factors may play a role in CAR T cell fatigue and AML dysfunction. Together, these findings imply that genome editing may be a useful tool for creating "universal" CAR T cells and that it can be used to mass produce healthy "off-the-shelf" T cells that are effective against a variety of targets.

9.5.2 Screening for Functional Genes

Numerous targeted medications have been created to treat a variety of disorders as a result of the notion of precision medicine. For instance, certain carcinogenic components (gene fragments or proteins) will bind to targeted medications created for recognized carcinogenic areas and cause the apoptosis of tumor cells without harming normal tissue cells. However, a clear disadvantage of this molecular targeting therapy is that in order for patients to respond to the targeted drug, there must be a specific mutation or change in gene expression. If not, drug resistance will continue to exist. Based on CRISPR/Cas9 technology, researchers have created libraries of gene mutations associated with particular functions, or mammalian genome-wide mutations, which are related to screening phenotypes through functional screening, PCR amplification, and deep sequencing analysis [206, 207]. For pharmacological screening or tailored pathway screening, the gRNA library is the perfect tool. The creation of gRNA libraries will be crucial for research into disease mechanisms, functional gene screening, and medication development. In order to investigate drug-gene interactions, functional genome screening using the CRISPR system could reveal changes in gene expression following cancer drug therapy. By adding small molecules as perturbations, this method could also identify novel targets for precise treatment and offer insights into disease development [208,

209]. Finding genotype-specific vulnerabilities is one of the main objectives of pooled CRISPR/Cas9 unbiased screening in cancer research, and AML was the first disease to be thoroughly examined with this technology [210]. By examining five frequently used AML cell lines and two solid tumor cell lines as controls, the authors used this platform to identify numerous well-known prospective targets for AML therapy, including BCL2, BRD4, MEN1, and DOT1L. Since then, numerous cancer cell lines have undergone extensive CRISPR/Cas9 screening in an effort to systematically find critical genes [211, 212]. As a result, around 1500 essential genes—five times more than the amount of genes previously discovered through shRNA screening—have been identified [100]. Long noncoding (lncRNA) loci can be found using CRISPR/Cas9, according to recent groundbreaking studies. These research searched for functional lncRNA loci that could affect cell proliferation [213, 214] and drug resistance [209, 215] using CRISPR-interference (CRISPRi) or CRISPR-activation (CRISPRa)-based libraries. Typically, a thorough sgRNA library was created to target the lncRNA transcription initiation site, and the library was then transduced into various cell lines. Following that, hundreds to thousands of lncRNAs supporting cell proliferation and drug resistance might be found using sequence analysis. The simultaneous mutation of two genes can result in an unanticipated phenotype that establishes the probable functional link between the genes, depending on the particular effects of each mutation [216]. The development of cancer therapies may be affected by this phenomenon, known as genetic interaction. For instance, in tumors with loss-of-function mutations in BRCA1 or BRCA2, an inhibitor of PARP1/2 (such as olaparib) could cause cell death by simultaneously disrupting both genes [217]. The CRISPR/Cas9 system is a useful tool for studying how cancer responds to chemotherapy by identifying gene interactions that work together. Researchers have developed a method using CRISPR technology, deep sequencing, and genetic analysis to study the interactions between two important drug targets, BCL2L1 and MCL1, in K562 leukemia cells. This approach allows for a better understanding of how these genes work together to enhance the effectiveness of chemotherapy [218]. The double-knockout screening method was used to identify 73 tumor genes in pairs. This method discovered synthetic lethal interactions between many known genes (such as BRCA-PARP) and unknown genes. Around 75% of these interactions were successfully replicated using combination drugs [219]. By combining pooled CRISPR screening with a perturbed drug, it is possible to identify genes that either work together effectively or provide resistance to the drug [220]. In one of the initial pooled CRISPR screens, researchers treated a large-scale knockout library of melanoma cells with the BRAF inhibitor vemurafenib. They then identified the genes that provided resistance to the drug [221].

9.5.3 Gene Diagnostic Tools

Genes that enhance the risk of cancer due to germline mutations are known as cancer predisposition genes [222]. For the purpose of preventing cancer, genetic diagnostics is essential for identifying such vulnerable genes. A CRISPR-based diagnostic technique called SHERLOCK (specific high-sensitivity enzymatic reporter UnLOCKing) has been developed to address this issue because low-frequency mutations are difficult to detect by sequencing [223]. Technically, the system is made up of the reporter signal and the RNA-guided endonuclease Cas13a, a member of the Cas family. The presence of Cas13a is a crucial component that successfully promotes trans-cleavage of non-specific single-stranded DNA (ssDNA). RNA cleavage results in the release of the reporter signal. When used to identify two cancer mutations, BRAF V600E and EGFR L858R, this strategy appeared to be a very sensitive detection tool [224]. There has also been the development of a different system known as DETECTR (DNA endonuclease-targeted CRISPR trans-reporter) [225]. Another enzyme, recombinase polymerase amplification (RPA), is employed as a detection tool to screen for viral infections in cancer and to amplify microsamples. Cas12a behaves similarly to Cas13a in this system. The technology seems to be a quick and affordable way to identify HPV 16/18 in lung carcinomas [226]. In order to explore breast cancer, lentivirus was utilized to infect healthy breast cells with the CRISPR nuclease-dead Cas9 (dCas9) system that had been coupled to a DNA methyltransferase effector. Using this technology, scientists have learned that the cyclin dependent kinase inhibitor 2A (CDKN2A) gene was a major contributor to carcinogenesis, causing unusually quick cell division that may 1 day serve as an early breast cancer diagnostic sign [227]. The BRCA1-delta11q optional splice isoform is a major contributor to PARPi and cisplatin treatment resistance in breast cancer, according to CRISPR/Cas9 gene editing as well as overexpression tests [228].

9.6 Section 6: Challenges in Therapeutic Targeting and Ethical Concerns

Despite the numerous advantages of genome editing, there are certain technical difficulties in converting these approaches to clinical disease therapy, particularly in terms of delivery, accuracy, and efficacy barriers. To meet these hurdles, researchers will need both meticulously crafted genome editing platforms in preclinical trials and in-depth understanding of the molecular basis of malignancies, particularly heterogeneous solid tumors.

9.6.1 Increasing the Specificity of Gene Correction

The capacity to modify the desired location of interest within the genome characterizes the accuracy of gene-editing technology. Off-target effects, or mutations in undesirable genomic loci, are inherently highly dangerous because they have the potential to cause genomic toxicity, genome instability, disruption of gene function, epigenetic changes, and even cancer [229–231]. Assays of utmost importance have been developed to evaluate the targeting specificities of ZFNs, TALENs, and Cas9 nucleases, such as in vitro selection libraries [232, 233], mismatch-detection nuclease assays [234], newly reported high-throughput profiling [235], next-generation sequencing (NGS) [236], and whole-genome sequencing (WGS) [237, 238]. This is because therapeutic gene targeting is strongly dependent on the creation of DSBs at specific target sites. As a result, the aforementioned investigations identified several variables that may influence the specificity of gene editing, which can be loosely split into two categories. First, the relative significance of each site that may vary between various sgRNA sequences may be determined by the intrinsic specificity encoded in the Cas9 protein. Second, the specificity is also influenced by how numerous efficient nuclease complexes are in comparison to the concentration of the target. CRISPR/Cas9 may have a higher potential for off-target effects in human cells when compared to ZFNs and TALENs [231]. Identical and highly similar DNA sequences can both be cleaved by Cas9-sgRNA when it binds to the target DNA, as previously mentioned. This might result in chromosomal rearrangements or off-target alterations [239, 240]. As a result, several strategies have been employed to reduce Cas9-mediated off-target effects and increase the cleavage specificity [231]. Both the structure and composition of gRNA can affect the level of off-target effects. The use of distinct target sequences with no resemblance to other sections of the genome has been reported to limit the off-target effects caused by Cas9. This technique is related to other methods for doing so [241]. Additionally, because this sgRNA structure is more sensitive to mismatches, the use of truncated and less-active sgRNAs that are shortened at the 5′ end by two to three nucleotides reduced undesirable mutagenesis at several off-target locations [230, 235]. Utilizing two nCas9 or RNA-guided FokI nucleases to create paired nicks rather than double-strand breaks is another way to lessen the impacts of off-target editing without reducing genome editing effectiveness [242, 243]. Reducing the amount of transfected DNA while improving specificity also leads to less cleavage at the target. As a result, it is important to strike a compromise between on-target cleavage efficiency and off-target consequences. Most recently, a SpCas9 with better genome-wide specificities was created by combining two distinct monomeric Streptococcus pyogenes variants. Functional mutations could also be combined to further boost specificity because they do not change the target range or size of the DNA needed to encode the desired Cas9 nuclease and a single gRNA. Alternative delivery strategies have also been devised to increase the editing process's specificity. When compared to plasmid-transfected delivery systems, direct distribution of recombinant Cas9 protein and in vitro generated sgRNA, either alone or in purified complexes, minimized off-target effects [88, 244]. Anti-CRISPR compounds, recently found CRISPR system inhibitors, may add to the fine control of genome editing strategy [245], which is being tried [246].

9.6.2 Nuclease Editing Efficiency Improvement

NHEJ has been found to be active during all stages of the cell cycle in a number of different cell types, including division and postmitosis [247, 248]. HDR, on the other hand, predominantly acts during the S/G2 phase and is thus mostly restricted to cells that are actively dividing, which restricts the range of treatments for the precise genomic modification of mitotic cells [249, 250]. Due to this distinction, disorders requiring gene insertion or genetic correction are more difficult to cure than those requiring gene inactivation. The main issue to date has been to increase HDR's effectiveness because NHEJ-mediated DSB repair can be used to encourage large amounts of gene disruption in most cell types. Notably, new research has shown innovative methods for enhancing genome editing effectiveness by blocking rival DNA repair processes, especially NHEJ-mediated DNA repair. A crucial enzyme (DNA ligase IV) in the NHEJ pathway was successfully targeted by Maruyama et al. using SCR7 to suppress NHEJ, boosting the efficacy of genome editing in cell lines and mice by up to 19-fold [251]. By silencing the genes for many essential NHEJ repair pathway components (KU70, KU80, or DNA ligase IV), Kuhn et al. enhanced the effectiveness of genome editing in a separate study by eliminating NHEJ activity in human and mouse cell lines [252]. With a smaller and less complex RNA-guided DNA nuclease than Cas9, an enhanced CRISPR system known as CRISPR/Cpf1 or CRISPR/Cas12a could target genomic areas that Cas9 cannot and produce multiplex gene disruption in vitro with frequencies of up to 45% [253]. Additionally, site-specific induction of DSBs and the insertion of new genetic information was accomplished with high HDR efficiency using timed delivery of Cas9-guide RNP (RNA ribonucleoprotein) complexes [254]. Gene site-directed mutation based on an HR mechanism is still inefficient, despite the CRISPR/Cas9 gene-editing technology improving the effectiveness of gene knockout and site-directed modification (including site-directed mutation and gene insertion). Base editor (BE) systems combining CRISPR/Cas9 and cytosine deaminase have been published one after another to increase the effectiveness of site-directed mutation [255–257]. The Cas9-cytidine deaminase and uracil glycosaminase inhibitor fusion protein can be directed at the desired site complementary to gRNA using this system without causing double-stranded DNA fragmentation. The amino group of pyrimidine (C) at the target site can be removed, converting it to uracil (U), which will then be replaced by thymidine (T) during DNA replication. In eukaryotic cells, the single-base C > T mutation is finally implemented accurately and effectively, resulting in single-base-pair replacements [258]. The BE method offers a crucial tool to the investigation and use of genome editing technology.

9.6.3 Optimizing the Delivery System

The creation of effective and secure techniques to deliver genetic editing elements to somatic cells in vivo as well as tumor cells ex vivo will be one of the major difficulties for the future deployment of gene-editing technologies. Viral and physical nonviral delivery mechanisms are both available. ZFNs, TALENs, and CRISPR have all been successfully delivered via nonviral physical delivery techniques like electroporation [259], hydrodynamic injection [260], and lipid nanoparticles [261] in a variety of cell lines and animal models. The therapeutic applications of those nonviral delivery techniques in vivo are constrained by their relatively ineffective delivery efficacy, notwithstanding their ease of use and safety [262]. Contrarily, viral vectors (such as retroviruses, lentivirus, adenovirus (AdV), and AAVs) have excellent delivery efficiency, and several of them have received clinical use approval [263, 264]. Despite the potential for unintended mutations and the existence of safety concerns, viral delivery systems have been shown to be the most efficient method for delivering plasmid-based nucleic acids to mammalian cells in vitro and in vivo [18, 265–267]. Other concerns impacting transport efficiency have also been brought to light recently, including the immunological vulnerability of host tumors and cells to Cas9 proteins [268] and the DSB P53 responses associated with genome editing [269]. To solve these issues, many new viral and nonviral systems have been created. By allaying issues about the security and immunogenicity of native cells in vivo, nonviral delivery techniques could broaden the scope of genome editing therapy. For instance, utilizing a PEG-PEI-cholesterol lipid polymer to transport plasmid DNA encoding a Cas9-sgRNA complex that targets the VEGF gene in osteosarcoma cells could result in a 50% gene knockout both in vitro and in vivo [270]. The use of lipid materials as vectors for genome-edited proteins was also examined by Zuris and colleagues. To increase negative charges on the surface of the proteins, they first fused Cas9 and TALEN into anionic GFP proteins. They then complexed these proteins with Lipofectamine 2000TM (a commercially available cationic lipofection reagent), which led to a 24% gene knockout of mouse embryonic stem cells in vitro and a 13% gene knockout of mouse cochlea hair cells in vivo. The compound had a 24% gene knockdown rate in mouse embryonic stem cells [261]. PEG-lipid-based lipid nanoparticles had high serum stability, according to Finn et al. After a single systemic injection, they induced a decline in serum protein levels of more than 97% and lasted for at least 12 months when used to deliver Cas9 mRNA and sgRNA targeting the mouse transthyretin gene in hepatocytes [271]. The transport of Cas9 and sgRNA plasmids with cationic alpha-helical polypeptides is predicted to increase gene editing effectiveness in vitro and in vivo, according to recent research by Cheng and Leong et al. Repeated intratumoral injections with this delivery method dramatically increased the survival of tumor-bearing mice by 67% targeted gene knockdown and >71% tumor growth suppression in a HeLa xenograft mouse model [272]. Moreover, compared to plasmid-based CRISPR/Cas9 and Cas9 mRNA/sgRNA, the Cas9 protein and sgRNA combination demonstrated greater efficiency. The delivery of purified Cas9 protein mediated by electroporation increased the editing efficiency to 79% [88, 273] because transgenic proteins degraded quickly and prevented long-lasting effects on the genome, as opposed to recombinant Cas9 proteins and sgRNA, which have been reported to achieve 16% editing efficiency in vitro through cell-penetrating peptide (CPP) [244, 274]. Hybrid virus vectors are made by combining different components of preexisting viruses to increase the precision and safety of viral-mediated gene-editing delivery. Point mutations and the incorporation of tiny molecules, synthetic polymers, and inorganic nanoparticles can modify the virus's structure [275]. The total transduction level of protease-activatable viruses (PAVs) was boosted by regulating the ratio of assembled wild-type viral capsid to protease-activatable subunits [276]. Asuri et al. developed a library of AAVs capsid genes with point mutations using error-prone polymerase chain reaction (EP-PCR), which led to a viral variation that was more effective at transferring genetic payloads to human stem cells [277]. Incorporating synthetic nonbiological elements like polymers and nanoparticles is another technique to further alter or improve the functional characteristics of viruses. PEG-5000 was added to adenoviral vectors by Hofherr et al. to create adeno-PEG-injected counterparts (Ad-PEG) [278].

PEGylation reduced thrombocytopenia and D-dimer formation after intravenous injection of adenovirus into mammalian blood. It also attenuated the interactions of the virus with platelets and endothelial cells. The right selection of various delivery methods and CRISPR/Cas9 variants also helps minimize off-target effects. Because bacterial expression sequences are removed when using minicircle DNA, it is more effective and immunogenic than plasmid DNA per-mass [279]. With great specificity, adenovirus-mediated codelivery of Cas9 and EGFR mutation-specific sgRNAs could effectively disrupt the oncogenic mutant allele [280]. A suitable carrier for the delivery of Cas9 RNP complexes and donor DNA has also been identified as nonviral polymers coupled to the gold nanoparticle hybridization system. This method may successfully reverse the disease phenotypes of muscle cells after intramuscular injection [281].

9.6.4 Ethical Concerns

The repercussions of modifying a human embryo are the biggest issue in relation to significant advancements in genome editing technologies. A ban on these experiments was suggested in 2015, but despite dissenting opinions from the scientific community, numerous research teams released the findings of their investigations on human embryos one after another. An objective lack of fundamental information, anticipating probable outcomes, a lack of legal regulation, and subjective anxieties about human selection, the rise of "invasive mutants," and the development of bioweapons are obstacles to scientific arguments about the benefits of such research [282]. A professional opinion on the subject of modifying the human embryonic genome was released in 2017 by the American Society for Human Genetics (ASHG) [283]. It declared that at this moment, germline gene editing that results in human pregnancy is inappropriate due to the type and quantity of unresolved scientific, moral, and policy problems. Additionally, their experts' views on in vitro germline genome editing are that, with the proper supervision and donor consent, there is no reason to prevent this editing on human embryos and gametes in order to facilitate research on potential future clinical uses of gene editing. Additionally, there should be no restrictions on making public funds available to support this research. Furthermore, the statement states that future clinical applications of human germline genome editing should not move forward unless, at the very least, there is (a) an overwhelming medical case, (b) an evidence supporting its clinical use, (c) a strong ethical case, and (d) an open, transparent process to gather and consider stakeholder input. It was reiterated that ASHG maintains the position statement wherein in vitro human germline genome editing is allowed while genome editing that involves human pregnancy is considered to be a misdemeanor (press release at ASHG website) when the report on the birth of twin sisters in China (2018) with edited genomes reached the authorities. This incident demonstrated that genome editing in humans is not yet suitable for clinical use and still has many unresolved legal and ethical questions.

References

1. Cring MR, Sheffield VC. Gene therapy and gene correction: targets, progress, and challenges for treating human diseases. Gene Ther. 2022;29(1):3–12.
2. Gonçalves GAR, Paiva RMA. Gene therapy: advances, challenges and perspectives. Einstein (São Paulo). 2017;15(3):369–75.
3. Mali S. Delivery systems for gene therapy. Indian J Hum Genet. 2013;19(1):3–8.
4. Dwivedi S, et al. Chapter 9—Gene therapy and gene editing in healthcare. In: Barh D, editor. Biotechnology in healthcare. New York: Academic Press; 2022. p. 147–75.
5. Sayed N, et al. Gene therapy: comprehensive overview and therapeutic applications. Life Sci. 2022;294:120375.
6. Rittiner J, et al. Therapeutic modulation of gene expression in the disease state: treatment strategies and approaches for the development of next-generation of the epigenetic drugs. Front Bioeng Biotechnol. 2022;10:1035543.
7. Savulescu J. Harm, ethics committees and the gene therapy death. J Med Ethics. 2001;27(3):148–50.
8. Hutchinson JA. Somatic cell-based therapy. Transplantation. 2015;99(6):1103–5.

9. Scheller EL, Krebsbach PH. Gene therapy: design and prospects for craniofacial regeneration. J Dent Res. 2009;88(7):585–96.
10. Petraitytė G, Preikšaitienė E, Mikštienė V. Genome editing in medicine: tools and challenges. Acta Med Litu. 2021;28(2):205–19.
11. Kim DY, et al. Efficient CRISPR editing with a hypercompact Cas12f1 and engineered guide RNAs delivered by adeno-associated virus. Nat Biotechnol. 2022;40(1):94–102.
12. Zu H, Gao D. Non-viral vectors in gene therapy: recent development, challenges, and prospects. AAPS J. 2021;23(4):78.
13. Shirley JL, et al. Immune responses to viral gene therapy vectors. Mol Ther. 2020;28(3):709–22.
14. Bulcha JT, et al. Viral vector platforms within the gene therapy landscape. Signal Transduct Target Ther. 2021;6(1):53.
15. Vargas JE, et al. Retroviral vectors and transposons for stable gene therapy: advances, current challenges and perspectives. J Transl Med. 2016;14(1):288.
16. Zhao Z, Anselmo AC, Mitragotri S. Viral vector-based gene therapies in the clinic. Bioeng Transl Med. 2022;7(1):e10258.
17. Milone MC, O'Doherty U. Clinical use of lentiviral vectors. Leukemia. 2018;32(7):1529–41.
18. Maggio I, et al. Adenoviral vector delivery of RNA-guided CRISPR/Cas9 nuclease complexes induces targeted mutagenesis in a diverse array of human cells. Sci Rep. 2014;4(1):5105.
19. Wang D, Tai PWL, Gao G. Adeno-associated virus vector as a platform for gene therapy delivery. Nat Rev Drug Discov. 2019;18(5):358–78.
20. Hudry E, Vandenberghe LH. Therapeutic AAV gene transfer to the nervous system: a clinical reality. Neuron. 2019;102(1):263.
21. Manservigi R, Argnani R, Marconi P. HSV recombinant vectors for gene therapy. Open Virol J. 2010;4:123–56.
22. Miyagawa Y, et al. Herpes simplex viral-vector design for efficient transduction of nonneuronal cells without cytotoxicity. Proc Natl Acad Sci U S A. 2015;112(13):E1632–41.
23. Wang C, et al. Emerging non-viral vectors for gene delivery. J Nanobiotechnol. 2023;21(1):272.
24. Cullis PR, Hope MJ. Lipid nanoparticle systems for enabling gene therapies. Mol Ther. 2017;25(7):1467–75.
25. Tenchov R, et al. Lipid nanoparticles—from liposomes to mRNA vaccine delivery, a landscape of research diversity and advancement. ACS Nano. 2021;15(11):16982–7015.
26. Balazs DA, Godbey W. Liposomes for use in gene delivery. J Drug Deliv. 2011;2011:326497.
27. Cheng Q, et al. Dendrimer-based lipid nanoparticles deliver therapeutic FAH mRNA to normalize liver function and extend survival in a mouse model of hepatorenal tyrosinemia type I. Adv Mater. 2018;30(52):1805308.
28. Chen CK, et al. Biodegradable polymers for gene-delivery applications. Int J Nanomedicine. 2020;15:2131–50.
29. Chen Z, et al. Recent advancements in polyethyleneimine-based materials and their biomedical, biotechnology, and biomaterial applications. J Mater Chem B. 2020;8(15):2951–73.
30. Yang J, et al. Surface-engineered dendrimers in gene delivery. Chem Rev. 2015;115(11):5274–300.
31. Filipe LC, et al. Exploring the structural properties of positively charged peptide dendrimers. J Phys Chem B. 2016;120(43):11323–30.
32. Dong S, et al. Engineered design of a mesoporous silica nanoparticle-based nanocarrier for efficient mRNA delivery in vivo. Nano Lett. 2023;23(6):2137–47.
33. Loh XJ, et al. Utilising inorganic nanocarriers for gene delivery. Biomater Sci. 2016;4(1):70–86.
34. Xia Q, et al. Size- and cell type-dependent cellular uptake, cytotoxicity and in vivo distribution of gold nanoparticles. Int J Nanomedicine. 2019;14:6957–70.
35. Chen L, et al. Recent progress in targeted delivery vectors based on biomimetic nanoparticles. Signal Transduct Target Ther. 2021;6(1):225.
36. Wallen M, et al. Exosomes as an emerging plasmid delivery vehicle for gene therapy. Pharmaceutics. 2023;15(7):1832.
37. Chen H, et al. Engineered exosomes as drug and RNA co-delivery system: new hope for enhanced therapeutics? Front Bioeng Biotechnol. 2023;11:1254356.
38. Banskota S, et al. Engineered virus-like particles for efficient in vivo delivery of therapeutic proteins. Cell. 2022;185(2):250–265.e16.
39. Schmidt-Wolf GD, Schmidt-Wolf IGH. Non-viral and hybrid vectors in human gene therapy: an update. Trends Mol Med. 2003;9(2):67–72.
40. Mukherjee A, et al. Lipid-polymer hybrid nanoparticles as a next-generation drug delivery platform: state of the art, emerging technologies, and perspectives. Int J Nanomedicine. 2019;14:1937–52.
41. Yu J, Li T, Zhu J. Gene therapy strategies targeting aging-related diseases. Aging Dis. 2023;14(2):398–417.
42. Petrich J, et al. Gene replacement therapy: a primer for the health-system pharmacist. J Pharm Pract. 2020;33(6):846–55.
43. Moshiri A. Animals models of inherited retinal disease. Int Ophthalmol Clin. 2021;61(3):113–30.
44. Ghosh S, Brown AM, Jenkins C, Campbell K. Viral vector systems for gene therapy: a comprehensive literature review of progress and biosafety challenges. Appl Biosaf. 2020;25(1):7–18.
45. Thomas CE, Ehrhardt A, Kay MA. Progress and problems with the use of viral vectors for gene therapy. Nat Rev Genet. 2003;4(5):346–58.
46. Gray SJ, et al. Optimizing promoters for recombinant adeno-associated virus-mediated gene expression in the peripheral and central nervous system

using self-complementary vectors. Hum Gene Ther. 2011;22(9):1143–53.
47. Qin S, et al. mRNA-based therapeutics: powerful and versatile tools to combat diseases. Signal Transduct Target Ther. 2022;7(1):166.
48. Kowalski PS, et al. Delivering the messenger: advances in technologies for therapeutic mRNA delivery. Mol Ther. 2019;27(4):710–28.
49. Bafleh WS, et al. Applications of advances in mRNA-based platforms as therapeutics and diagnostics in reproductive technologies. Front Cell Dev Biol. 2023;11:1198848.
50. Castro-Muñoz LJ, et al. Modulating epigenetic modifications for cancer therapy (Review). Oncol Rep. 2023;49(3):59.
51. Kelly TK, De Carvalho DD, Jones PA. Epigenetic modifications as therapeutic targets. Nat Biotechnol. 2010;28(10):1069–78.
52. Lu Y, et al. Epigenetic regulation in human cancer: the potential role of epi-drug in cancer therapy. Mol Cancer. 2020;19(1):79.
53. Li K, et al. Interrogation of enhancer function by enhancer-targeting CRISPR epigenetic editing. Nat Commun. 2020;11(1):485.
54. Simó-Riudalbas L, Esteller M. Targeting the histone orthography of cancer: drugs for writers, erasers and readers. Br J Pharmacol. 2015;172(11):2716–32.
55. Ryan FJ, et al. Colonic microbiota is associated with inflammation and host epigenomic alterations in inflammatory bowel disease. Nat Commun. 2020;11(1):1512.
56. Jones PA, Issa JP, Baylin S. Targeting the cancer epigenome for therapy. Nat Rev Genet. 2016;17(10):630–41.
57. Majchrzak-Celińska A, Warych A, Szoszkiewicz M. Novel approaches to epigenetic therapies: from drug combinations to epigenetic editing. Genes (Basel). 2021;12(2):208.
58. Khalil AM. The genome editing revolution: review. J Genet Eng Biotechnol. 2020;18(1):68.
59. Blaese RM, et al. T lymphocyte-directed gene therapy for ADA-SCID: initial trial results after 4 years. Science. 1995;270(5235):475–80.
60. Rouet P, Smih F, Jasin M. Introduction of double-strand breaks into the genome of mouse cells by expression of a rare-cutting endonuclease. Mol Cell Biol. 1994;14(12):8096–106.
61. Shukla V, et al. The landscape of cellular and gene therapy products: authorization, discontinuations, and cost. Hum Gene Ther Clin Dev. 2019;30(3):102–13.
62. Musunuru K, et al. In vivo CRISPR base editing of PCSK9 durably lowers cholesterol in primates. Nature. 2021;593(7859):429–34.
63. Wright WD, Shah SS, Heyer W-D. Homologous recombination and the repair of DNA double-strand breaks. J Biol Chem. 2018;293(27):10524–35.
64. Cox DB, Platt RJ, Zhang F. Therapeutic genome editing: prospects and challenges. Nat Med. 2015;21(2):121–31.
65. Silva G, et al. Meganucleases and other tools for targeted genome engineering: perspectives and challenges for gene therapy. Curr Gene Ther. 2011;11(1):11–27.
66. Prieto J, et al. Understanding the indirect DNA readout specificity of I-CreI Meganuclease. Sci Rep. 2018;8(1):10286.
67. Gaj T, et al. Genome-editing technologies: principles and applications. Cold Spring Harb Perspect Biol. 2016;8(12):a023754.
68. Kleinstiver BP, et al. Monomeric site-specific nucleases for genome editing. Proc Natl Acad Sci U S A. 2012;109(21):8061–6.
69. Maeder ML, Gersbach CA. Genome-editing technologies for gene and cell therapy. Mol Ther. 2016;24(3):430–46.
70. Khan SH. Genome-editing technologies: concept, pros, and cons of various genome-editing techniques and bioethical concerns for clinical application. Mol Ther Nucleic Acids. 2019;16:326–34.
71. Deng D, et al. Structural basis for sequence-specific recognition of DNA by TAL effectors. Science. 2012;335(6069):720–3.
72. Boch J, et al. Breaking the code of DNA binding specificity of TAL-type III effectors. Science. 2009;326(5959):1509–12.
73. Mussolino C, et al. TALENs facilitate targeted genome editing in human cells with high specificity and low cytotoxicity. Nucleic Acids Res. 2014;42(10):6762–73.
74. Huo Z, et al. Generation of a heterozygous p53 R249S mutant human embryonic stem cell line by TALEN-mediated genome editing. Stem Cell Res. 2019;34:101360.
75. Packer MS, Rees HA, Liu DR. Phage-assisted continuous evolution of proteases with altered substrate specificity. Nat Commun. 2017;8(1):956.
76. Gaj T, et al. Protein delivery using Cys2-His2 zinc-finger domains. ACS Chem Biol. 2014;9(8):1662–7.
77. Kim YG, Cha J, Chandrasegaran S. Hybrid restriction enzymes: zinc finger fusions to Fok I cleavage domain. Proc Natl Acad Sci U S A. 1996;93(3):1156–60.
78. Paschon DE, et al. Diversifying the structure of zinc finger nucleases for high-precision genome editing. Nat Commun. 2019;10(1):1133.
79. Segal DJ, et al. Evaluation of a modular strategy for the construction of novel polydactyl zinc finger DNA-binding proteins. Biochemistry. 2003;42(7):2137–48.
80. Gaj T, et al. Targeted gene knockout by direct delivery of zinc-finger nuclease proteins. Nat Methods. 2012;9(8):805–7.
81. Gaj T, et al. A comprehensive approach to zinc-finger recombinase customization enables genomic targeting in human cells. Nucleic Acids Res. 2013;41(6):3937–46.
82. Jiang W, et al. RNA-guided editing of bacterial genomes using CRISPR-Cas systems. Nat Biotechnol. 2013;31(3):233–9.

83. Hsu PD, Lander ES, Zhang F. Development and applications of CRISPR-Cas9 for genome engineering. Cell. 2014;157(6):1262–78.
84. Anders C, et al. Structural basis of PAM-dependent target DNA recognition by the Cas9 endonuclease. Nature. 2014;513(7519):569–73.
85. Makarova KS, et al. An updated evolutionary classification of CRISPR-Cas systems. Nat Rev Microbiol. 2015;13(11):722–36.
86. Li H, et al. Applications of genome editing technology in the targeted therapy of human diseases: mechanisms, advances and prospects. Signal Transduct Target Ther. 2020;5(1):1.
87. Fu Y, et al. Improving CRISPR-Cas nuclease specificity using truncated guide RNAs. Nat Biotechnol. 2014;32(3):279–84.
88. Kim S, et al. Highly efficient RNA-guided genome editing in human cells via delivery of purified Cas9 ribonucleoproteins. Genome Res. 2014;24(6):1012–9.
89. Cao J, et al. An easy and efficient inducible CRISPR/Cas9 platform with improved specificity for multiple gene targeting. Nucleic Acids Res. 2016;44(19):e149.
90. Gapinske M, et al. CRISPR-SKIP: programmable gene splicing with single base editors. Genome Biol. 2018;19(1):107.
91. Gschweng E, De Oliveira S, Kohn DB. Hematopoietic stem cells for cancer immunotherapy. Immunol Rev. 2014;257(1):237–49.
92. Vogelstein B, et al. Cancer genome landscapes. Science. 2013;339(6127):1546–58.
93. Sanchez-Garcia I, Grutz G. Tumorigenic activity of the BCR-ABL oncogenes is mediated by BCL2. Proc Natl Acad Sci U S A. 1995;92(12):5287–91.
94. Do TU, et al. Zinc finger nuclease induced DNA double stranded breaks and rearrangements in MLL. Mutat Res. 2012;740(1–2):34–42.
95. Tanaka A, et al. A novel therapeutic molecule against HTLV-1 infection targeting provirus. Leukemia. 2013;27(8):1621–7.
96. Herrmann F, et al. p53 gene repair with zinc finger nucleases optimised by yeast 1-hybrid and validated by Solexa sequencing. PLoS One. 2011;6(6):e20913.
97. Kanu GA, et al. Gold nanoparticle-mediated gene therapy. Cancers (Basel). 2022;14(21):5366.
98. Wang J, et al. TALENs-mediated gene disruption of FLT3 in leukemia cells: using genome-editing approach for exploring the molecular basis of gene abnormality. Sci Rep. 2015;5(1):18454.
99. Tufail M. Genome editing: an essential technology for cancer treatment. Med Omics. 2022;4:100015.
100. Wang T, et al. Gene essentiality profiling reveals gene networks and synthetic lethal interactions with oncogenic Ras. Cell. 2017;168(5):890–903.e15.
101. Birsoy K, et al. An essential role of the mitochondrial electron transport chain in cell proliferation is to enable aspartate synthesis. Cell. 2015;162(3):540–51.
102. Dwivedi S, Sharma P. Stem cell biology: a new hope in regenerations and replenishments therapy. Indian J Clin Biochem. 2018;33(4):369–71.
103. Hazafa A, et al. CRISPR/Cas9: a powerful genome editing technique for the treatment of cancer cells with present challenges and future directions. Life Sci. 2020;263:118525.
104. Mintz RL, et al. CRISPR/Cas9-mediated mutagenesis to validate the synergy between PARP1 inhibition and chemotherapy in BRCA1-mutated breast cancer cells. Bioeng Transl Med. 2020;5(1):e10152.
105. Tian X, et al. CRISPR/Cas9—an evolving biological tool kit for cancer biology and oncology. NPJ Precis Oncol. 2019;3:8.
106. Pont M, et al. Applications of CRISPR technology to breast cancer and triple negative breast cancer research. Cancers (Basel). 2023;15(17):4364.
107. Karn V, et al. CRISPR/Cas9 system in breast cancer therapy: advancement, limitations and future scope. Cancer Cell Int. 2022;22(1):234.
108. Böiers C, et al. A human IPS model implicates embryonic B-myeloid fate restriction as developmental susceptibility to B acute lymphoblastic leukemia-associated ETV6-RUNX1. Dev Cell. 2018;44(3):362–377.e7.
109. Hnisz D, et al. Activation of proto-oncogenes by disruption of chromosome neighborhoods. Science. 2016;351(6280):1454–8.
110. Rahman S, et al. Activation of the LMO2 oncogene through a somatically acquired neomorphic promoter in T-cell acute lymphoblastic leukemia. Blood. 2017;129(24):3221–6.
111. Dimitri A, Herbst F, Fraietta JA. Engineering the next-generation of CAR T-cells with CRISPR-Cas9 gene editing. Mol Cancer. 2022;21(1):78.
112. Poirot L, et al. Multiplex genome-edited T-cell manufacturing platform for "off-the-shelf" adoptive T-cell immunotherapies. Cancer Res. 2015;75(18):3853–64.
113. Barroso I, McCarthy MI. The genetic basis of metabolic disease. Cell. 2019;177(1):146–61.
114. Dwivedi S, et al. Diseases and molecular diagnostics: a step closer to precision medicine. Indian J Clin Biochem. 2017;32:374–98.
115. Sharma P, Dwivedi S. Nutrigenomics and nutrigenetics: new insight in disease prevention and cure. Indian J Clin Biochem. 2017;32(4):371–3.
116. Nakagawa Y, et al. Hyperlipidemia and hepatitis in liver-specific CREB3L3 knockout mice generated using a one-step CRISPR/Cas9 system. Sci Rep. 2016;6(1):27857.
117. Claussnitzer M, et al. FTO obesity variant circuitry and adipocyte browning in humans. N Engl J Med. 2015;373(10):895–907.
118. Wang X, et al. Characterization of novel cytochrome P450 2E1 knockout rat model generated by CRISPR/Cas9. Biochem Pharmacol. 2016;105:80–90.
119. Naylor J, et al. Use of CRISPR/Cas9-engineered INS-1 pancreatic β cells to define the pharmacol-

ogy of dual GIPR/GLP-1R agonists. Biochem J. 2016;473(18):2881–91.
120. Chen S, Yu X, Guo D. CRISPR-Cas targeting of host genes as an antiviral strategy. Viruses. 2018;10(1):40.
121. Ebina H, et al. Harnessing the CRISPR/Cas9 system to disrupt latent HIV-1 provirus. Sci Rep. 2013;3(1):2510.
122. Hu W, et al. RNA-directed gene editing specifically eradicates latent and prevents new HIV-1 infection. Proc Natl Acad Sci U S A. 2014;111(31):11461–6.
123. Moody CA, Laimins LA. Human papillomavirus oncoproteins: pathways to transformation. Nat Rev Cancer. 2010;10(8):550–60.
124. Ding W, et al. Zinc finger nucleases targeting the human papillomavirus E7 oncogene induce E7 disruption and a transformed phenotype in HPV16/18-positive cervical cancer cells. Clin Cancer Res. 2014;20(24):6495–503.
125. Hu Z, et al. TALEN-mediated targeting of HPV oncogenes ameliorates HPV-related cervical malignancy. J Clin Invest. 2015;125(1):425–36.
126. Zhen S, et al. In vitro and in vivo synergistic therapeutic effect of cisplatin with human papillomavirus16 E6/E7 CRISPR/Cas9 on cervical cancer cell line. Transl Oncol. 2016;9(6):498–504.
127. Cradick TJ, et al. Zinc-finger nucleases as a novel therapeutic strategy for targeting hepatitis B virus DNAs. Mol Ther. 2010;18(5):947–54.
128. Ramanan V, et al. CRISPR/Cas9 cleavage of viral DNA efficiently suppresses hepatitis B virus. Sci Rep. 2015;5(1):10833.
129. Wang J, Quake SR. RNA-guided endonuclease provides a therapeutic strategy to cure latent herpesviridae infection. Proc Natl Acad Sci U S A. 2014;111(36):13157–62.
130. Cai S, et al. The research advances and applications of genome editing in hereditary eye diseases. Zhonghua Yan Ke Za Zhi. 2017;53(5):386–91.
131. Björk S, et al. REEPs are membrane shaping adapter proteins that modulate specific g protein-coupled receptor trafficking by affecting ER cargo capacity. PLoS One. 2013;8(10):e76366.
132. Arno G, et al. Mutations in REEP6 cause autosomal-recessive retinitis pigmentosa. Am J Hum Genet. 2016;99(6):1305–15.
133. Bowes C, et al. Retinal degeneration in the rd mouse is caused by a defect in the β subunit of rod cGMP-phosphodiesterase. Nature. 1990;347(6294):677–80.
134. Keeler CE. The geotropic reaction of rodless mice in light and in darkness. J Gen Physiol. 1928;11(4):361–8.
135. Wu W-H, et al. CRISPR repair reveals causative mutation in a preclinical model of retinitis pigmentosa. Mol Ther. 2016;24(8):1388–94.
136. Bassuk AG, et al. Precision medicine: genetic repair of retinitis pigmentosa in patient-derived stem cells. Sci Rep. 2016;6(1):19969.
137. Lee W-H, Murphree AL, Benedict WF. Expression and amplification of the N-myc gene in primary retinoblastoma. Nature. 1984;309(5967):458–60.
138. Naert T, et al. CRISPR/Cas9 mediated knockout of rb1 and rbl1 leads to rapid and penetrant retinoblastoma development in Xenopus tropicalis. Sci Rep. 2016;6(1):35264.
139. Tu J, et al. Generation of human embryonic stem cell line with heterozygous RB1 deletion by CRIPSR/Cas9 nickase. Stem Cell Res. 2018;28:29–32.
140. Hollands H, et al. Do findings on routine examination identify patients at risk for primary open-angle glaucoma? The rational clinical examination systematic review. JAMA. 2013;309(19):2035–42.
141. Alward WL, et al. Clinical features associated with mutations in the chromosome 1 open-angle glaucoma gene (GLC1A). N Engl J Med. 1998;338(15):1022–7.
142. Stone EM, et al. Identification of a gene that causes primary open angle glaucoma. Science. 1997;275(5300):668–70.
143. Kim BS, et al. Targeted disruption of the myocilin gene (Myoc) suggests that human glaucoma-causing mutations are gain of function. Mol Cell Biol. 2001;21(22):7707–13.
144. Joe MK, et al. Accumulation of mutant myocilins in ER leads to ER stress and potential cytotoxicity in human trabecular meshwork cells. Biochem Biophys Res Commun. 2003;312(3):592–600.
145. Liu Y, Vollrath D. Reversal of mutant myocilin non-secretion and cell killing: implications for glaucoma. Hum Mol Genet. 2004;13(11):1193–204.
146. Yam GH-F, et al. Aggregated myocilin induces russell bodies and causes apoptosis: implications for the pathogenesis of myocilin-caused primary open-angle glaucoma. Am J Pathol. 2007;170(1):100–9.
147. Jain A, et al. CRISPR-Cas9-based treatment of myocilin-associated glaucoma. Proc Natl Acad Sci U S A. 2017;114(42):11199–204.
148. Liao H, et al. Development of allele-specific therapeutic siRNA in Meesmann epithelial corneal dystrophy. PLoS One. 2011;6(12):e28582.
149. Courtney D, et al. CRISPR/Cas9 DNA cleavage at SNP-derived PAM enables both in vitro and in vivo KRT12 mutation-specific targeting. Gene Ther. 2016;23(1):108–12.
150. Shima M, et al. Factor VIII–mimetic function of humanized bispecific antibody in hemophilia A. N Engl J Med. 2016;374(21):2044–53.
151. Park C-Y, et al. Functional correction of large factor VIII gene chromosomal inversions in hemophilia A patient-derived iPSCs using CRISPR-Cas9. Cell Stem Cell. 2015;17(2):213–20.
152. Guan Y, et al. CRISPR/Cas9-mediated somatic correction of a novel coagulator factor IX gene mutation ameliorates hemophilia in mouse. EMBO Mol Med. 2016;8(5):477–88.
153. Rees DC, Williams TN, Gladwin MT. Sickle-cell disease. Lancet. 2010;376(9757):2018–31.
154. Dever DP, et al. CRISPR/Cas9 β-globin gene targeting in human haematopoietic stem cells. Nature. 2016;539(7629):384–9.

155. Park SH, et al. Highly efficient editing of the β-globin gene in patient-derived hematopoietic stem and progenitor cells to treat sickle cell disease. Nucleic Acids Res. 2019;47(15):7955–72.
156. Anzalone AV, et al. Search-and-replace genome editing without double-strand breaks or donor DNA. Nature. 2019;576(7785):149–57.
157. Khosravi MA, et al. Targeted deletion of BCL11A gene by CRISPR-Cas9 system for fetal hemoglobin reactivation: a promising approach for gene therapy of beta thalassemia disease. Eur J Pharmacol. 2019;854:398–405.
158. Ye L, et al. Genome editing using CRISPR-Cas9 to create the HPFH genotype in HSPCs: an approach for treating sickle cell disease and β-thalassemia. Proc Natl Acad Sci U S A. 2016;113(38):10661–5.
159. Long C, et al. Postnatal genome editing partially restores dystrophin expression in a mouse model of muscular dystrophy. Science. 2016;351(6271):400–3.
160. Ryu S-M, et al. Adenine base editing in mouse embryos and an adult mouse model of Duchenne muscular dystrophy. Nat Biotechnol. 2018;36(6):536–9.
161. Kuo CY, et al. Site-specific gene editing of human hematopoietic stem cells for X-linked hyper-IgM syndrome. Cell Rep. 2018;23(9):2606–16.
162. De Ravin SS, et al. CRISPR-Cas9 gene repair of hematopoietic stem cells from patients with X-linked chronic granulomatous disease. Sci Transl Med. 2017;9(372):eaah3480.
163. Pavel-Dinu M, et al. Gene correction for SCID-X1 in long-term hematopoietic stem cells. Nat Commun. 2019;10(1):1634.
164. Goodwin M, et al. CRISPR-based gene editing enables FOXP3 gene repair in IPEX patient cells. Sci Adv. 2020;6(19):eaaz0571.
165. Chen Y, et al. Generation of cynomolgus monkey chimeric fetuses using embryonic stem cells. Cell Stem Cell. 2015;17(1):116–24.
166. Chen Y, et al. Functional disruption of the dystrophin gene in rhesus monkey using CRISPR/Cas9. Hum Mol Genet. 2015;24(13):3764–74.
167. Xu L, et al. CRISPR-mediated genome editing restores dystrophin expression and function in mdx mice. Mol Ther. 2016;24(3):564–9.
168. Bengtsson NE, et al. Muscle-specific CRISPR/Cas9 dystrophin gene editing ameliorates pathophysiology in a mouse model for Duchenne muscular dystrophy. Nat Commun. 2017;8(1):14454.
169. Nelson CE, et al. In vivo genome editing improves muscle function in a mouse model of Duchenne muscular dystrophy. Science. 2016;351(6271):403–7.
170. Pai S-Y, et al. Transplantation outcomes for severe combined immunodeficiency, 2000–2009. N Engl J Med. 2014;371(5):434–46.
171. Lombardo A, et al. Gene editing in human stem cells using zinc finger nucleases and integrase-defective lentiviral vector delivery. Nat Biotechnol. 2007;25(11):1298–306.
172. Flisikowska T, et al. Efficient immunoglobulin gene disruption and targeted replacement in rabbit using zinc finger nucleases. PLoS One. 2011;6(6):e21045.
173. Wang Y, et al. Generation of knockout rabbits using transcription activator-like effector nucleases. Cell Regen. 2014;3:1–9.
174. Yan Q, et al. Generation of multi-gene knockout rabbits using the Cas9/gRNA system. Cell Regen. 2014;3(1):12.
175. Zhou J, et al. One-step generation of different immunodeficient mice with multiple gene modifications by CRISPR/Cas9 mediated genome engineering. Int J Biochem Cell Biol. 2014;46:49–55.
176. Ott de Bruin LM, Volpi S, Musunuru K. Novel genome-editing tools to model and correct primary immunodeficiencies. Front Immunol. 2015;6:250.
177. Chen Y, et al. Modeling Rett syndrome using TALEN-edited MECP2 mutant cynomolgus monkeys. Cell. 2017;169(5):945–955.e10.
178. Chen J, et al. Effects of genetic correction on the differentiation of hair cell-like cells from iPSCs with MYO15A mutation. Cell Death Differ. 2016;23(8):1347–57.
179. Jiang W, et al. Production of Wilson disease model rabbits with homology-directed precision point mutations in the ATP7B gene using the CRISPR/Cas9 system. Sci Rep. 2018;8(1):1332.
180. Kurome M, et al. 361 growth hormone receptor mutant pigs produced by using the clustered regularly interspaced short palindromic repeats (CRISPR) and CRISPR-associated systems in in vitro-produced zygotes. Reprod Fertil Dev. 2015;27(1):269.
181. Tseng W-C, et al. Modeling Niemann-Pick disease type C1 in zebrafish: a robust platform for in vivo screening of candidate therapeutic compounds. Dis Model Mech. 2018;11(9):dmm034165.
182. Abdul-Wahab A, Qasim W, McGrath JAJ. Gene therapies for inherited skin disorders. Semin Cutan Med Surg. 2014;33(2):83–90.
183. Gao Z, et al. Extinction of all infectious HIV in cell culture by the CRISPR-Cas12a system with only a single crRNA. Nucleic Acids Res. 2020;48(10):5527–39.
184. Zhu W, et al. The CRISPR/Cas9 system inactivates latent HIV-1 proviral. DNA. 2015;12:1–7.
185. Cyranoski DJN. CRISPR gene-editing tested in a person for the first time. Nature. 2016;539(7630):479.
186. Lazaro C, et al. Neurofibromatosis type 1 due to germ-line mosaicism in a clinically normal father. N Engl J Med. 1994;331(21):1403–7.
187. Wilen CB, et al. Engineering HIV-resistant human CD4+ T cells with CXCR4-specific zinc-finger nucleases. PLoS Pathog. 2011;7(4):e1002020.
188. Van Lunzen J, et al. Transfer of autologous gene-modified T cells in HIV-infected patients with advanced immunodeficiency and drug-resistant virus. Mol Ther. 2007;15(5):1024–33.
189. Voit RA, et al. Generation of an HIV resistant T-cell line by targeted "stacking" of restriction factors. Mol Ther. 2013;21(4):786–95.

190. Xu L, et al. CRISPR-edited stem cells in a patient with HIV and acute lymphocytic leukemia. N Engl J Med. 2019;381(13):1240–7.
191. Choi K-D, et al. Hematopoietic and endothelial differentiation of human induced pluripotent stem cells. Stem Cells. 2009;27(3):559–67.
192. Dolan G, et al. Haemophilia B: where are we now and what does the future hold? Blood Rev. 2018;32(1):52–60.
193. Bauer DE, Orkin SH. Hemoglobin switching's surprise: the versatile transcription factor BCL11A is a master repressor of fetal hemoglobin. Curr Opin Genet Dev. 2015;33:62–70.
194. Moore CT, et al. Personalised genome editing—the future for corneal dystrophies. Prog Retin Eye Res. 2018;65:147–65.
195. Cho GY, et al. CRISPR-mediated ophthalmic genome surgery. Curr Ophthalmol Rep. 2017;5:199–206.
196. Kirkwood JM, et al. Immunotherapy of cancer in 2012. CA Cancer J Clin. 2012;62(5):309–35.
197. Rein LA, et al. Applications of gene editing technologies to cellular therapies. Biol Blood Marrow Transplant. 2018;24(8):1537–45.
198. June CH, Sadelain M. Chimeric antigen receptor therapy. N Engl J Med. 2018;379(1):64–73.
199. Sadelain M, Brentjens R, Rivière I. The promise and potential pitfalls of chimeric antigen receptors. Curr Opin Immunol. 2009;21(2):215–23.
200. Davenport AJ, et al. CAR T cells inflict sequential killing of multiple tumor target cells. Cancer Immunol Res. 2015;3(5):483–94.
201. Zhao J, et al. Universal CARs, universal T cells, and universal CAR T cells. J Hematol Oncol. 2018;11:132.
202. Torikai H, et al. A foundation for universal T-cell based immunotherapy: T cells engineered to express a CD19-specific chimeric-antigen-receptor and eliminate expression of endogenous TCR. Blood. 2012;119(24):5697–705.
203. Torikai H, et al. Toward eliminating HLA class I expression to generate universal cells from allogeneic donors. Blood. 2013;122(8):1341–9.
204. Liu X, et al. CRISPR-Cas9-mediated multiplex gene editing in CAR-T cells. Cell Res. 2017;27(1):154–7.
205. Ren J, et al. A versatile system for rapid multiplex genome-edited CAR T cell generation. Oncotarget. 2017;8(10):17002.
206. Wong AS, et al. Multiplexed barcoded CRISPR-Cas9 screening enabled by CombiGEM. Proc Natl Acad Sci U S A. 2016;113(9):2544–9.
207. Jason S, Yusa K. Genome-wide CRISPR-Cas9 screening in mammalian cells. Methods. 2019;164:29–35.
208. Luo J. CRISPR/Cas9: from genome engineering to cancer drug discovery. Trends Cancer. 2016;2(6):313–24.
209. Bester AC, et al. An integrated genome-wide CRISPRa approach to functionalize lncRNAs in drug resistance. Cell. 2018;173(3):649–664.e20.
210. Tzelepis K, et al. A CRISPR dropout screen identifies genetic vulnerabilities and therapeutic targets in acute myeloid leukemia. Cell Rep. 2016;17(4):1193–205.
211. Munoz DM, et al. CRISPR screens provide a comprehensive assessment of cancer vulnerabilities but generate false-positive hits for highly amplified genomic regions. Cancer Discov. 2016;6(8):900–13.
212. Hart T, et al. High-resolution CRISPR screens reveal fitness genes and genotype-specific cancer liabilities. Cell. 2015;163(6):1515–26.
213. Liu SJ, et al. CRISPRi-based genome-scale identification of functional long noncoding RNA loci in human cells. Science. 2017;355(6320):eaah7111.
214. Zhu S, et al. Genome-scale deletion screening of human long non-coding RNAs using a paired-guide RNA CRISPR–Cas9 library. Nat Biotechnol. 2016;34(12):1279–86.
215. Esposito R, et al. Hacking the cancer genome: profiling therapeutically actionable long non-coding RNAs using CRISPR-Cas9 screening. Cancer Cell. 2019;35(4):545–57.
216. Boettcher M, et al. Dual gene activation and knockout screen reveals directional dependencies in genetic networks. Nat Biotechnol. 2018;36(2):170–8.
217. Yuan Z, et al. Olaparib hydroxamic acid derivatives as dual PARP and HDAC inhibitors for cancer therapy. Bioorg Med Chem. 2017;25(15):4100–9.
218. Han K, et al. Synergistic drug combinations for cancer identified in a CRISPR screen for pairwise genetic interactions. Nat Biotechnol. 2017;35(5):463–74.
219. Shen JP, et al. Combinatorial CRISPR–Cas9 screens for de novo mapping of genetic interactions. Nat Methods. 2017;14(6):573–6.
220. Baliou S, et al. CRISPR therapeutic tools for complex genetic disorders and cancer. Int J Oncol. 2018;53(2):443–68.
221. Shalem O, Sanjana NE, Zhang F. High-throughput functional genomics using CRISPR–Cas9. Nat Rev Genet. 2015;16(5):299–311.
222. Rahman N. Mainstreaming genetic testing of cancer predisposition genes. Clin Med (Lond). 2014;14(4):436.
223. Gootenberg JS, et al. Nucleic acid detection with CRISPR-Cas13a/C2c2. Science. 2017;356(6336):438–42.
224. Abudayyeh OO, et al. RNA targeting with CRISPR–Cas13. Nature. 2017;550(7675):280–4.
225. Chen JS, et al. CRISPR-Cas12a target binding unleashes indiscriminate single-stranded DNase activity. Science. 2018;360(6387):436–9.
226. Chertow DS. Next-generation diagnostics with CRISPR. Science. 2018;360(6387):381–2.
227. Yang H, et al. Break breast cancer addiction by CRISPR/Cas9 genome editing. J Cancer. 2018;9(2):219.
228. Wang Y, et al. The BRCA1-Δ11q alternative splice isoform bypasses germline mutations and promotes therapeutic resistance to PARP inhibition and cisplatin. Cancer Res. 2016;76(9):2778–90.

229. Cathomen T, Joung JK. Zinc-finger nucleases: the next generation emerges. Mol Ther. 2008;16(7):1200–7.
230. Wen WS, et al. CRISPR-Cas9 systems: versatile cancer modelling platforms and promising therapeutic strategies. Int J Cancer. 2016;138(6):1328–36.
231. Fu Y, et al. High-frequency off-target mutagenesis induced by CRISPR-Cas nucleases in human cells. Nat Biotechnol. 2013;31(9):822–6.
232. Guilinger JP, et al. Broad specificity profiling of TALENs results in engineered nucleases with improved DNA-cleavage specificity. Nat Methods. 2014;11(4):429–35.
233. Pattanayak V, et al. Revealing off-target cleavage specificities of zinc-finger nucleases by in vitro selection. Nat Methods. 2011;8(9):765–70.
234. Vouillot L, Thélie A, Pollet N. Comparison of T7E1 and surveyor mismatch cleavage assays to detect mutations triggered by engineered nucleases. G3 (Bethesda). 2015;5(3):407–15.
235. Pattanayak V, et al. High-throughput profiling of off-target DNA cleavage reveals RNA-programmed Cas9 nuclease specificity. Nat Biotechnol. 2013;31(9):839–43.
236. Seeger C, Sohn JA. Complete spectrum of CRISPR/Cas9-induced mutations on HBV cccDNA. Mol Ther. 2016;24(7):1258–66.
237. Gabriel R, et al. An unbiased genome-wide analysis of zinc-finger nuclease specificity. Nat Biotechnol. 2011;29(9):816–23.
238. Osborn MJ, et al. Evaluation of TCR gene editing achieved by TALENs, CRISPR/Cas9, and megaTAL nucleases. Mol Ther. 2016;24(3):570–81.
239. Hsu PD, et al. DNA targeting specificity of RNA-guided Cas9 nucleases. Nat Biotechnol. 2013;31(9):827–32.
240. Mali P, et al. CAS9 transcriptional activators for target specificity screening and paired nickases for cooperative genome engineering. Nat Biotechnol. 2013;31(9):833–8.
241. Cho SW, et al. Analysis of off-target effects of CRISPR/Cas-derived RNA-guided endonucleases and nickases. Genome Res. 2014;24(1):132–41.
242. Xie S, et al. sgRNAcas9: a software package for designing CRISPR sgRNA and evaluating potential off-target cleavage sites. PLoS One. 2014;9(6):e100448.
243. Sander JD, et al. ZiFiT (Zinc Finger Targeter): an updated zinc finger engineering tool. Nucleic Acids Res. 2010;38(Suppl 2):W462–8.
244. Suresh B, et al. Cell-penetrating peptide-mediated delivery of Cas9 protein and guide RNA for genome editing. Methods Mol Biol. 2017;1507:81–94.
245. Dong L, et al. An anti-CRISPR protein disables type V Cas12a by acetylation. Nat Struct Mol Biol. 2019;26(4):308–14.
246. Shin J, et al. Disabling Cas9 by an anti-CRISPR DNA mimic. Sci Adv. 2017;3(7):e1701620.
247. Chang HH, et al. Non-homologous DNA end joining and alternative pathways to double-strand break repair. Nat Rev Mol Cell Biol. 2017;18(8):495–506.
248. Shrivastav M, De Haro LP, Nickoloff JA. Regulation of DNA double-strand break repair pathway choice. Cell Res. 2008;18(1):134–47.
249. Ciccia A, Elledge SJ. The DNA damage response: making it safe to play with knives. Mol Cell. 2010;40(2):179–204.
250. Chapman JR, Taylor MR, Boulton SJ. Playing the end game: DNA double-strand break repair pathway choice. Mol Cell. 2012;47(4):497–510.
251. Maruyama T, et al. Increasing the efficiency of precise genome editing with CRISPR-Cas9 by inhibition of nonhomologous end joining. Nat Biotechnol. 2015;33(5):538–42.
252. Chu VT, et al. Increasing the efficiency of homology-directed repair for CRISPR-Cas9-induced precise gene editing in mammalian cells. Nat Biotechnol. 2015;33(5):543–8.
253. Shah SZ, et al. Advances in research on genome editing CRISPR-Cas9 technology. J Ayub Med Coll Abbottabad. 2018;31(1):108–22.
254. Lin S, et al. Enhanced homology-directed human genome engineering by controlled timing of CRISPR/Cas9 delivery. Elife. 2014;3:e04766.
255. Kim K, et al. Highly efficient RNA-guided base editing in mouse embryos. Nat Biotechnol. 2017;35(5):435–7.
256. Zafra MP, et al. Optimized base editors enable efficient editing in cells, organoids and mice. Nat Biotechnol. 2018;36(9):888–93.
257. Grünewald J, et al. CRISPR DNA base editors with reduced RNA off-target and self-editing activities. Nat Biotechnol. 2019;37(9):1041–8.
258. Zong Y, et al. Efficient C-to-T base editing in plants using a fusion of nCas9 and human APOBEC3A. Nat Biotechnol. 2018;36(10):950–3.
259. Hur JK, et al. Targeted mutagenesis in mice by electroporation of Cpf1 ribonucleoproteins. Nat Biotechnol. 2016;34(8):807–8.
260. Gori JL, et al. Delivery and specificity of CRISPR/Cas9 genome editing technologies for human gene therapy. Hum Gene Ther. 2015;26(7):443–51.
261. Zuris JA, et al. Cationic lipid-mediated delivery of proteins enables efficient protein-based genome editing in vitro and in vivo. Nat Biotechnol. 2015;33(1):73–80.
262. Mout R, et al. In vivo delivery of CRISPR/Cas9 for therapeutic gene editing: progress and challenges. Bioconjug Chem. 2017;28(4):880–4.
263. Yin H, Kauffman KJ, Anderson DG. Delivery technologies for genome editing. Nat Rev Drug Discov. 2017;16(6):387–99.
264. Kotterman MA, Schaffer DV. Engineering adeno-associated viruses for clinical gene therapy. Nat Rev Genet. 2014;15(7):445–51.
265. Feng M, et al. Stable in vivo gene transduction via a novel adenoviral/retroviral chimeric vector. Nat Biotechnol. 1997;15(9):866–70.

266. Koike-Yusa H, et al. Genome-wide recessive genetic screening in mammalian cells with a lentiviral CRISPR-guide RNA library. Nat Biotechnol. 2014;32(3):267–73.
267. Paulk NK, et al. Adeno-associated virus gene repair corrects a mouse model of hereditary tyrosinemia in vivo. Hepatology. 2010;51(4):1200–8.
268. Charlesworth CT, et al. Identification of preexisting adaptive immunity to Cas9 proteins in humans. Nat Med. 2019;25(2):249–54.
269. Ihry RJ, et al. p53 inhibits CRISPR–Cas9 engineering in human pluripotent stem cells. Nat Med. 2018;24(7):939–46.
270. Liang C, et al. Tumor cell-targeted delivery of CRISPR/Cas9 by aptamer-functionalized lipopolymer for therapeutic genome editing of VEGFA in osteosarcoma. Biomaterials. 2017;147:68–85.
271. Finn JD, et al. A single administration of CRISPR/Cas9 lipid nanoparticles achieves robust and persistent in vivo genome editing. Cell Rep. 2018;22(9):2227–35.
272. Wang H-X, et al. Nonviral gene editing via CRISPR/Cas9 delivery by membrane-disruptive and endosomolytic helical polypeptide. Proc Natl Acad Sci U S A. 2018;115(19):4903–8.
273. Ma Y, et al. Increasing the efficiency of CRISPR/Cas9-mediated precise genome editing in rats by inhibiting NHEJ and using Cas9 protein. RNA Biol. 2016;13(7):605–12.
274. Ramakrishna S, et al. Gene disruption by cell-penetrating peptide-mediated delivery of Cas9 protein and guide. RNA. 2014;24(6):1020–7.
275. Guenther CM, et al. Synthetic virology: engineering viruses for gene delivery. Wiley Interdiscip Rev Nanomed Nanobiotechnol. 2014;6(6):548–58.
276. Ho ML, et al. Efficiency of protease-activatable virus nanonodes tuned through incorporation of wild-type capsid subunits. Cell Mol Bioeng. 2014;7:334–43.
277. Asuri P, et al. Directed evolution of adeno-associated virus for enhanced gene delivery and gene targeting in human pluripotent stem cells. Mol Ther. 2012;20(2):329–38.
278. Hofherr SE, et al. Polyethylene glycol modification of adenovirus reduces platelet activation, endothelial cell activation, and thrombocytopenia. Hum Gene Ther. 2007;18(9):837–48.
279. Kay MA, He C-Y, Chen Z-Y. A robust system for production of minicircle DNA vectors. Nat Biotechnol. 2010;28(12):1287–9.
280. Koo T, et al. Selective disruption of an oncogenic mutant allele by CRISPR/Cas9 induces efficient tumor regression. Nucleic Acids Res. 2017;45(13):7897–908.
281. Lee K, et al. Nanoparticle delivery of Cas9 ribonucleoprotein and donor DNA in vivo induces homology-directed. DNA Repair. 2017;1(11):889–901.
282. Rothschild J. Ethical considerations of gene editing and genetic selection. J Gen Fam Med. 2020;21(3):37–47.
283. Ormond KE, et al. Human germline genome editing. Am J Hum Genet. 2017;101(2):167–76.

RNA Therapies in Cancer

10

Fatemeh Khatami, Setareh Zahedian, Leila Zareian, Rahil Mashhadi, Mandana Hasanzad, Zeinab Ahadi, and Seyed Mohamamd Kazem Aghamir

What Will We Learn in This Chapter?
In this chapter, we will learn about the Ribonucleic acid (RNA)-based therapy in cancers, as a new molecule for cancer treatment. RNA-based therapy in cancers has a narrow therapeutic window depending on the patient's genome and personalized-dependent efficacy. We will present the molecular structure of RNA, mRNA therapy, antisense oligonucleotide (ASO), RNAi (RNA interference), RNA-based immune checkpoint inhibitors, and self-delivering RNAi (sd-rxRNA). We will also discuss the clinical evidence, challenges, and implications of implementing RNA-based cancer therapy in routine practice.

Rationale and Importance
Over the past several decades, RNA-based vaccines have evolved from a theoretical concept to a clinical reality. RNA-based therapy in cancers is a treatment that regulates tumor cell apoptosis and treats cancer by affecting the metabolism of mRNAs in cells. There are several advantages of RNA-based cancer therapy medication over traditional vaccine techniques, including their high potency, rapid development, low-cost manufacturing, and safe administration. We introduced the discovery and development of RNA therapy and discussed its classification, mechanisms, advantages, and challenges. Moreover, we highlighted how RNA therapy works in killing tumor cells as well as the progress it has made in related research.

F. Khatami (✉)
Urology Research Center, Tehran University of Medical Sciences, Tehran, Iran

Personalized Medicine Research Center, Endocrinology and Metabolism Clinical Sciences Institute, Tehran University of Medical Sciences, Tehran, Iran

S. Zahedian · L. Zareian · R. Mashhadi · Z. Ahadi · S. M. K. Aghamir
Urology Research Center, Tehran University of Medical Sciences, Tehran, Iran
e-mail: lzareian@farabi.tums.ac.ir; mkaghamir@tums.ac.ir

M. Hasanzad
Personalized Medicine Research Center, Endocrinology and Metabolism Clinical Sciences Institute, Tehran University of Medical Sciences, Tehran, Iran

10.1 Introduction

The treatment of cancer has undergone significant developments over the course of several decades. Chemotherapy, targeted therapies, and radiation therapy have consistently remained the primary methods of treating advanced solid tumors [1]. However, there is a growing interest and acceptance of various innovative therapies in the field.

M. Hasanzad (ed.), *A Glimpse at Medicine in the Future*,
https://doi.org/10.1007/978-981-97-4974-4_10

RNA therapy, an intervention that modulates cellular proteins and ameliorates ailments through its impact on cellular mRNA metabolism, has garnered significant attention in the realm of research on complicated diseases, such as inherited disorders and malignancies [2, 3]. This kind of therapy is a therapeutic approach that relies on the level of RNA and has the capacity to impact the metabolic pathways of messenger RNAs (mRNA) through the utilization of oligonucleotides that are combined with them via base pairing [3]. In the past decade, RNA therapy has exhibited remarkable activity across numerous significant domains and has emerged as a prominent area of investigation within the field of immunotherapy [3, 4]. To date, multiple RNA-based medications have been approved, along with numerous undergoing phase III investigations [5]. More investigations on RNA therapy showed a diversity of genes involved in different cellular pathways can be inhibited by RNA-based drugs. The aim of utilizing these drugs is to target several important main sources of multiple gene diseases, like tumors. In addition, RNA-based drugs can effectively abate tumor resistance to drugs, and in advanced tumors, they are able to halt tumor growth. RNA therapy, with its remarkable specificity, a great variety of targets, and beneficial drug characteristics, exhibits unparalleled efficacy in the treatment of tumors [3].

RNA-based gene therapy necessitates the functional presence of therapeutic RNA within specific cellular targets while concurrently evading any undesirable immune reactions [6].

RNA therapeutics in the treatment of tumors involve various mechanisms. These mechanisms encompass the inhibition of tumor cell proliferation, prevention of malignant tumor cell metastasis, induction of tumor cell apoptosis, disruption of tumor cell expression, inhibition of tumor cell angiogenesis, reconstruction of the tumor microenvironment, reprogramming of tumor cells, and reduction of tumor cell resistance to drugs [3].

Therapeutic RNA drugs can be briefly classified into antisense oligonucleotide (ASO) therapy, small interfering RNA (siRNAs) therapy, and microRNA (miRNAs) therapy, which are explained more in the following chapters. These molecules have the ability to directly target mRNAs and noncoding RNAs (ncRNAs) by means of Watson-Crick base pairing [7].

Antisense oligonucleotides (ASO) are synthetic RNA (or DNA) sequences that are single-stranded and extensively modified. These sequences are specifically designed to bind selectively to RNA molecules that encode the gene of interest through complementary base-pairing. ASOs have undergone examination in various diseases, including cancer.

Antisense oligonucleotides (ASO) are single-stranded, highly modified, synthetic RNA (or DNA) sequences designed to bind selectively to RNA through its complementary base-pairing, which encodes the gene of interest, and have been tested in several disorders, including cancer [8].

Small interfering RNA (siRNA) has been extensively examined as a viable therapeutic intervention for both viral diseases and cancer, with the ultimate objective of inhibiting numerous pathogenic genes [9]. These molecules are connected to the process of gene silencing. Furthermore, when it comes to malignant cells, siRNA possess the ability to modulate the level of expression of specific proteins which have a crucial role in their growth [10]. SiRNA is generated through a two-step process, namely the initial and operative stages. During the initial stage, a lengthy double-stranded RNA (ranging from 500 to 200 base pairs) is fragmented into 23–21 nucleotide segments by Dicer, thereby yielding siRNA. SiRNA serves as a functional entity. The double-stranded RNA can be derived from the viral genome, bacterial DNA, or synthesized RNA obtained through the utilization of bioinformatics data [11].

MicroRNAs (miRNA) are a class of small non-coding RNA (ncRNA) (with around 22 nucleotides) in cells that modulate the expression of some specific genes by impeding the conversion of mRNAs and facilitating their subsequent degradation [3, 12]. In recent years, the examination of miRNA biology within the realm of cancer has paved the way for novel diagnostic and therapeutic prospects [12]. One potential approach for therapeutically manipulating the tumor microenvironment involves the modulation of microRNA (miRNA) levels. MiRNAs are small, noncoding RNA molecules that can travel back and forth between malignant cells and the surrounding tumor microenvironmental cells. These specific RNA transcripts possess two distinct characteristics: (a) their expression is highly specific to particular types of tumors, and (b) they have a role in the initial regulation of immune responses. Consequently, miRNAs represent promising candidates for employment in cancer therapy. Numerous miRNAs exhibit abnormal expression patterns in human cancer cells, thus presenting new avenues for cancer treatment [13].

Several of these molecules hinder the translation of mRNA, leading to a decrease in protein activity. Conversely, other molecules enhance protein activity by inhibiting the suppression of miRNA. Granted that these molecules have become the primary focus of research in the field of RNA drugs, resulting in the generation of numerous RNA drugs for the various types of tumors. To date, RNA drugs with anti-tumor properties have demonstrated promising results in clinical trials. Additionally, nowadays there is a burgeoning number of pharmaceutical companies which focus on RNA-based anti-tumor therapy [3].

It should be mentioned that if long non-coding RNAs (lncRNAs) are located within the nucleus, the implementation of RNA interference (RNAi) will not be feasible for the suppression of them. In such instances, the utilization of antisense oligonucleotides (ASOs) represents a more appropriate approach [14]. As an example, the oncological RAS and MYC proteins, which have traditionally been considered untargetable by small-molecule drugs, have presented a persistent challenge. However, recent studies have demonstrated a burgeoning body of evidence suggesting that targeting RNAs could potentially yield novel therapeutic interventions against RAS, MYC, and other molecular sites. Moreover, the manipulation of RNA holds great promise in influencing aberrant non-coding RNAs, thereby modifying the profiles of altered proteins [2]. A variety of small molecules can attach themselves to RNA molecules.

10.2 Antisense Oligonucleotide

In 1978, antisense oligonucleotides (ASOs) were invented for the first time to inhibit viral replication of Rous sarcoma virus (RSV) in vitro [15]. It took around 20 years to introduce the first ASO drug, which has the FDA (US Food and Drug Administration) approval. In the light of that, more research has been done to utilize ASO for other human diseases [16]. For example, recent research has proven the silencing of five essential upregulated transcripts has a significant impact on the survival and migration of breast cancer cell lines. It also can target LncRNA LLNLR-299G3. One by antisense oligonucleotide in squamous cell carcinoma cells (ESCC) (in vitro and in vivo), which can halt these tumor cells malignancy [17, 18].

The importance of ASOs in oncology goes back to their excellent features, including being water-soluble, specific, non-toxic, uncharged, and stable under physiologic conditions, and not being cleaved by endogenous nucleases. It can be employed for various types of cancer cells, including lung, liver, oral squamous, and breast cancer cells, to inhibit the primary and metastatic tumor growth or raise the sensitivity of cells to chemical therapy [19–21]. It can eliminate the need for genetic manipulation and inhibit protein expression on a dose-dependent basis [15, 22]. AO-based approaches differ from antibodies and small molecules in that they direct their focus toward the RNA level, specifically by employing a synthetic AO to specifically bind to target genes involved in the pathologies associated with cancer. This binding enables the modulation of gene expression, particularly in relation to pro-oncogenic genes [23, 24]. Conversely, only a mere number of the human genome (0.05%) is affected by approved therapeutic substances, specifically therapies in which protein would be considered as a target, like antibodies and small-molecule chemicals. This limited impact is due to the fact that the majority of sequences in DNA sequences in the genome of humans do not produce protein-coding transcripts [25].

ASO therapy is known as a novel tool of transcript-targeted therapy which holds great potential for the field of personalized medicine [15, 26]. Indeed, ASO is a synthetic single-stranded deoxyribonucleotide with 16–22 bases which has a complementary sequence to a specific mRNA target or non-coding RNA and through different mechanisms can alter (reduce, restore, or inhibit) the protein expression of genes that cause or are associated with diseases like cancer and genetic disorders [16, 27–29]. ASOs can be classified into two categories, namely, occupancy-mediated degradation (or enzymatic RNA degradation, which relies on a specific enzyme) and occupancy-only (or steric block mechanism, which is a binding process that does not need the help of any enzyme) models, which are distinguished by their distinct post-hybridization mechanisms [30, 31].

Furthermore, in a variety of studies, ASOs are utilized in combinational therapies like radiation therapy, chemotherapy, and photodynamic therapy [22, 32]. In general, the mechanism of ASO can differ depending on the chemistry of ASO, its binding sequence, and its target. The mechanism of action for ASO includes: (1) The activation of RNase H endonuclease, which degrades the RNA-DNA heteroduplex, is induced by the RNase-dependent mechanism, resulting in a noteworthy decrease in the translation of the target gene [15, 28]; (2) inhibiting 5′ cap formation [28]; (3) altering the splicing process (splice switching) through altering the open reading frame [15, 28]; (4) acting as a steric hindrance and inhibiting RNA loading on ribosomes, which results in halting mRNA translation [15, 33].

ASO therapies face a considerable obstacle in delivering the drug to the intended location of action. ASO drugs, upon injection, need to navigate through the bloodstream, traverse biological obstacles, and undergo cellular internalization. Having entered the cell, one important challenge for ASOs is evading degradation by the lysosome. They must prevent being captured by secretory vesicles which involve in cellular exportation [34]. Numerous approaches have been and continue to be improved in order to optimize the stability, targeting, and trafficking of ASO. These strategies encompass chemical alternations, bioconjugating it to different parts, along with the utilization of delivery vehicles [35, 36].

It is apparent that ASOs necessitate a strong affinity and stability with the target transcripts while concurrently maintaining resistance to nuclease activity. To accomplish this, diverse chemical modifications have been executed on ASOs. These modifications serve the purpose of safeguarding the sequence integrity and facilitating the attainment of their intended impact. The initial in vivo implementations of ASOs demonstrated restricted clinical promise. In fact, the unaltered backbone of phosphoribose leads to a high vulnerability of ASO's to be degraded swiftly by endonucleases and exonucleases [32, 37]. Nevertheless, a substantial quantity of chemical alternations has currently been administered to ASOs, leading to enhanced its properties with regard to pharmacology (Fig. 10.1). According to the nature of these chemical modifications, ASOs are categorized into first, second, and third generations.

A sulfur can be replaced by a non-binding oxygen atom in the backbone of ASO through phosphorothioate backbone incorporation. This can significantly enhance the resistance of cells to the nuclease enzyme and augment serum proteins binding to ASO. Granted that, ASO's half-lives prolong in serum [38, 39], while still facilitating their utilization in applications involving the downregulation of target RNA.

Furthermore, ASO's effectiveness and safety can be enhanced by an alternation of ribose sugar at the 2′ location, which makes an additional category of ASOs. This includes 2′-*O*-methyl (2′-OMe) and 2′-*O*-methoxy-ethyl (2′-MOE), that are extensively investigated within this particular group.

Another generation of antisense agents is composed of phosphorothioate and segments consisting of oligoribonucleotide in their backbone, which are substituted at the 2′-O position,

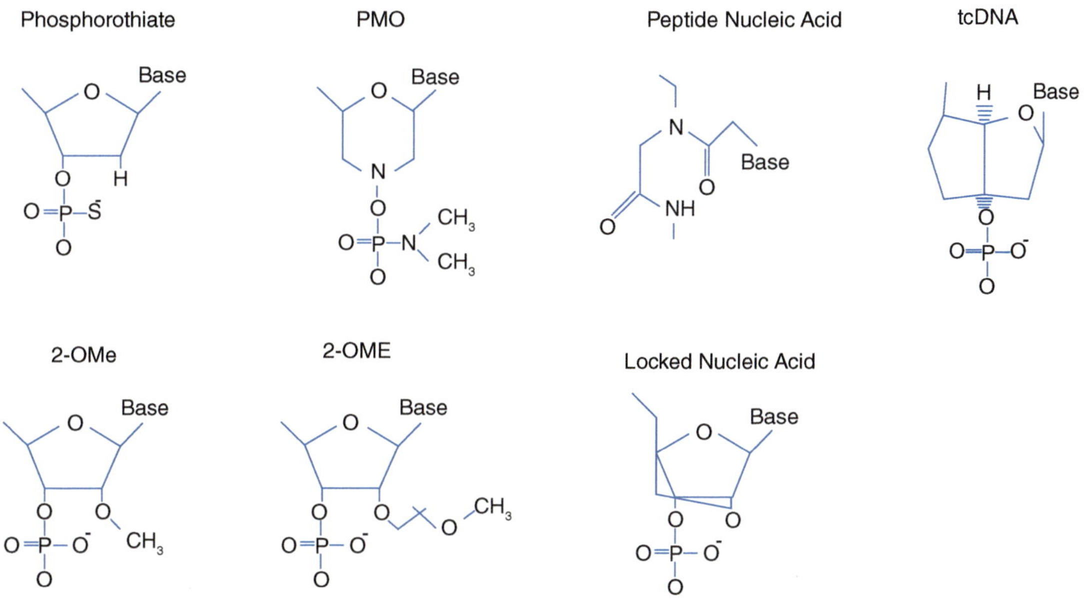

Fig. 10.1 In recent decades, several chemical changes have been implemented on the phosphodiester backbone to enhance the ASO features related to pharmacology, tolerance, and binding affinity to the target. Phosphorothioate DNA, phosphorodiamidate morpholino (PMO), and peptide nucleic acid design all provide resistance against nucleases and exhibit improved cellular uptake, resulting in an increased potency of the ASO. Tricyclo-DNAs (tcDNA) are DNA analogs that are structurally constrained, leading to higher potency and enhanced tissue uptake when administered systemically. Various ribose substitutions, such as 2′-*O*-methyl (2′-OMe), 2′-*O*-methoxyethyl (2′-MOE), and locked nucleic acid, are commonly combined to enhance stability, improve binding to the intended target, and totally reduce the toxicity associated with unmodified designs

have exhibited an enhanced affinity for hybridization with target RNA [26–28]. Furthermore, these modified agents have demonstrated an increased resistance against the degradation caused by nucleases [28, 29] and have displayed a reduced immunostimulatory activity when compared to counterparts that did not undergo modifications. It is important to note that the naturally occurring backbone of phosphoribose does not exist in other ASOs. As an example, the replacement of the deoxyribose part by a morpholine ring and the replacement of charged phosphodiester inter-subunit linkage by a phosphorodiamidate linkage which is uncharged can make an oligomer called phosphorodiamidate morpholino oligomers (PMO) [40]. Oligonucleotides like these exhibit a high level of resistance against degradation caused by nucleases and proteases and commonly play a role in modulating splicing and halting translation processes. Spinal muscular atrophy (SMA) and Duchenne muscular dystrophy (DMD) (recently approved) serve as exemplars of this generation of ASOs, as explained above [26].

10.3 RNAi (RNA Interference)

RNA interference (RNAi) is a process in which a specific RNA is degraded through the introduction of double-stranded (ds) RNAs in cells and suppresses concerned gene expression upon identification of infection or genetic irregularity [3, 31]. This process serves as an inherent defense mechanism against viral infections and the insertion of transposable elements [41, 42]. Cancer constitutes one of the primary focal points for therapeutic interventions based on RNA interference [43].

There exist a variety of methodologies to elicit RNA interference (RNAi), encompassing the utilization of synthetic molecules, RNAi vectors, and in vitro cleavage. Synthetic molecules encompass standard small interfering RNAs (siRNA) as well as specifically designed RNAi sequences formulated through programs established by biological enterprises to cater to the requirements of their clientele. RNAi vectors can be segregated into two categories, namely small hairpin RNAs (shRNA) and microRNAs (miRNAs).

Dicer is the primary enzyme responsible for processing long double-stranded RNA (dsRNA) in the in vitro environment (Fig. 10.2).

10.3.1 mRNA Therapy

Messenger ribonucleic acid (mRNA) has garnered considerable interest within the medical field. mRNA is found as a plausible strategy of cancer treatment, owing to diverse treatment methodologies such as protein replacement therapies, gene editing, and cell engineering [44]. mRNA-based therapeutics can be classified into three fundamental categories, namely preventive vaccines, therapeutic vaccines, and protein-encoding therapies [45] which, on the other hand, based on the constituents that elicit an immune response within the human body, can be classified as conventional tumor cell-based vaccines, peptide- or protein-based vaccines generated through synthesis, and vaccines founded on nucleic acids (specifically DNA and mRNA) [46]. Self-amplifying mRNA (saRNA) vaccines and non-replicating mRNA (also known as conventional mRNA) vaccines are two main varieties of mRNA vaccines. Both types of RNA exhibit 5 fundamental components of mRNA: a 50 cap, a 50 untranslated region (UTR), an open

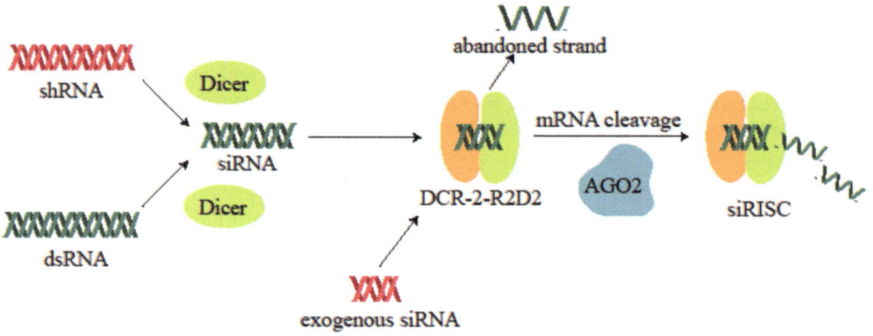

Fig. 10.2 RNA therapy through RNA-induced silencing complex (RISC) formation

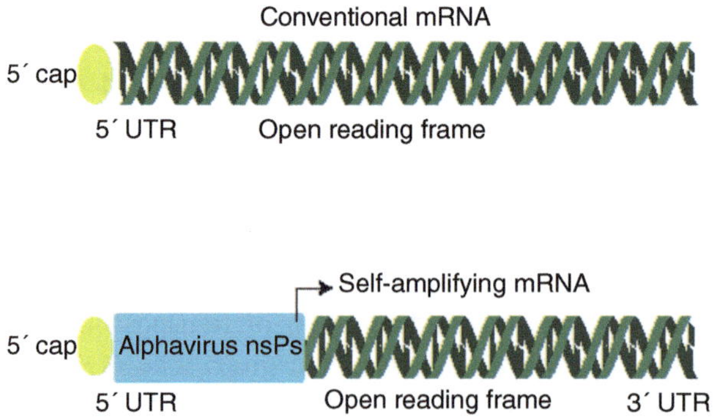

Fig. 10.3 Key structures of in vitro transcribed mRNA (IVT mRNA) encompass various components: (a) A conventional mRNA is comprised of a 50 cap, a 50 untranslated region (UTR), an open reading frame (ORF), a 30 UTR, and a 30 poly(A) tail. (b) On the other hand, self-amplifying mRNA incorporates a 50 cap, a 50 UTR, an ORF, a 30 UTR, a 30 poly(A) tail, as well as an additional alphavirus sequence encoding nonstructural proteins (nsPs)

reading frame (ORF) responsible for encoding an antigen, a 30 UTR, and a poly(A) tail (a. Fig. 10.3). Furthermore, apart from these five components, a saRNA possesses a sequence of gene that encodes the RNA replicase complex (b. Fig. 10.3) [46].

The success of these vaccines during the Coronavirus pandemic has captured the attention of the general public. In the recent past, many researchers have focused on mRNA as a groundbreaking and powerful approach for the treatment of cancer [45]. Cancer vaccines, in a broad sense, can be categorized according to their various applications and functions. Also, they can be categorized into four primary classifications, namely cell-based vaccines (including dendritic cells (DCs) or T cells), virus-based vaccines, bacteria-based vaccines, and molecular vaccines, based on disparities among antigen-carrying vectors [47].

Classical cancer-preventative vaccines are designed to specifically target the viruses that have the potential to induce certain types of cancers [48], whereas therapeutic vaccines have the capability to activate immune responses against preexisting tumors [49]. Compared to conventional vaccines, nucleic acid-based vaccines possess the benefits of a more condensed manufacturing timeline, the proficient generation of safeguarding immunological reactions, simplified industrialization, comparatively economical pricing, and minimal adverse effects [46, 50].

Various methods for mRNA cancer vaccines, such as dendritic cell vaccines and different forms of directly injectable mRNA, have been utilized in numerous clinical trials focused on cancer treatment. Some of these trials have yielded encouraging outcomes, demonstrating antigen-specific T cell responses and extended periods of disease-free survival in certain instances [50].

The delivery of mRNA to specific organs and cells may present a formidable challenge due to the inherent instability of its unadorned structure and the limited ability of cells to uptake it. Indeed, traditional physical methods like electroporation (EP) and gene gun have been found to have detrimental effects on cells [51, 52]. On the other hand, viruses have proven to be effective in the delivery of mRNA into cells, although their safety is not entirely guaranteed due to the possibility of integration into the host genome [53]. In order to strike a balance between safety and efficiency, numerous non-viral nanodelivery systems have been formulated, encompassing lipid nanoparticles (LNPs), polymers, hybrid NPs, peptides, gold nanoparticle (AuNP)-DNA conjugates, and mRNA-loaded exosomes, all of which exhibit distinctive merits [46].

The fundamental element of cancer vaccines consists of an mRNA (known as in vitro transcribed mRNA or IVT mRNA), which is a single-stranded polynucleotide possessing identical structure and biological functionality to endogenous mRNA.

This IVT mRNA is capable of encoding either a tumor-associated antigen (TAA) or a tumor-specific antigen (TSA). TAAs exhibit high expression levels in actively proliferating tumor cells, while normal cells also generate small quantities of TAAs. However, vaccines based on TAAs may elicit undesired immune reactions in healthy tissues [54]. On the other hand, TSAs, also referred to as neoantigens, are solely expressed in tumor cells and not in normal cells at any developmental stage. The mechanisms underlying the anti-cancer effects of TAA and TSA mRNA vaccines remain indistinguishable, despite certain disparities between them [55].

In light of this, many studies focus on the alteration of mRNA and advancement of mRNA delivery (such as utilizing nanoparticles) [44]. Different methods have been employed to solve the problem of its instability, such as enhancing the structure of the 5′ cap and manipulating the length of the 3′ poly (A) tail, as well as manipulating the regulatory elements found within the 5′ and 3′ untranslated regions [50].

10.4 RNA and Cancer Immunotherapy

Under conditions of homeostasis, there exists an equilibrium between signaling that promotes inflammation and signaling that inhibits inflammation. This equilibrium is upheld by immune checkpoints, which consist of a collection of pathways that either hinder or stimulate the activity of immune cells [56]. The balance of these immune checkpoints is disrupted by malignant cells, which actively encourage a state of immunosuppression that facilitates evasion of the immune system and the growth of tumors [56, 57]. The way in which cancer cells achieve this disruption is by recruiting regulatory T cells (Tregs), reducing the expression of tumor antigens, inducing tolerance or apoptosis in T cells, and producing cytokines that suppress the immune response [57]. Consequently, a highly immunosuppressive tumor microenvironment (TME) is formed [58], which is distinct and specific to each tumor. In an endeavor to counteract the effects of this immunosuppression, immune checkpoint inhibitors function by obstructing the effects of certain inhibitory pathways [56, 57, 59].

The revelation of immune checkpoints has presented a fresh pathway within the realm of immunology for the treatment of cancer. Immune checkpoints hold significant importance in the regulation of the magnitude of immune reactions primarily in T lymphocytes through the implementation of inhibitory or co-stimulatory mechanisms [56]. The well-defined immune checkpoints encompass a wide array of entities, which include but are not limited to cytotoxic T-lymphocyte-associated protein 4 (CTLA-4), Programmed cell death protein 1 (PD-1), Lymphocyte-activation gene 3 (LAG3), T-cell immunoreceptor with Ig and ITIM domains (TIGIT), B- and T-lymphocyte attenuator (BTLA), Indoleamine-2,3-dioxygenase 1 (IDO1), and T cell immunoglobulin and mucin-domain containing-3 (TIM-3). The delicate equilibrium between immune checkpoint proteins that exert inhibitory effects and those that provide co-stimulation is of utmost importance in the maintenance of an individual's self-tolerance, thereby effectively thwarting the onset of autoimmunity [60]. As an example, the activation of lymphocyte gene-3 (LAG-3, CD223) is facilitated by its interaction with the major histocompatibility complex (MHC) class II. This molecule is expressed by various immune cells, such as activated T cells, natural killer (NK) cells, B cells, and dendritic cells (DCs) [61, 62]. The precise mechanism by which LAG-3 operates is not yet fully understood; however, its engagement with MHC class II leads to the downregulation of T cell cytokine production, as well as the contraction of CD4 and CD8 T cell populations. Furthermore, LAG-3 promotes the adoption of a Treg phenotype to prevent tissue damage and autoimmunity [62]. Within the tumor microenvironment (TME), specifically in tumor-infiltrating lymphocytes (TILs), LAG-3 is overexpressed, resulting in cellular dysfunction, immune exhaustion, and creating a favorable environment for tumor growth [63]. Consequently, the blockade of LAG-3 facilitates immune activation against malignant cells while augmenting the effects of other immune checkpoint inhibitors (ICIs), such as anti-PD-1 agents, and potentially other forms of immunotherapy. However, the combination of LAG-3 inhibitors with other ICIs may lead to an increased occurrence and severity of adverse events (AEs) [62, 64].

The human genome harbors approximately 20,000 genes responsible for coding proteins, constituting a mere fraction of the entirety of the genome, amounting to less than 2% [65]. The identification of non-coding RNAs (ncRNAs) coincided with the revelation that the vast majority of the genome's "dark matter" is actively transcribed [65]. ncRNAs have a strong correlation with the advancement and progression of cancer owing to their capacity to regulate the expression of genes, protein translation, and pathways related to growth [66]. In the year 1998, the Nobel Prize winners in Physiology or Medicine of 2006 discovered that non-coding transcripts possess the capability to impede mRNAs post-transcriptionally [67]. The non-coding RNAs possess the ability to regulate a plethora of cellular mechanisms ranging from cellular proliferation and specialization to the immune response and pharmacological resistance [66]. In the context of cancer, they exhibit the capacity to control the microenvironment of tumors, as well as the spread of cancer cells and resistance to drugs

[68]. The capability of ncRNAs to induce resistance to cancer drugs is currently a subject of great importance, especially regarding emerging therapeutic approaches like immune checkpoint inhibitors (ICIs) [66].

10.4.1 sd-rxRNA

A novel category of double-stranded, hydrophobic, asymmetric RNAi compounds has been recently discovered. These compounds, known as "self-delivering rxRNAs" (sd-rxRNA), have undergone significant modifications and are employed to suppress immune checkpoints by targeting and eliminating particular RNAs prior to their translation into proteins [3, 69]. They possess a small duplex region consisting of less than 15 base pairs and feature a fully phosphorothioate single-stranded tail. Remarkably, sd-rxRNA compounds possess the ability to effectively target numerous immunosuppressive targets, achieving a high knockout efficiency, all without the need for a delivery vehicle [3, 69]. sd-rxRNAs can be utilized alone or in combination with other therapies, such as chemotherapy [3, 70–72].

siRNA is a macromolecule with a high negative charge. It has a similar chemical structure to mRNA, making it water-soluble and unstable, just like mRNA. The cell membrane acts as a barrier, keeping mRNA inside the cells for cellular functions and preventing the entry of external substances, including exogenous siRNA. The negative charge of the siRNA molecule causes it to be repelled by the negatively charged cell membrane. Additionally, the hydrophilic nature of the siRNA backbone is incompatible with the hydrophobic interior of the lipid bilayer. As a result, siRNAs, including modified versions like 20-F or 20-OMe siRNA, cannot cross the membrane or be taken up by cells on their own. To enter cells, siRNA needs to be delivered with the help of siRNA delivery carriers. There are different technologies to deliver sd-rxRNAs to cells, such as hydrophobic modulation and cholesterol tagging technology [73].

10.4.2 Development of Tumor RNA Vacancy

The wide-ranging etiologies of cancer malignancies propose that patient care should be customized to provide a potent remedy for every individual. Furthermore, the objective is to intervene at the molecular level and manipulate cellular entities and ongoing mechanisms to combat ailments. For a considerable duration, the concept of utilizing the body's innate immune system for cancer treatment has captivated researchers due to its apparent straightforwardness and potential efficacy [74, 75]. The principal objective of cancer immunotherapies is to augment the host's anti-tumor immunity and alter the tumor's inhibitory microenvironment. In this manner, these therapies endeavor to impede the expansion of the patient's tumor and protract their lifespan [76]. Additionally, self-replicating RNA (srRNA) viral vaccine platforms have gained significant attention from many researchers. These vaccines are derived from distinct, single-stranded RNA viruses that contain RNA sequences encoding specific target antigens, as well as replication genes that possess the ability to significantly amplify RNA messages post-infection. Additionally, the encoded replica genes activate the innate immune response, rendering srRNA vectors highly suitable for anti-tumor vaccination [77]. Multiple categories of single-stranded RNA viruses have been employed for the purpose of self-replicating vaccine platforms, encompassing both positive-strand viruses such as alphaviruses and flaviviruses, as well as negative-strand viruses like Measles viruses and rhabdoviruses [78, 79]. In the context of negative-strand viruses, the generation of engineered replicons necessitates the utilization of reverse genetics and packaging cell lines, whereas the generation of positive-stranded viruses can be accomplished through the application of in vitro transcription (IVT) techniques, as well as the utilization of intermediate DNA vectors [77]. To produce srRNA vectors from these viruses, the nonstructural replica genes remain unaffected while the target antigen(s) are substituted for structural genes. These replicons contain the genes encoding the non-structural proteins that constitute the viral replicas complex, the cis-active replication sequences located at the 5′- and 3′-ends, as well as a subgenomic promoter that naturally directs the expression of the encoded antigen [80]. Vaccine replicons lack the capacity to form virions that would enable cell-to-cell infection, but they can generate multiple RNA copies in the cytoplasm, resulting in heightened antigen expression [81]. Self-replicating RNA-based vaccines in cancer therapies encompass:

1. 'Naked' srRNA cancer vaccines:

   Over two decades ago, the initial attempts to employ self-replicating RNA vaccines to combat cancer involved the utilization of intramuscular vaccination. This method entailed the administration of naked srRNA molecules that encoded a LacZ model antigen, followed by the implantation of a LacZ-expressing colorectal carcinoma known as CT26.CL25. A comprehensive investigation of this approach revealed that the direct injection of srRNA, in varying doses ranging from 0.1 to 100 μg, was capable of inducing anti-LacZ responses and providing defense against the metastatic challenge posed by CT26.CL25.
2. DNA-launched srRNA Replicon (DREP) cancer vaccines:

   The production of srRNA vectors necessitated the utilization of DNA templates, which had already been recognized as a modality for cancer vaccination [82, 83]. The studies provided evidence that the inhibition of programmed cell death, known as apoptosis, by delivering plasmids expressing Bcl-XL, diminished the effectiveness of TRP-1 DREP vaccination in combating tumors [84]. Subsequent investigations using mice lacking the Type I interferon receptor (IFNAR) demonstrated the significance of Type I interferon activation in the DREP-induced adaptive immune response specific to TRP-1 and its anti-tumor effects [85]. These investigations into the mechanisms established the critical role of apoptosis induction and interferon responses in the activation of anti-tumor immunity through DREP vaccines [86].
3. Viral replicon particles (VRP) cancer vaccines:

   The utilization of cancer vaccines composed of srRNA derived from unadorned RNA and DNA-encoded intermediates has also instigated the advancement of viral replicon particles (VRP) through the aid of helper plasmids [87–89]. These particular vaccines encompass the inherent structural proteins that possess the potential to enhance the efficiency of the vaccine by potentially assisting in the process of trafficking and/or stimulating the innate immune response [77].

In general, RNA, when administered in isolation or employed for the transfection of dendritic cells, possesses a multitude of merits as a vaccine, encompassing feasibility, applicability, safety, and efficacy in terms of engendering immune responses [90]. A study in 2020 indicated that RNA-LPX vaccination is a highly effective immunotherapeutic modality in individuals afflicted with melanoma who have previously received CPI treatment, thus indicating the broad applicability of non-mutated, shared antigens specific to tumorous lesions as viable candidates for cancer immunization [91]. Moreover, neoantigen-based personalized mRNA vaccines, particularly tumor-associated antigens, have emerged as novel therapeutic approaches that demonstrate significant clinical advantages in the treatment of solid tumors. A study in 2021 showed DNA topoisomerase II alpha (TOP2A), neutrophil cytosol factor 4 (NCF4), formin-like protein 1 (FMNL1), and docking protein 3 (DOK3) as potential effective neoantigens for kidney renal clear cell carcinoma (KIRC) mRNA vaccine development, and patients with renal cell carcinoma immune subtype 2 (RIS2) tumor might benefit more from mRNA vaccination [92].

Recently, investigations showed interleukin 1 (IL-1) and interleukin 1 receptor antagonist (IL-1ra) are key regulators of the inflammatory response to RNA vaccines since the activation of RNA vaccine-associated innate signaling is significantly influenced by certain lipids utilized in vaccine formulations that incorporate *N*1-methylpseudouridine-modified RNA, thus augmenting the effect in an unexpected manner. This serves to reduce the activation of Toll-like receptor signaling [93].

10.4.3 Adjuvant Cellular Immunotherapy (C-Met-CAR T Cells)

In recent years, the field of hematological tumor treatment has experienced a significant breakthrough with the emergence of CAR-T cells. The technology of CAR-T cells represents an advanced form of immunotherapy for the effective treatment of recurrent or refractory tumors [94]. This innovative approach involves the utilization of genetically modified T cells that have been specifically engineered to target antigens associated with tumors, thereby exerting potent antitumor effects [95]. The majority of CAR-T cells currently accessible are derived from T cells collected from patients, subsequently transformed into CAR-T cells [94]. The most significant benefit lies in their ability to adhere directly to the exterior of malignant cells without significant limitations imposed by the major histocompatibility complex (MHC), thereby triggering apoptosis of neoplastic cells [96]. CAR-T cells are composed of three primary constituents: an extracellular segment (SCFV fragment) that identifies a tumor antigen, a transmembrane segment (CD8), and an intracellular segment (costimulatory molecules) that facilitates T lymphocyte activation (Fig. 10.4) [97].

CAR-T cells are increasingly being employed to address solid tumors, including urologic neo-

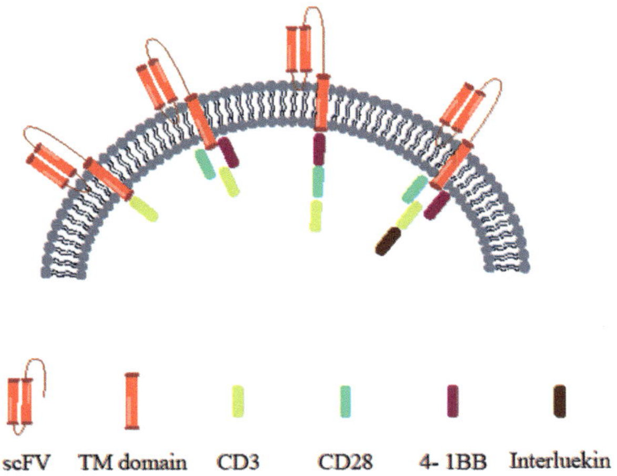

Fig. 10.4 The schematic diagram depicting the arrangement of CAR-T cells across the first to fourth generations is presented. Upon infiltration of tumor tissues during tumor therapy, SCFV fragments selectively attach to analogous antigens located on the external surface of tumor cells. Consequently, the activation signal progresses via the transmembrane domain toward the intracellular costimulatory domain, thereby instigating T cell activation through the activation of the costimulatory molecule, which ultimately results in the eradication of the tumor cells. The fundamental constituents of CAR T cells encompass the SCFV fragment (extracellular domain), CD8 (transmembrane domain), and CD3z (intracellular domain). Successive generations of CAR-T cells have been enhanced to promote T cell proliferation and augment the capacity to eliminate tumors by elevating the presence of intracellular costimulatory molecules. In contrast, the initial generation of CAR-T cells solely comprised the CD3z costimulation domain within the intracellular domain. The cytotoxicity of CAR-T cells was significantly augmented in the second generation through the incorporation of a costimulatory molecule (either CD28 or 4-1BB). To further enhance T cell activation, the intracellular domain of third-generation CAR-T cells contains two costimulatory molecules. In the fourth generation of CAR-T cells, the intracellular co-expression of cytokines is increased compared to the second generation. These co-expressed cytokines play a key role in promoting the proliferation and differentiation of CAR T cells, as well as attracting peripheral immune cells to effectively target and eliminate tumor cells

plasms. Urologic malignancies predominantly encompass malignancies of the prostate, bladder, kidney, adrenal gland, penis, testicle, and ureter [98, 99]. Notably, prostate cancer presents numerous relatively specific targets. These neoplasms exhibit a slower progression compared to tumors found in other bodily organs, thereby allowing for ample opportunity for the application of CAR-T cell therapy. Moreover, urologic neoplasms exhibit the expression of tumor-associated antigens, including prostate stem cell antigen (PSCA), prostate-specific membrane antigen (PSMA), and epidermal growth factor receptor (EGFR). Consequently, the selection of CAR-T cells for urologic neoplasms offers an inherent advantage.

It can be deduced that CAR-T cell technology possesses inherent advantages when it comes to urologic neoplasms. Extensive research has been conducted on the use of CAR-T cells in the treatment of urologic neoplasms, leading to preliminary accomplishments in this area [99]. The targeted entities that have been chosen for preclinical investigations on CAR-T cells in the management of urologic neoplasms largely encompass CAIX, PSMA, PSCA, and other common targets associated with solid tumors. The design of the CAR primarily corresponds to the second generation, and the chosen costimulatory molecules are predominantly CD28 and 4-1BB. A relatively limited number of third-generation and fourth-generation CAR-T cells have been employed for the therapeutic intervention of urologic neoplasms [99].

Challenges in CAR T cell immunotherapy encompass: (1) The identification of tumor-specific targets within solid tumors, such as CD19 in hematological tumor cells, poses a significant challenge. Granted that only tumor-associated antigens can be utilized, potentially resulting in the emergence of CAR-T cells that focus on extratumoral cytotoxicity. (2) The drug exhibits a favorable diffusion effect within the bloodstream, facilitating direct contact between the drug and tumor cells. However, solid tumors are characterized by a dense stromal component and an immunosuppressive microenvironment. This poses difficulties for CAR-T cells to effectively infiltrate the tumor tissue and establish contact with the tumor cells. Moreover, even if CAR-T cells manage to infiltrate the tumor tissue, they must overcome the obstacle of immunosuppression in order to exert their tumor-killing function [99, 100]. To address these challenges, there is an ongoing investigation into the structural optimization and target screening of CAR-T cells [99].

Other anti-tumor methods, when used in conjunction with CAR-T cells, have also become a focal point of scientific investigation. For instance, in 2022, a study showed the tumor burden was significantly diminished when OAV-decorin was administered in combination with CAIX-CAR. Furthermore, the extracellular matrix (ECM) composition was altered due to the inhibition of collagen fiber distribution. Additionally, there was a decrease in TGF-b expression, an increase in IFN-a secretion, and an augmentation in the generation of CAR-T cells. The collective treatment approach also extended the survival of the murine subjects. These findings also serve to validate the efficacy of oncolytic adenovirus and CAIXCAR-T cells in combatting solid tumors [101].

Tumor-associated antigens assume a crucial function in the implementation of CAR-T cell technology for cancer therapy, and urologic neoplasms demonstrate a comparatively elevated quantity of distinct targets in comparison to other organ tumors within the physique. AIX, PMSA, and PSCA-directed chimeric antigen receptor (CAR) T cells have exhibited cytotoxic effects on tumors in numerous preclinical investigations, while clinical investigations have shown that PMSA-CAR-T cells can be well-tolerated and possess commendable antineoplastic properties [102]. Also, it is confirmed that tumor-associated target antigens that are frequently expressed in solid tumors are also present in urologic neoplasms. These antigens include MUC1, EGFR, VEGFR2, EpCAM, C-Met, NKG2DL, MUC1, etc. Consequently, the utilization of CAR-T cells in urologic neoplasms possesses a distinct advantage compared to other systemic tumors when it comes to the selection of target antigens [99].

c-Met, a receptor located on the cell membrane, is the product of the Met gene. It possesses independent phosphatase activity and falls under the category of tyrosine kinase receptors, specifically within the superfamily. Primarily, its expression occurs in epithelial cells [103]. c-Met, an entity derived from the proto-oncogene Met, holds significant importance in the oncogenic process, specifically pertaining to cancer progression [104, 105]. Activated c-Met can stimulate cellular proliferation, invasion, viability, angiogenesis, and locomotion. In non-small cell lung cancer (NSCLC) samples, an increased level of c-Met protein expression is commonly observed [106]. Numerous investigations have demonstrated that c-Met is excessively expressed in 60% of cases, while phosphorylated c-Met is elevated in 40–100% of cases [107, 108]. Based on our previously documented investigations, the humanized c-Met antibody is capable of effectively impeding invasion and inducing apoptosis in NSCLC cells both in vivo and in vitro.

10.4.4 Personalized RNA Vaccine Against Cancers

In recent years, the predominant concerns in modern oncology revolve primarily around the advancement of individualized medical care [109]. Developing preclinical or co-clinical strategies that predict patient response is essential due to the variability of drug responses among patients [110]. Different personalized vaccines have shown varying efficacy for a wide range of cancers in recent years [111–114]. Utilizing mRNA vaccines and molecular biomarkers gained many researchers attention [109]. Indeed, RNA therapeutics represent a flourishing class of medications that have the potential to revolutionize the treatment of numerous diseases and bring personalized medicine into reality [115]. mRNA vaccines are gaining recognition as promising contenders for precision cancer treatment due to their distinct advantages compared to other vaccine formats [116]. The capacity to swiftly modify the arrangement of the mRNA construct enables personalized treatments or adjustments to be made in response to a changing pathogen [115]. In 2017, the initial utilization of personalized mRNA vaccine in humans was documented for the treatment of melanoma. This vaccination successfully triggered targeted immune response, resulting in a reduction in the rate of metastasis and an extension of patients' progression-free survival [117]. Apart from addressing tumor heterogeneity by encoding personalized proteins based on the tumor's genetic expression profile, mRNA vaccines fulfill the criteria for precision therapy by emphasizing accurate targeting, high efficacy, safety, and cost-effectiveness [116].

The experience gained from devising tactics to combat SARS-CoV-2 mutations can be utilized in the development and implementation of a customized mRNA vaccine for a variety of cancers [116].

It should be noted that precise targeting is crucial for personalized therapy. In line with this concept, mRNA serves as a blueprint for protein synthesis and utilizes the machinery within host cells to produce vaccines. This unique feature enables the post-translational modification of protein products, including proper folding to ensure optimal functionality [45, 50]. Moreover, this approach facilitates the production of correctly folded and assembled multimeric proteins that cannot be generated in bioreactors. Additionally, it allows transmembrane and intracellular proteins to be transported to their specific cellular locations. Consequently, mRNA vaccines generate protein products with inherent characteristics, thereby ensuring accurate targeting [116].

The production of mRNA cancer vaccines typically involves in vitro transcription (IVT) followed by 5′ capping and polyadenylation at the 3′ end, which mimics the natural process of mRNA maturation in eukaryotic cells' cytoplasm. IVT is a relatively straightforward process, but the challenge lies in generating high-quality therapeutic mRNAs that do not trigger severe inflammation. Recent advancements in capping and tailing techniques, incorporation of modified nucleosides [118], optimization of coding sequences [119], and meticulous purification of IVT mRNA have significantly addressed the issues of inflammation and innate immunity. These techniques are instrumental in reducing the presence of exogenous mRNAs, thereby minimizing the inflammatory response and enhancing mRNA translation [114].

Presently, there are two personalized mRNA cancer vaccines, namely Moderna vaccine mRNA-4157 and BioNTech vaccine BNT122 (RO7198457), which are undergoing phase II clinical trials. mRNA-4157 underwent a specific screening process and contained 20–34 neoantigens on a single mRNA molecule, depending on the cancer mutations of the patient. The interim results of a phase I trial indicated that mRNA-4157, either as a monotherapy or in combination with the PD-1 inhibitor Keytruda (NCT03313778), was well tolerated at all tested doses and stimulated a T cell response specific to the neoantigens. Encouraged by the positive outcomes of the phase I trial, a phase II clinical trial (NCT03897881) is currently underway to assess whether postoperative adjuvant therapy with mRNA-4157 and pembrolizumab improves recurrence-free survival compared to pembrolizumab alone in patients with complete resection of cutaneous melanoma and a high risk of recurrence. BioNTech is also investigating the effectiveness and safety of RO7198457 in combination with a PD-L1 antibody for the treatment of various cancers. Phase II clinical trials for melanoma, NSCLC, and CRC were initiated in the first half of 2021. Personalized mRNA vaccines offer a promising avenue for precise tumor treatment. With the existing clinical data and numerous ongoing trials, personalized mRNA vaccines have the potential to serve as adjuvant therapy for immunotherapy. However, the efficacy data primarily demonstrate individualized immune responses, and these tumor-specific immune responses rarely translate into tumor reduction. Nonetheless, mRNA cancer vaccines encoding tumor-specific antigens can enhance tumor immunogenicity and increase the sensitivity of tumor cells to immune checkpoint inhibitors, thereby making the clinical response rate significant [114].

10.5 Conclusion

The use of RNA as an anticancer is a new approach. However, transferring of RNA oligonucleotides' inhibition mechanism into clinical trials almost took time to become a reality. RNA therapy mostly comprises Antisense oligonucleotide (ASO), RNA interference (RNAi), and mRNA therapy. Varieties of mechanisms of RNA therapy work on target RNAs expression to cure diseases. RNA therapy destroys tumor cells by inhibiting proliferation and metastasis and inducing apoptosis.

References

1. DeVita VT Jr, Rosenberg SA. Two hundred years of cancer research. N Engl J Med. 2012;366(23):2207–14.
2. Bumcrot D, Manoharan M, Koteliansky V, Sah DW. RNAi therapeutics: a potential new class of pharmaceutical drugs. Nat Chem Biol. 2006;2(12):711–9.
3. Liang X, Li D, Leng S, Zhu X. RNA-based pharmacotherapy for tumors: from bench to clinic and back. Biomed Pharmacother. 2020;125:109997.
4. Mullard A. FDA approves landmark RNAi drug. Nat Rev Drug Discov. 2018;17(9):613–4.
5. Wang F, Zuroske T, Watts JK. RNA therapeutics on the rise. Nat Rev Drug Discov. 2020;19(7):441–2.
6. Paunovska K, Loughrey D, Dahlman JE. Drug delivery systems for RNA therapeutics. Nat Rev Genet. 2022;23(5):265–80.
7. Crooke ST, Witztum JL, Bennett CF, Baker BF. RNA-targeted therapeutics. Cell Metab. 2018;27(4):714–39.
8. Rossor AM, Reilly MM, Sleigh JN. Antisense oligonucleotides and other genetic therapies made simple. Pract Neurol. 2018;18(2):126–31.
9. Farrow B, Rychahou P, Murillo C, O'Connor KL, Iwamura T, Evers BM. Inhibition of pancreatic cancer cell growth and induction of apoptosis with novel therapies directed against protein kinase A. Surgery. 2003;134(2):197–205.
10. Chadar R, Kesharwani P. Nanotechnology-based siRNA delivery strategies for treatment of triple negative breast cancer. Int J Pharm. 2021;605:120835.
11. Agrawal N, Dasaradhi P, Mohmmed A, Malhotra P, Bhatnagar RK, Mukherjee SK. RNA interference: biology, mechanism, and applications. Microbiol Mol Biol Rev. 2003;67(4):657–85.
12. Shah V, Shah J. Recent trends in targeting miRNAs for cancer therapy. J Pharm Pharmacol. 2020;72(12):1732–49.

13. Smolle MA, Calin HN, Pichler M, Calin GA. Noncoding RNA s and immune checkpoints—clinical implications as cancer therapeutics. FEBS J. 2017;284(13):1952–66.
14. Tripathi V, Shen Z, Chakraborty A, Giri S, Freier SM, Wu X, et al. Long noncoding RNA MALAT1 controls cell cycle progression by regulating the expression of oncogenic transcription factor B-MYB. PLoS Genet. 2013;9(3):e1003368.
15. Barresi V, Musmeci C, Rinaldi A, Condorelli DF. Transcript-targeted therapy based on RNA interference and antisense oligonucleotides: current applications and novel molecular targets. Int J Mol Sci. 2022;23(16):8875.
16. Rinaldi C, Wood MJ. Antisense oligonucleotides: the next frontier for treatment of neurological disorders. Nat Rev Neurol. 2018;14(1):9–21.
17. Tilli TM, Carels N, Tuszynski JA, Pasdar M. Validation of a network-based strategy for the optimization of combinatorial target selection in breast cancer therapy: siRNA knockdown of network targets in MDA-MB-231 cells as an in vitro model for inhibition of tumor development. Oncotarget. 2016;7(39):63189.
18. Tian L, Huang Y, Zhang B, Song Y, Yang L, Chen Q, et al. Targeting LncRNA LLNLR-299G3.1 with antisense oligonucleotide inhibits malignancy of esophageal squamous cell carcinoma cells in vitro and in vivo. Oncol Res. 2023;31(4):463.
19. Shimojo M, Kasahara Y, Inoue M, Tsunoda S-I, Shudo Y, Kurata T, et al. A gapmer antisense oligonucleotide targeting SRRM4 is a novel therapeutic medicine for lung cancer. Sci Rep. 2019;9(1):7618.
20. Harada T, Matsumoto S, Hirota S, Kimura H, Fujii S, Kasahara Y, et al. Chemically modified antisense oligonucleotide against ARL4C inhibits primary and metastatic liver tumor growth. Mol Cancer Ther. 2019;18(3):602–12.
21. Sun Y, Yan L, Guo J, Shao J, Jia R. Downregulation of SRSF3 by antisense oligonucleotides sensitizes oral squamous cell carcinoma and breast cancer cells to paclitaxel treatment. Cancer Chemother Pharmacol. 2019;84:1133–43.
22. Ambady P, Wu YJ, Kersch CN, Walker JM, Holland S, Muldoon LL, et al. Radiation enhances the delivery of antisense oligonucleotides and improves chemo-radiation efficacy in brain tumor xenografts. Cancer Gene Ther. 2022;29(5):533–42.
23. Le BT, Raguraman P, Kosbar TR, Fletcher S, Wilton SD, Veedu RN. Antisense oligonucleotides targeting angiogenic factors as potential cancer therapeutics. Mol Ther Nucleic Acids. 2019;14:142–57.
24. Dean NM, Bennett CF. Antisense oligonucleotide-based therapeutics for cancer. Oncogene. 2003;22(56):9087–96.
25. Falese JP, Donlic A, Hargrove AE. Targeting RNA with small molecules: from fundamental principles towards the clinic. Chem Soc Rev. 2021;50(4):2224–43.
26. Quemener AM, Bachelot L, Forestier A, Donnou-Fournet E, Gilot D, Galibert MD. The powerful world of antisense oligonucleotides: from bench to bedside. Wiley Interdiscip Rev RNA. 2020;11(5):e1594.
27. Li M, Ding X, Zhang Y, Li X, Zhou H, Yang L, et al. Antisense oligonucleotides targeting lncRNA AC104041.1 induces antitumor activity through Wnt2B/β-catenin pathway in head and neck squamous cell carcinomas. Cell Death Dis. 2020;11(8):672.
28. Di Fusco D, Dinallo V, Marafini I, Figliuzzi MM, Romano B, Monteleone G. Antisense oligonucleotide: basic concepts and therapeutic application in inflammatory bowel disease. Front Pharmacol. 2019;10:305.
29. Bajan S, Hutvagner G. RNA-based therapeutics: from antisense oligonucleotides to miRNAs. Cells. 2020;9(1):137.
30. Crooke ST, Baker BF, Crooke RM, Liang X-H. Antisense technology: an overview and prospectus. Nat Rev Drug Discov. 2021;20(6):427–53.
31. Zhu Y, Zhu L, Wang X, Jin H. RNA-based therapeutics: an overview and prospectus. Cell Death Dis. 2022;13(7):644.
32. Dias N, Stein CA. Antisense oligonucleotides: basic concepts and mechanisms. Mol Cancer Ther. 2002;1(5):347–55.
33. Boiziau C, Kurfurst R, Cazenave C, Roig V, Thuong NT, Toulmé J-J. Inhibition of translation initiation by antisense oligonucleotides via an RNase-H independent mechanism. Nucleic Acids Res. 1991;19(5):1113–9.
34. Linnane E, Davey P, Zhang P, Puri S, Edbrooke M, Chiarparin E, et al. Differential uptake, kinetics and mechanisms of intracellular trafficking of next-generation antisense oligonucleotides across human cancer cell lines. Nucleic Acids Res. 2019;47(9):4375–92.
35. Gagliardi M, Ashizawa AT. The challenges and strategies of antisense oligonucleotide drug delivery. Biomedicines. 2021;9(4):433.
36. Thomas OS, Weber W. Overcoming physiological barriers to nanoparticle delivery—are we there yet? Front Bioeng Biotechnol. 2019;7:415.
37. Eder PS, DeVine RJ, Dagle JM, Walder JA. Substrate specificity and kinetics of degradation of antisense oligonucleotides by a 3′ exonuclease in plasma. Antisense Res Dev. 1991;1(2):141–51.
38. De Clercq E, Eckstein F, Merigan T. Interferon induction increased through chemical modification of a synthetic polyribonucleotide. Science. 1969;165(3898):1137–9.
39. Rifai A, Brysch W, Fadden K, Clark J, Schlingensiepen K-H. Clearance kinetics, biodistribution, and organ saturability of phosphorothioate oligodeoxynucleotides in mice. Am J Pathol. 1996;149(2):717.

40. Egholm M, Buchardt O, Christensen L, Behrens C, Freier SM, Driver DA, et al. PNA hybridizes to complementary oligonucleotides obeying the Watson–Crick hydrogen-bonding rules. Nature. 1993;365(6446):566–8.
41. Elbashir SM, Harborth J, Lendeckel W, Yalcin A, Weber K, Tuschl T. Duplexes of 21-nucleotide RNAs mediate RNA interference in cultured mammalian cells. Nature. 2001;411(6836):494–8.
42. Kim DH, Rossi JJ. Strategies for silencing human disease using RNA interference. Nat Rev Genet. 2007;8(3):173–84.
43. Mansoori B, Shotorbani SS, Baradaran B. RNA interference and its role in cancer therapy. Adv Pharm Bull. 2014;4(4):313.
44. Zhong Y, Du S, Dong Y. mRNA delivery in cancer immunotherapy. Acta Pharm Sin B. 2023; 13:1348.
45. Barbier AJ, Jiang AY, Zhang P, Wooster R, Anderson DG. The clinical progress of mRNA vaccines and immunotherapies. Nat Biotechnol. 2022;40(6):840–54.
46. Wang Y, Zhang R, Tang L, Yang L. Nonviral delivery systems of mRNA vaccines for cancer gene therapy. Pharmaceutics. 2022;14(3):512.
47. Kantoff PW, Higano CS, Shore ND, Berger ER, Small EJ, Penson DF, et al. Sipuleucel-T immunotherapy for castration-resistant prostate cancer. N Engl J Med. 2010;363(5):411–22.
48. McAllister RM. Viruses and cancer—a review. Calif Med. 1965;102(5):344.
49. Gredner T, Behrens G, Stock C, Brenner H, Mons U. Cancers due to infection and selected environmental factors: estimation of the attributable cancer burden in Germany. Dtsch Arztebl Int. 2018;115(35–36):586.
50. Pardi N, Hogan MJ, Porter FW, Weissman D. mRNA vaccines—a new era in vaccinology. Nat Rev Drug Discov. 2018;17(4):261–79.
51. Mahvi DM, Sheehy MJ, Yang NS. DNA cancer vaccines: a gene gun approach. Immunol Cell Biol. 1997;75(5):456–60.
52. Van Meirvenne S, Straetman L, Heirman C, Dullaers M, De Greef C, Van Tendeloo V, et al. Efficient genetic modification of murine dendritic cells by electroporation with mRNA. Cancer Gene Ther. 2002;9(9):787–97.
53. Thomas CE, Ehrhardt A, Kay MA. Progress and problems with the use of viral vectors for gene therapy. Nat Rev Genet. 2003;4(5):346–58.
54. Chen L, Liu W-q, Lei J-h, Guan F, Li M-j, Song W-j, et al. Chronic Schistosoma japonicum infection reduces immune response to vaccine against hepatitis B in mice. PLoS One. 2012;7(12):e51512.
55. Li Y, Wang M, Peng X, Yang Y, Chen Q, Liu J, et al. mRNA vaccine in cancer therapy: current advance and future outlook. Clin Transl Med. 2023;13(8):e1384.
56. Pardoll DM. The blockade of immune checkpoints in cancer immunotherapy. Nat Rev Cancer. 2012;12(4):252–64.
57. Vinay DS, Ryan EP, Pawelec G, Talib WH, Stagg J, Elkord E, et al. Immune evasion in cancer: mechanistic basis and therapeutic strategies. Semin Cancer Biol. 2015;35(Suppl):S185–98.
58. Whiteside T. The tumor microenvironment and its role in promoting tumor growth. Oncogene. 2008;27(45):5904–12.
59. Chen L, Flies DB. Molecular mechanisms of T cell co-stimulation and co-inhibition. Nat Rev Immunol. 2013;13(4):227–42.
60. Lim S, Phillips JB, Madeira da Silva L, Zhou M, Fodstad O, Owen LB, et al. Interplay between immune checkpoint proteins and cellular metabolism. Cancer Res. 2017;77(6):1245–9.
61. Marin-Acevedo JA, Dholaria B, Soyano AE, Knutson KL, Chumsri S, Lou Y. Next generation of immune checkpoint therapy in cancer: new developments and challenges. J Hematol Oncol. 2018;11:1–20.
62. Goldberg MV, Drake CG. LAG-3 in cancer immunotherapy. Curr Top Microbiol Immunol. 2011;344:269–78.
63. Anderson AC, Joller N, Kuchroo VK. Lag-3, Tim-3, and TIGIT: co-inhibitory receptors with specialized functions in immune regulation. Immunity. 2016;44(5):989–1004.
64. Andrews LP, Marciscano AE, Drake CG, Vignali DA. LAG 3 (CD 223) as a cancer immunotherapy target. Immunol Rev. 2017;276(1):80–96.
65. Stein LD. End of the beginning. Nature. 2004;431(7011):915–6.
66. Shek D, Read SA, Akhuba L, Qiao L, Gao B, Nagrial A, et al. Non-coding RNA and immune-checkpoint inhibitors: friends or foes? Immunotherapy. 2020;12(7):513–29.
67. Fire A, Xu S, Montgomery MK, Kostas SA, Driver SE, Mello CC. Potent and specific genetic interference by double-stranded RNA in Caenorhabditis elegans. Nature. 1998;391(6669):806–11.
68. Wang W-T, Han C, Sun Y-M, Chen T-Q, Chen Y-Q. Noncoding RNAs in cancer therapy resistance and targeted drug development. J Hematol Oncol. 2019;12(1):1–15.
69. Byrne M, Tzekov R, Wang Y, Rodgers A, Cardia J, Ford G, et al. Novel hydrophobically modified asymmetric RNAi compounds (sd-rxRNA) demonstrate robust efficacy in the eye. J Ocul Pharmacol Ther. 2013;29(10):855–64.
70. Zhou Y-J, Wan W-J, Tong Y, Chen M-T, Wang D-D, Wang Y, et al. Stimuli-responsive nanoparticles for the codelivery of chemotherapeutic agents doxorubicin and siPD-L1 to enhance the antitumor effect. J Biomed Mater Res B Appl Biomater. 2020;108(4):1710–24.
71. Ligtenberg MA, de Coaña YP, Shmushkovich T, Yoshimoto Y, Truxova I, Yang Y, et al. Self-delivering RNAi targeting PD-1 improves

tumor-specific T cell functionality for adoptive cell therapy of malignant melanoma. Mol Ther. 2018;26(6):1482–93.
72. Lian S, Xie R, Ye Y, Xie X, Li S, Lu Y, et al. Simultaneous blocking of CD47 and PD-L1 increases innate and adaptive cancer immune responses and cytokine release. EBioMedicine. 2019;42:281–95.
73. Zheng Y, Tai W. Insight into the siRNA transmembrane delivery—from cholesterol conjugating to tagging. Wiley Interdiscip Rev Nanomed Nanobiotechnol. 2020;12(3):e1606.
74. Pilch YH, de Kernion JB, Skinner DG, Ramming KP, Schick PM, Fritze D, et al. Immunotherapy of cancer with "immune" RNA: a preliminary report. Am J Surg. 1976;132(5):631–7.
75. Steele G Jr, Wang BS, Richie J, Wilson RE, Ervin T, Yankee R, et al. In vivo effect and parallel in vitro lymphocyte-mediated tumor cytolysis after phase I xenogeneic immune RNA treatment of patients with widespread melanoma or metastatic renal cell carcinoma. Cancer Res. 1980;40(7):2377–82.
76. Miao L, Zhang Y, Huang L. mRNA vaccine for cancer immunotherapy. Mol Cancer. 2021;20(1):1–23.
77. Dailey GP, Crosby EJ, Hartman ZC. Cancer vaccine strategies using self-replicating RNA viral platforms. Cancer Gene Ther. 2023;30(6):794–802.
78. Fuller DH, Berglund P. Amplifying RNA vaccine development. N Engl J Med. 2020;382(25):2469–71.
79. Lundstrom K. Self-amplifying RNA viruses as RNA vaccines. Int J Mol Sci. 2020;21(14):5130.
80. Perri S, Greer CE, Thudium K, Doe B, Legg H, Liu H, et al. An alphavirus replicon particle chimera derived from venezuelan equine encephalitis and sindbis viruses is a potent gene-based vaccine delivery vector. J Virol. 2003;77(19):10394–403.
81. Pushko P, Parker M, Ludwig GV, Davis NL, Johnston RE, Smith JF. Replicon-helper systems from attenuated venezuelan equine encephalitis virus: expression of heterologous genes in vitro and immunization against heterologous pathogensin vivo. Virology. 1997;239(2):389–401.
82. Weber LW, Bowne WB, Wolchok JD, Srinivasan R, Qin J, Moroi Y, et al. Tumor immunity and autoimmunity induced by immunization with homologous DNA. J Clin Invest. 1998;102(6):1258–64.
83. Lee S-H, Danishmalik SN, Sin J-I. DNA vaccines, electroporation and their applications in cancer treatment. Hum Vaccin Immunother. 2015;11(8):1889–900.
84. Leitner WW, Hwang LN, Bergmann-Leitner ES, Finkelstein SE, Frank S, Restifo NP. Apoptosis is essential for the increased efficacy of alphaviral replicase-based DNA vaccines. Vaccine. 2004;22(11–12):1537–44.
85. Leitner WW, Bergmann-Leitner ES, Hwang LN, Restifo NP. Type I interferons are essential for the efficacy of replicase-based DNA vaccines. Vaccine. 2006;24(24):5110–8.
86. van de Wall S, Ljungberg K, Ip PP, Boerma A, Knudsen ML, Nijman HW, et al. Potent therapeutic efficacy of an alphavirus replicon DNA vaccine expressing human papilloma virus E6 and E7 antigens. Onco Targets Ther. 2018;7(10):e1487913.
87. Tews BA, Meyers G. Self-replicating RNA. In: RNA vaccines: methods and protocols. Berlin: Springer; 2017. p. 15–35.
88. Lundstrom K. Replicon RNA viral vectors as vaccines. Vaccine. 2016;4(4):39.
89. Ljungberg K, Liljeström P. Self-replicating alphavirus RNA vaccines. Expert Rev Vaccines. 2015;14(2):177–94.
90. Bringmann A, Held SAE, Heine A, Brossart P. RNA vaccines in cancer treatment. J Biomed Biotechnol. 2010;2010:623687.
91. Sahin U, Oehm P, Derhovanessian E, Jabulowsky RA, Vormehr M, Gold M, et al. An RNA vaccine drives immunity in checkpoint-inhibitor-treated melanoma. Nature. 2020;585(7823):107–12.
92. Xu H, Zheng X, Zhang S, Yi X, Zhang T, Wei Q, et al. Tumor antigens and immune subtypes guided mRNA vaccine development for kidney renal clear cell carcinoma. Mol Cancer. 2021;20(1):1–7.
93. Tahtinen S, Tong A-J, Himmels P, Oh J, Paler-Martinez A, Kim L, et al. IL-1 and IL-1ra are key regulators of the inflammatory response to RNA vaccines. Nat Immunol. 2022;23(4):532–42.
94. Huang R, Li X, He Y, Zhu W, Gao L, Liu Y, et al. Recent advances in CAR-T cell engineering. J Hematol Oncol. 2020;13(1):1–19.
95. Chen DS, Mellman I. Elements of cancer immunity and the cancer–immune set point. Nature. 2017;541(7637):321–30.
96. Lim WA, June CH. The principles of engineering immune cells to treat cancer. Cell. 2017;168(4):724–40.
97. Gill S, June CH. Going viral: chimeric antigen receptor T-cell therapy for hematological malignancies. Immunol Rev. 2015;263(1):68–89.
98. Bray F, Ferlay J, Soerjomataram I, Siegel RL, Torre LA, Jemal A. Global cancer statistics 2018: GLOBOCAN estimates of incidence and mortality worldwide for 36 cancers in 185 countries. CA Cancer J Clin. 2018;68(6):394–424.
99. Zhang Z, Li D, Yun H, Liu W, Chai K, Tong J, et al. CAR-T cells in the treatment of urologic neoplasms: present and future. Front Oncol. 2022;12:915171.
100. Wagner J, Wickman E, DeRenzo C, Gottschalk S. CAR T cell therapy for solid tumors: bright future or dark reality? Mol Ther. 2020;28(11):2320–39.
101. Zhang C, Fang L, Wang X, Yuan S, Li W, Tian W, et al. Oncolytic adenovirus-mediated expression of decorin facilitates CAIX-targeting CAR-T therapy against renal cell carcinoma. Mol Ther Oncolytics. 2022;24:14–25.

102. Junghans RP, Ma Q, Rathore R, Gomes EM, Bais AJ, Lo AS, et al. Phase I trial of anti-PSMA designer CAR-T cells in prostate cancer: possible role for interacting interleukin 2-T cell pharmacodynamics as a determinant of clinical response. Prostate. 2016;76(14):1257–70.
103. Kim K-H, Kim H. Progress of antibody-based inhibitors of the HGF–cMET axis in cancer therapy. Exp Mol Med. 2017;49(3):e307.
104. Zhuang P-H, Xu L, Gao L, Lu W, Ruan L-T, Yang J. Correlations of microvascular blood flow of contrast-enhanced ultrasound and HGF/c-Met signaling pathway with clinicopathological features and prognosis of patients with hepatocellular carcinoma. Onco Targets Ther. 2017;10:847–57.
105. Salgia R. MET in lung cancer: biomarker selection based on scientific rationale. Mol Cancer Ther. 2017;16(4):555–65.
106. Miranda O, Farooqui M, Siegfried JM. Status of agents targeting the HGF/c-Met axis in lung cancer. Cancers. 2018;10(9):280.
107. Peschard P, Park M. From Tpr-Met to Met, tumorigenesis and tubes. Oncogene. 2007;26(9):1276–85.
108. Danilkovitch-Miagkova A, Zbar B. Dysregulation of Met receptor tyrosine kinase activity in invasive tumors. J Clin Invest. 2002;109(7):863–7.
109. Kolenda T, Guglas K, Baranowski D, Sobocińska J, Kopczyńska M, Teresiak A, et al. cfRNAs as biomarkers in oncology—still experimental or applied tool for personalized medicine already? Rep Pract Oncol Radiother. 2020;25(5):783–92.
110. Gilles M-E, Hao L, Huang L, Rupaimoole R, Lopez-Casas PP, Pulver E, et al. Personalized RNA medicine for pancreatic cancer. Clin Cancer Res. 2018;24(7):1734–47.
111. Chiocchetti A, Cappellano G, Dianzani U. To each his own: a personalized vaccine for metastatic melanoma. Gland Surg. 2019;8(4):329.
112. Shibata H, Zhou L, Xu N, Egloff AM, Uppaluri R. Personalized cancer vaccination in head and neck cancer. Cancer Sci. 2021;112(3):978–88.
113. Cafri G, Gartner JJ, Zaks T, Hopson K, Levin N, Paria BC, et al. mRNA vaccine-induced neoantigen-specific T cell immunity in patients with gastrointestinal cancer. J Clin Invest. 2020;130(11):5976–88.
114. Deng Z, Tian Y, Song J, An G, Yang P. mRNA vaccines: the dawn of a new era of cancer immunotherapy. Front Immunol. 2022;13:887125.
115. Damase TR, Sukhovershin R, Boada C, Taraballi F, Pettigrew RI, Cooke JP. The limitless future of RNA therapeutics. Front Bioeng Biotechnol. 2021;9:161.
116. Huang X, Zhang G, Tang T-Y, Gao X, Liang T-B. Personalized pancreatic cancer therapy: from the perspective of mRNA vaccine. Mil Med Res. 2022;9(1):53.
117. Sahin U, Derhovanessian E, Miller M, Kloke B-P, Simon P, Löwer M, et al. Personalized RNA mutanome vaccines mobilize poly-specific therapeutic immunity against cancer. Nature. 2017;547(7662):222–6.
118. Hoerr I, Obst R, Rammensee HG, Jung G. In vivo application of RNA leads to induction of specific cytotoxic T lymphocytes and antibodies. Eur J Immunol. 2000;30(1):1–7.
119. Thess A, Grund S, Mui BL, Hope MJ, Baumhof P, Fotin-Mleczek M, et al. Sequence-engineered mRNA without chemical nucleoside modifications enables an effective protein therapy in large animals. Mol Ther. 2015;23(9):1456–64.

Ethical Concerns in Novel Medicine 11

Navid Ravan, Mandana Hasanzad, Bagher Larijani, and Hamidreza Namazi

What Will You Learn in This Chapter?
We investigated the complicated ethical issues underlying the development and use of novel medical technology. One of the most important things we obtained was the significance of considering the prospective advantages of novel treatments against the ethical ideals of autonomy, beneficence, nonmaleficence, and fairness.

N. Ravan
Pharmaceutical Sciences Research Center, The Institute of Pharmaceutical Sciences, Tehran University of Medical Sciences, Tehran, Iran

M. Hasanzad
Personalized Medicine Research Center, Endocrinology and Metabolism Clinical Sciences Institute, Tehran University of Medical Sciences, Tehran, Iran

B. Larijani
Endocrinology and Metabolism Research Center, Endocrinology and Metabolism Clinical Sciences Institute, Tehran University of Medical Sciences, Tehran, Iran

H. Namazi (✉)
Medical Ethics and History of Medicine Research Center, Tehran University of Medical Sciences, Tehran, Iran

Department of Medical Ethics, School of Medicine, Tehran University of Medical Sciences, Tehran, Iran

Department of History of Medicine, Tehran University of Medical Sciences, Tehran, Iran

Rational and Importance
The protection of individual rights and autonomy is very important. Novel medical interventions may involve significant risks and uncertainties, and considering ethical principles such as informed consent allows individuals to make autonomous decisions about their healthcare. And also, ethical scrutiny helps mitigate potential harms and ensures the safety of patients and research participants. By identifying and addressing ethical issues early in the development process, we can minimize adverse effects and maximize the benefits of new medical technologies. Therefore, addressing ethical concerns promotes public trust and confidence in healthcare systems and scientific research.

11.1 New Approaches to (Bio)ethics and a Pace Further for Addressing Ethical Issues of Novel Medicine

Ethics is a branch of philosophy that focuses on the moral principles and values of human action. In common sense, ethics is tied to judging human behaviors and actions and categorizing them as good or bad. However, ethics as a field of inquiry is a reflection based on human action as a wise and autonomous entity. Ethics can be applied to any realm of human action.

M. Hasanzad (ed.), *A Glimpse at Medicine in the Future*,
https://doi.org/10.1007/978-981-97-4974-4_11

As one of the oldest human activities, medicine is also an area for ethical considerations and reflections.

Medical ethics, also known as bioethics in a broader sense, investigates the ethical issues and implications of medicine. The history of investigating ethical subjects in medicine goes back to the Hippocratic Oath [1]. However, the term medical ethics was first used in the nineteenth century by Thomas Percival [1]. In this chapter, we use bioethics instead of medical ethics to address a broader set of issues like what is rooted in biomedical sciences.

As biomedical sciences evolve progressively, new ethical issues come into play increasingly. We believe that to address the issues in novel medicine and future medicine, we need to rethink bioethics and consider it a critical, reflective, and argumentative field. Moreover, we support philosophical bioethics instead of mere code writing.

In this chapter, we first highlight fields that are a step further and complementary to medicine. Then, with this framework, we investigate the philosophical and ethical challenges of some pioneer fields in novel medicine. We cannot address all the ethical issues of novel medicine in one single chapter. Thus we aim to introduce a framework considering new trends in bioethics and some examples of the implications of these new approaches in the philosophical and ethical investigation of different fields of novel medicine.

11.2 Philosophical Bioethics Versus Bioethics as a Common Morality

Bioethics is practiced from at least two main perspectives. While some bioethicists consider bioethics as a philosophical field that needs philosophical reflections, others believe that bioethics doesn't need theory [2, 3]. Bioethics without theory is not always, but usually rooted in a common morality approach to bioethics. The common morality approach is based on recognizing moral values and virtues and advises medical practitioners to act morally—just like all people who should act and live morally [2]. Philosophy-free bioethics led to increasing code writing for every circumstance in medicine and medical encounters [4].

The overload of codes in medicine is what can be called codification of medicine. This can cause the over-ethicalization of medicine. Over-ethicalization occurs when "everything" in medicine is codified. This leads to moral overload and moral fatigue in medical practitioners. Moral overload would cause burnout as an individual and defensive medicine as a professional consequence. Defensive medicine is an unwanted way of practicing medicine by which a medical practitioner only acts in a way to prevent blaming or punishing. It especially occurs when ethics change into laws and codes of ethics are redefined as legal necessities [5].

Philosophy-free bioethics, however, cannot be blamed or considered a useless superficial attempt. Besides, philosophical bioethics is not superior only because it is philosophical, and being philosophical is presumed a value. Instead, we claim that merely clinging to bioethics codes is insufficient for emerging ethical issues of novel medicine. For novel medicine, we need an always-renewing and critical framework of ethics.

We believe that for addressing ethical issues in novel medicine, we need a return to the reflective roots of modern ethics. As Immanuel Kant in *Groundwork for Metaphysics of Morals* mentioned, acting based on the superior principle of ethics is tied to the freedom of human beings. This means that only the one who is liberated from extrinsic determinants of action acts in compliance with the intrinsic principle of ethics, which is acceptable by all rational agents. This intrinsic principle is: "act only in accordance with that maxim through which you at the same time can will that it become a universal law" [6].

It may seem like a simple principle, but this idea is importantly tied to freedom and rationality [6, 7]. Moreover, as a groundwork for the modern philosophy of ethics, it refuses any extrinsic ethical rule. However, many different approaches toward ethics have been introduced after Kant; it is important to still adhere to this immanent reflective ethics, which is appreciating reason and freedom as important values in the contemporary world.

To clarify, it is worth noting that having one superior ethical principle in Kant's philosophy of ethics does not mean that we couldn't recognize ethical values and intercultural diversities. Instead, Kant attempted to introduce ethics based on reason in the field of philosophy of ethics and not involve pragmatic judgment of single actions or values. He also doesn't deny the need for ethical trade-offs. So we can still obey the Kantian framework parallel with considering different ethical values.

11.3 Medicalization, Pharmaceuticalization, and the Moral Expansion of Medicine

Medicalization is a process by which what was not previously a medical problem is defined as a medical issue that requires medical observation and intervention [8–10]. For example, in the past decades, sadness has been defined as a depressive disorder [8] and shyness as a social anxiety disorder [11]. Addiction is no longer a crime but rather a disease. Different stages of life, such as birth [12], menopause [13–15], old age [8], and death [16], have entered the field of medicine.

Medicalization involves not only the transformation of everyday conditions into treatable diseases but also the expansion of medical considerations of health maintenance and lifestyle [8]. For example, although nutrition and physical activity are not treatable problems, they are nowadays within the framework of medical considerations, so even ordinary people use medicalized words to describe their behaviors in this field (such as intake and burning of calories or body mass index).

In addition, enhancement has also entered the field of medical considerations [8, 17]. Cosmetic surgeries are an example of the medicalization of beauty, and increasing the consumption of food supplements to increase physical and mental strength, or sports supplements to increase the strength and beauty of muscles and body organs, is an example of the expansion of medicalization to issues related to improvement. Medicalization studies researchers have often tried to show the negative aspects and adverse consequences of medicalization in the aforementioned cases.

To explain the different aspects of medicalization, in the following, the medicalization of birth will be expanded. The example of the medicalization of birth is important because it shows how a wide range of everyday situations have entered the realm of medicine. In the last few decades, from the medicalization of sexual behavior, intercourse, and contraception to the medicalization of pregnancy, childbirth, and postpartum care of the mother and neonate, all have entered the realm of medicine. For example, pregnancy and childbirth were non-medical processes until about a century ago. The pregnant person was taken care of by the family members at home, and the delivery was done in the private home of the people. But today, in middle- and high-income countries, a pregnant person visits the doctor several times during pregnancy, and there are many medical considerations regarding pregnancy. Finally, delivery and subsequent care of the mother and neonate are done in the hospital under the supervision of the medical institution.

Similar to our point of view, critics may object that ironically, the medicalization of birth has prevented many congenital malformations or maternal and fetal deaths during pregnancy and childbirth. According to the statistics of the World Health Organization, the maternal mortality rate has decreased by 44% from 1990 to 2015 [18] and the infant mortality rate has increased from 5 million to 2.4 million from 1990 to 2020 [19]. Critics will generally say that medicalization of various problems of daily life can give consistency and completeness to the symptoms or problems of the patient, make the problem more understandable, and facilitate the management of the disease by the patient [8]. As Talcott Parsons, in his classic formulation of the sick role, shows that medical diagnosis and consideration of a condition can justify and accept a person's problems [20]. For example, although the decriminalization of addiction and defining it as a medical condition can lead to the neglect of its social aspects, on the other hand, it can help the social acceptance of addiction and encourage more individual and social efforts to solve the problem of addiction [8].

To show the relevance of medicalization with ethical issues in novel medicine, we review the history of medicalization theories and critical perspectives. The first critics of medicalization like Irving Zola and Ivan Illich were sociologists. They criticized medicalization as a negative process that harnesses human autonomy and destroys society [21–23]. Feminism approach also came into play, but at the beginning, they also presumed medicalization was a negative process that evokes gender disparities [24].

In 2007, *Lancet* published a special issue [25] about medicalization, which could be considered a turn in medicalization studies. Nikolas Rose, in his article *Beyond Medicalisation* in this issue, claimed that medicalization is increasingly becoming a neutral concept [26]. Therefore, the role of medicalization studies is changing from blaming doctors and the institution of medicine for making people sick to critically reflecting on the essence of medicalization and its consequences in a concrete way [26].

This special issue attracts attention to the concept of medicalization from inside the medical community and shows the way to rethink medicalization. Based on this rethinking, new approaches toward medicalization highlight the role of technoscience [27], pharmaceutical companies [28–30], and society and individuals [8, 30] for medicalization instead of only blaming medical doctors.

In this manner, biomedicalization, which emphasized the importance of technoscience in the expansion of medicine, and pharmaceuticalization, which focused on the role of pharmaceutical companies in this process, emerged. Pharmaceuticalization is the process of defining everyday life as problems that should be treated by medications [28–30].

(Bio)medicalization and pharmaceuticalization can be worsened by many trends in novel medicine, as we will show, P4 medicine, AI, and many other trends. The ethical problem of these processes is that patients become increasingly dependent on medicine to cope with every single problem. This can harm the autonomy of the patient. Moreover, the personal lived narratives of patients would be replaced by medical narratives that can lead to existential problems and the dehumanization of medicine.

Furthermore, concerns related to justice should be taken into consideration. As Clarke and colleagues mentioned, we face stratified biomedicalization, which means that rich people are medicalized more than poor people and high-income countries more than low-income countries [10]. In other words, biomedicalization results in the disparity of resource allocation. Moreover, it has been argued that medicalization can lead to gender discrimination too.

Howard Brody has put drug-centered care in front of patient-centered care to portray an ethical problem of the expansion of medicine and medical narratives (e.g., serotonin theory for depression). He claims that drug-centered care is a condition in which the patient is neglected and the benefits of pharmaceutical companies and other stakeholders are prioritized. This leads to the destruction of patient-centered care that acknowledges the patient as an autonomous entity and respects her dignity [31].

New pathways in medicalization studies are becoming more reflective and considering all parties and stakeholders share in the process of medicalization and pharmaceuticalization. For instance, Emilia Kaczmarek stated that:

> *The pharmaceutical industry functions in a culture which has already been medicalized. Therefore, its actions should be seen in the context of other broad social changes that may be observed in Western societies since the twentieth century, such as secularization, consumerism, the cult of youth, healthism, or population aging* [30].

This is interestingly correlated with the idea of Michel Foucault who believes that there is no black institution of power. But instead, power determines the interactions of all parties in the society [32, 33]. It is an important insight for ethical and philosophical investigation of different aspects of medicalization and parallel processes evoked by novel medicine and trends for the future.

Another insightful approach toward medicalization is scrutinizing it in a broader context of "the moral expansion of medicine," or a more exact phrase, "the expansion of the subject matter of the moral imperative of medicine," which Bjørn Hofmann has introduced and elaborated [34]. This enabled him to investigate the expansion of medicine as a neutral concept that can be even more beneficial.

With a radical investigation, he suggested that there is an epistemic difference between present outcomes and events future probable events in favor of the present. Besides, he argues that there is an asymmetry in negative and positive conditions, i.e., disease and health, in favor of negative ones (disease). With this viewpoint, he claims that we should prioritize the treatment of present disease over the enhancement of future well-being [34].

To sum up, medicalization and correlated concepts are inevitable results of novel medicine. To investigate the ethical and philosophical implications of novel medicine, we should take a neutral and concrete position for analyzing medicalization.

11.4 Fraud or Epistemic Corruption? Traditional Research Ethics vs. Meta-Research

Meta-research is a field of inquiry that examines research itself as a subject [35]. Meta-research is interdisciplinary, critical, and theoretical. Reflection on science and the way we reach scientific facts (=research) has a long history. However, meta-research is a young term. So, the question is: "With this long history, what kind of novelty meta-research brings for us?". To answer that question, we briefly review the history of thinking critically about science and research.

Similar to all reflective attempts, thinking about science has a beginning in ancient Greece. The way that humankind can reach the knowledge of nature is one of the topics that can be found in Plato's and Aristotle's works. In this era, the acquisition of knowledge from nature is valuable in a teleological context. With the rise of modern philosophy, teleology disappeared and gave its place to a neutral approach to nature and science [36]. Philosophy of science is a leading field that emerged in the nineteenth century. The philosophy of science is a philosophical reflection on science and its issues.

In 2005, John Ioannidis wrote an important article, "Why most published research findings are false?" [37]. It is believed that this article gave birth to the field of meta-research. Only 11 years after this article, a study published in Nature asked 1576 scientists from different fields about the existence and the magnitude of the replication crisis. Ninety percent said that a replication crisis exists to some extent [38]. This shows that only in 11 years, attention to concerns of meta-research has been increased considerably. The reason is that meta-research emerged within the realm of science and was introduced by a well-known scientist in epidemiology. This is also why meta-research is different from previous schools of research on research.

Meta-research and traditional research ethics represent the classic dichotomy of structure and agency. Traditional research ethics focus on intentional fraud, misconduct, conflict of interests, and similar deliberate corruption in research. This approach implicitly accuses researchers or supporting institutions of research for impaired research results. Conversely, meta-research seeks for structural issues in research. For instance, meta-research focuses on rethinking the *P*-value [39] or norms of designing and performing randomized clinical trials [40]. Thus, it highlights the structural impairment and corruption.

Epistemic corruption is a term that Sergio Sismondo introduced to point to the structural impairment in research. Epistemic corruption occurs "when a knowledge system importantly loses integrity, ceasing to provide the kinds of trusted knowledge expected of it" [40]. Considering epistemic corruption is a pace further than classic research ethics, which simply tries to give some advice to researchers.

Ethical issues in novel medicine with no doubt can be addressed by new frameworks of ethical considerations. Although ways to prevent fraud and misconduct are still necessary, meta-research and parallel fields of inquiry are needed to prevent new problems in novel medicine.

11.5 Ethical and Philosophical Issues of Technology in Medicine and Medicine as a Technology

Novel medicine is heavily tied to technology. The relationship between technology and medicine is controversial and has been scrutinized from different perspectives. Before starting the section, it is worth clarifying that technology can be seen in different ways.

From a traditional perspective, technology can be seen as hardware and tools. With this perspective, stethoscopes and CT scanners are technologies in medicine. However, technology is not limited to tools. It can be seen as a rule. In this approach, technology is the rationality that is used to change the world by humans. Technology can also be seen as a socio-technical system. In this approach, technology is fully dependent on its human context. In the medical setting, the hospital could be an example of technology from a socio-technical viewpoint [41].

Technology in a broad sense that we have accepted is the rationality of changing the world by human intention. Thus, medicine itself can also be seen as a technology. While there is no consensus agreement between philosophers of medicine whether medicine is art, science, technology, phronesis, or a blend of them, considering medicine itself as a philosophy sheds light on some important aspects of medicine. For instance, this perspective helps us to acquire insights from the philosophy of technology and reflections on the essence of technology to deepen our reflection on the essence of medicine and its ethical implications.

11.5.1 Technology Shapes the Experience of Health, Illness, and Disease

Some philosophers of technology like Peter Verbeek and Don Ihde believe that technology mediates between humans and the world [42–44]. In other words, we perceive and interpret the world through the lens of technology. Technology can also affect the human life world [43]. This means that technology determines how we experience the world and how we *live*. Our bodies are deeply connected to technology, and there is no way to escape from it.

The use of technology in medicine can alter the way that we face the world. Eyeglasses are a metaphoric example that shows we see the world through the lens of technology. But all technological devices and processes in medicine do so. The use of medical imaging, for example, changes the way that the patient perceives her body and sufferings, and the use of psychopharmaceuticals in the way that a patient grasps, interprets, and copes with her environment. Hofmann and Svenaeus have categorized the ways that technology shapes our experience of illness into six categories, including: creating awareness of disease, revealing risk factors of a developing disease, affecting the experience of an already existing illness, redefining a condition as disease instead of unfortunate, influencing illness experience via changing sociocultural values and norms, and influencing the experience of being healthy in contrast to being ill [45] .

Technology can also change the perception of ability and vulnerability. Even simple medical devices such as wheelchairs and eyeglasses can considerably change the perception of the person from the world and positively affect her quality of life [46]. From the perspective of the phenomenology of medicine, the lived experience of the patient plays an important role in medical effectiveness. So, we should take the lived experience into account.

Hofmann also argued that technological values are dominant in contemporary medicine [47]. Technology in his view is not a neutral tool but a value-laden concept. Therefore, he claims that some of the important characteristics of contemporary medicine are determined by underlying technoscientific basis. For example, technology is the desire for manipulation of nature, and it reflects in medicine in the way that medicine is becoming more interventive and focuses on treatment rather than caring and prevention. Moreover, the "abstraction from person's" values of technology results in emancipation from the subjective patient and her lived experience [47].

11.5.2 Technology Lead to the Expansion of Medicine: Biomedicalization

Clarke and colleagues in their 2003 book introduced biomedicalization as a new approach to the expansion of medicine other than medicalization [48]. They believe that classis medicalization critics ignored the importance of the role of contemporary technoscience in the expansion of medicine to other areas of human life [10, 27, 48]. Based on their viewpoint, it could be claimed that novel technologies are accelerating the expansion of medicine outside its current realm.

For example, they show that new biological and genetic technologies lead to the molecularization of human life. Molecularization is a process by which every human condition is reduced to a molecular level [10]. Even though molecularization is based on scientific research, it can substitute the lived experience and personal narration of people of their conditions.

These reflections show the necessity of considering the philosophical implications of technological medicine. Medicine is becoming more technological through novel developments; it is necessary to take the philosophy of technology into account to provide insights and inspirations for philosophical bioethics concerning novel medicine.

11.6 Ethical and Philosophical Issues of P4 Medicine

P4 medicine is a novel trend in contemporary medicine which contains four leading components: personalized, participatory, preventive, and predictive medicine. P4 medicine is also interrelated with systems medicine, which is a pioneer field of medicine originating from systems biology [49]. Vogt and colleagues have scrutinized P4 and systems medicine regarding medicalization theories. They believe that the technological holism based on P4 and systems medicine would medicalize health and life itself. In this manner, health itself is medicalized, and the focus is on health itself rather than the one who is healthy. When the subject matter of medicalization changes from disease to health, the medicalization process becomes limitless because health can be defined in such a broad sense that it can never be fully achieved [49]. However, P4 medicine has different ethical aspects other than medicalization criticisms. In the following sections, we briefly mention the ethical and philosophical implications of P4 medicine.

11.6.1 Personalized Medicine

Personalized medicine is an idea for medicine to treat each patient in a unique and personalized way. It is believed that each person with a special race, gender, and genome profile in a given environment and time would have a unique response to a given treatment or medication. Therefore, to have the most effective medications, we need to design personalized treatments for each patient [49, 50]. Although it seems interesting initially, it may have some important ethical implications that we will investigate in this section.

Personalized medicine is expensive. So we may face the problem of resource allocation for health. Even though personalized medicine could be more effective, the question is whether governments can afford it. Furthermore, the problem of equity and justice is considerable. Besides, the ethical considerations regarding resource allocation would be noticeable with the development of personalized medicine.

On the other side, although personalized medicine differs from patient-centered care and does not necessarily lead to it, the emphasis on the patient and personalization would be an opportunity for fulfilling patient-centered medicine and noticing each patient's uniqueness. This can cause more respect for the patient's dignity.

11.6.2 Participatory Medicine

The idea of participatory medicine is based on the expansion of the responsibility of health maintenance to families, communities, and societal institutions [49]. This may have positive implications, but in the context of P4 medicine, it might have undesirable consequences.

The expectation from all individuals to maintain their health and improve it would gradually alter the health to a super-value that everybody is responsible for. Healthism is a term used for this process [51, 52]. Healthism would bring about a biological citizenship [10] which biopolitics researchers have argued under the term biopower. In this condition, individuals are reduced to their health/sick bodies. This can destroy the first-hand experience of health and illness. The dominance of medical language and narratives in public discourse, along with healthism, results in the loss of lived narratives of patients.

Furthermore, the expectation from all individuals to actively seek maximum health status would cause health anxiety. Health anxiety occurs when a person is always anxious for her health and therefore unsatisfied because she cannot reach the maximum level of health. Producing health anxiety is opposed to the moral responsibility of medicine to benefit individuals and prevent harm.

11.6.3 Predictive and Preventive Medicine

Preventive medicine is an approach to medicine that focuses more on prevention rather than treatment [49]. It may seem clear that preventing disease is less expensive and less invasive than letting them develop and treating them. But preventive medicine under the P4 medicine approach is also a subject for ethical and philosophical reflection.

One of the notable counterarguments of preventive medicine is the problem of overdiagnosis. Overdiagnosis is the unnecessary use of diagnostic tools to detect those problems that would not result in any harm, injury, or death in diagnosed patients [53]. Overdiagnosis inevitably leads to overtreatment of disease. Overtreatment may put patients at risk of adverse effects of treatments, and governments are under the burden of providing budget and facilities for unnecessary care and treatment. Overdiagnosis would also cause patients to have health anxiety, which we explored its ethical implications in the previous section.

11.7 Ethical and Philosophical Issues of the Use of Artificial Intelligence (AI) in Medicine

Artificial intelligence is the product of human attempts to reproduce natural intelligence. Some of the important performances of AI are pattern recognition, learning, natural language understanding, reasoning, etc. [54]. Similar to almost all fields of human knowledge, AI can be used in medicine. It can be an assistant for medical practitioners to enhance their performance. Medical practitioners can also be replaced by AI [55]. Putting all fears, hopes, and hypes aside, in this section, we will scrutinize the ethical implications of using AI in medicine.

11.7.1 Explicability, Transparency, and Responsibility

One of the most important concerns related to the use of AI in medicine is "Who is responsible for AI decisions?". It is questionable whether AI can be seen as a legal entity and be punished for its probable faults, and if yes, what is the meaning and benefit of this punishment. It would be argued that the developer company may be responsible for AI decisions. But this only works when the process of decision-making by AI is transparent and explainable [56].

Explicability is a crucial problem of AI. Explicability is defined as the ability to explain the AI decision-making processes logically for a human entity. If we are not able to do so, we will be blind to the ethical basis of decisions made by AI [56]. Critics may argue that the process is complex and it's a basic characteristic of AI. But complexity isn't opposed to explicability. A process may be so complex that we can't do it like AI, but it can be explained simply. Conversely, a simple process could be unexplainable.

Another concern about the use of AI in medicine is the protection of personal data and data privacy. As AI uses big data, it is important to have a transparent process of using health information (i.e., electronic health records) and responsible use of those data based on the ethical principle of confidentiality of medical records [56].

11.7.2 Reproducing and Proliferating Bias

AI uses previous data to produce new knowledge. In medicine, there is a lot of data in health systems that can be used by AI. Although it is a piece of good news for AI, the bad news is that medical data are highly in danger of unwanted biases. It is still among the hot topics in medicine that women and gender minorities as well as indigenous people and color people are neglected in contemporary medical research. This means that existing data are gender- and race-biased, and these data are used by AI.

Furthermore, one of the concerns in meta-research is the reproducibility crisis. Reproducibility crisis refers to the issue that most published research, especially medical research, cannot reach the previous result if it is repeated. This means that current available medical big data cannot be fully trustable. The problem would worsen if AI reproduces and proliferates biases and doubtful medical data [56–58].

The GIGO (garbage-in, garbage-out) metaphor addresses this issue. This refers to a condition that one puts garbage in a system and the output would be garbage, at least with some modifications. When the AI is working by using biased, flawed, and doubtful data, we couldn't expect it to repair this set of data [56–58].

Moreover, medicalization would be enforced by AI. When the input of AI is medicalized, AI would take it as a fact and increase it. So the ethical and philosophical issues of medicalization as we mentioned in the previous section would worsen.

11.7.3 The Right to Refuse AI Medical Decisions

Thomas Ploug and Soren Holm have emphasized the necessity of having the right to refuse diagnostics and treatment planning by AI. Considering the informed consent right, it seems that all patients have this right to refuse treatment by AI. However, they argue that there is also a need for an alternative to AI. That means a patient not only has the right to refuse diagnosis and treatment by AI but also the right to choose an alternative, i.e., by a human doctor [59].

They believe that patients may want to refuse AI for many reasons. Regarding the ethical importance of respecting the autonomy of the patient and the value of democracy, we should provide and maintain the right to refuse AI even in the case that the patient cannot explain the reason rationally [59].

As we mentioned, AI might reproduce biases and proliferate discrimination against gender and racial minorities. So the right to refuse AI decisions and have an alternative should be provided for every single patient. Furthermore, it should be considered that we should preserve the right of the next generation to have this choice.

11.7.4 AI in the Light of the Philosophy of Technology

Some researchers investigated the use of AI in medicine in light of the philosophy of technology. Considering both medicine and AI as technologies would bring about insights for analyzing the effect of AI on medicine. Kyle E. Karches has investigated AI in medicine in light of Heidegger and Borgmann's philosophy of technology [60].

Martin Heidegger, in his distinguished book *The Question Concerning Technology,* introduced Gestell as the essence of technology. *Gestell* shows nature as a ready subject for manipulation and exploitation. For instance, *Gestell* leads us to see a tree as some piece of lumber [60, 61]. Using this philosophical framework, Karches believes that using AI in medicine would bring about seeing health as a commodity and healthcare as a production line. Thus, it may be harmful to both doctors and patients.

Philosophy of technology reflections on AI can also provide some invaluable insights into the use and effectiveness of AI in the healthcare context. Some critics of current trends in AI believe that artificial intelligence is a human attempt to reproduce natural intelligence. As the view of human intelligence is heavily based on the Cartesian theory of knowledge, we have missed some important aspects of intelligence in AI [60].

Descartes made a deep distinction between the mind as a thinking substance and the body as an extended substance [62]. It seems this distinction is also reflected in the contemporary view of AI [60]. Developers of AI seem to believe that they can produce intelligence without the body. But philosophers of technology in the phenomenological school like Dreyfus usually argue that we cannot achieve intelligence without having sensation, emotion, imagination, and socialization [60, 63]. These characteristics are rooted in the embodiment that contemporary AI is deprived of. As Maurice Merleau-Ponty stated, we cannot achieve maximum grasp without having body and bodily perception [60, 64, 65].

11.7.5 Gift of Time: The Time for Empathy and Improving the Doctor–Patient Relationship

A positive implication of AI would be the fact that using AI can emancipate medical practitioners from many works, such as pattern recognition in radiography, and that AI can perform better than a human agent. Given that AI would not replace medical doctors completely, this can save time and energy as Eric Topol states it is a gift of time [66]. This can be an opportunity to deepen the humanistic aspect of the patient–doctor relationship and increase the empathy of medical practitioners.

11.8 Conclusion

Medicine as a field that has an imperative to benefit patients and avoid harm has a deep relation to ethics. Ethical reflection on medicine is not a new field. However, novel medicine brings about new concerns that need philosophical and ethical investigations. We do not claim that we can address all ethical aspects of all fields in novel medicine. Rather, we attempted to provide a framework for this reflection. Such a framework would be fulfilled by rethinking the basics of bioethics and acknowledging new turns in contemporary bioethics. We do believe that we cannot achieve an efficient investigation without an updated bioethics framework. Thus, we provided a reflective philosophical bioethics framework to address ethical issues in novel medicine and investigate some pioneer fields of novel medicine using this framework.

Acknowledgments Authors are grateful to express their thanks to Dr. Mehran Baskhah, who helped in framing the idea of this chapter in the initial steps.

Conflicts of Interests The authors declare to have no conflicts of interest.

References

1. Jonsen AR. A short history of medical ethics. Oxford: Oxford University Press; 2000.
2. Rhodes R. Why not common morality? J Med Ethics. 2019;45:770.
3. Holm S. Bioethics without theory? Camb Q Healthc Ethics. 2024;33(2):159–66.
4. Stempsey WE. Medical humanities and philosophy: is the universe expanding or contracting? Med Health Care Philos. 2007;10:373–83.

5. Tancredi LR, Barondess JA. The problem of defensive medicine. Science. 1978;200(4344):879–82.
6. Kant I, Schneewind JB. Groundwork for the metaphysics of morals. New Haven: Yale University Press; 2002.
7. Kant I. Critique of practical reason. Indianapolis: Hackett Publishing; 2002.
8. Conrad P. The medicalization of society: on the transformation of human conditions into treatable disorders. Baltimore: Johns Hopkins University Press; 2007.
9. Kaczmarek E. How to distinguish medicalization from over-medicalization? Med Health Care Philos. 2019;22(1):119–28.
10. Clarke AE, Jeske M, Mamo L, Shim JK. Biomedicalization revisited. In: The Wiley Blackwell companion to medical sociology. New York: Wiley; 2021. p. 125–49.
11. Scott S. The medicalisation of shyness: from social misfits to social fitness. Social Health Illn. 2006;28(2):133–53.
12. Ballard K, Elston MA. Medicalisation: a multidimensional concept. Soc Theory Health. 2005;3(3):228–41.
13. Bell SE. Sociological perspectives on the medicalization of menopause. Ann N Y Acad Sci. 1990;592(1):173–8.
14. Bell SE. Changing ideas: the medicalization of menopause. Soc Sci Med. 1987;24(6):535–42.
15. McCrea FB, The politics of menopause: The "discovery" of a deficiency disease. Soc Probl. 1983;31(1):111–23.
16. Goh SAE, Cynthia R. Medicalization of dying: are we turning the corner? J Palliat Med. 2012;15(7):728–9.
17. Poli Neto P, Caponi SNC. The 'medicalization' of beauty. Interface. 2007;11:569–84.
18. World Health Organization. 2015. https://www.who.int/news/item/12-11-2015-maternal-deaths-fell-44-since-1990-un#:~:text=Maternal%20deaths%20fell%2044%25%20since%201990%20%E2%80%93%20UN.
19. World Health Organization. 2021. https://www.who.int/news-room/fact-sheets/detail/levels-and-trends-in-child-mortality-report-2021#:~:text=Globally%202.4%20million%20children%20died,up%20from%2040%25%20in%201990.
20. Parsons T. The social system. New York, NY: Free Press; 1951. p. 575.
21. Zola IK. Medicine as an institution of social control. Sociol Rev. 1972;20(4):487–504.
22. Illich I. Medical nemesis: Citeseer. London: Calder & Boyars; 1975.
23. Illich I. The medicalization of life. J Med Ethics. 1975;1(2):73–7.
24. Riessman CK. Women and medicalization: a new perspective. Soc Policy. 1983;14(1):3–18.
25. Metzl JM, Herzig RM. Medicalisation in the 21st century: introduction. Lancet. 2007;9562(369):697–8.
26. Rose N. Beyond medicalisation. Lancet. 2007;369(9562):700–2.
27. Clarke AE, Mamo L, Fosket JR, Fishman JR, Shim JK. Biomedicalization: technoscience, health, and illness in the US. Durham: Duke University Press; 2010.
28. Abraham JJS. Pharmaceuticalization of society in context: theoretical, empirical and health dimensions. Sociology. 2010;44(4):603–22.
29. Bell SE, Figert AE. Medicalization and pharmaceuticalization at the intersections: looking backward, sideways and forward. Soc Sci Med. 2012;75(5):775–83.
30. Kaczmarek E. Promoting diseases to promote drugs: the role of the pharmaceutical industry in fostering good and bad medicalization. Br J Clin Pharmacol. 2022;88(1):34–9.
31. Brody H. Patient-centered care or drug-centered care: the influence of pharmaceutical marketing on medical science and public health. In: Philosophical issues in pharmaceutics. Berlin: Springer; 2017. p. 109–24.
32. Foucault M. The history of sexuality. 1st American ed. New York: Pantheon Books; 1978.
33. Lupton D. Foucault and the medicalisation critique. Foucault, health and medicine. London: Routledge; 2002. p. 94–110.
34. Hofmann B. Managing the moral expansion of medicine. BMC Med Ethics. 2022;23(1):1–13.
35. Ioannidis JP. Meta-research: why research on research matters. PLoS Biol. 2018;16(3):e2005468.
36. Russell B. History of western philosophy. London: Routledge; 2004.
37. Ioannidis JP. Why most published research findings are false. PLoS Med. 2005;2(8):e124.
38. Baker M. Reproducibility crisis. Nature. 2016;533(26):353–66.
39. Ioannidis JP. The proposal to lower P value thresholds to .005. JAMA. 2018;319(14):1429–30.
40. Sismondo S. Epistemic corruption, the pharmaceutical industry, and the body of medical science. Front Res Metr Anal. 2021;6:614013.
41. Dusek V. What is technology? Defining or characterizing technology. In: Philosophy of technology: an introduction. Oxford: Blackwell; 2006. p. 26–38.
42. Verbeek P-P. Moralizing technology: understanding and designing the morality of things. Chicago: University of Chicago Press; 2011.
43. Ihde D. Technology and the lifeworld: from garden to earth. Bloomington: Indiana University Press; 1990.
44. Verbeek P-P. Toward a theory of technological mediation. In: Technoscience and postphenomenology: the Manhattan papers. Lanham: Lexington Books; 2015. p. 189.
45. Hofmann B, Svenaeus F. How medical technologies shape the experience of illness. Life Sci Soc Policy. 2018;14(1):1–11.
46. Hansson SO. Philosophy of medical technology. In: Philosophy of technology and engineering sciences. Amsterdam: Elsevier; 2009. p. 1275–300.
47. Hofmann B. On the value-ladenness of technology in medicine. Med Health Care Philos. 2001;4(3):335–45.

48. Clarke AE, Shim JK, Mamo L, Fosket JR, Fishman JR. Biomedicalization: technoscientific transformations of health, illness, and US biomedicine. Am Soc Rev. 2003;68:161–94.
49. Vogt H, Hofmann B, Getz L. The new holism: P4 systems medicine and the medicalization of health and life itself. Med Health Care Philos. 2016;19(2):307–23.
50. Jain KK. Personalized medicine. Curr Opin Mol Ther. 2002;4(6):548–58.
51. Cheek J. Healthism: a new conservatism? Qual Health Res. 2008;18(7):974–82.
52. Crawford R. Healthism and the medicalization of everyday life. Int J Health Serv. 1980;10(3):365–88.
53. Hofmann B. Medicalization and overdiagnosis: different but alike. Med Health Care Philos. 2016;19(2):253–64.
54. Sadegh-Zadeh K. Handbook of analytic philosophy of medicine. Berlin: Springer; 2012.
55. Loh E. Medicine and the rise of the robots: a qualitative review of recent advances of artificial intelligence in health. BMJ Leader. 2018;2:59.
56. Buruk B, Ekmekci PE, Arda B. A critical perspective on guidelines for responsible and trustworthy artificial intelligence. Med Health Care Philos. 2020;23(3):387–99.
57. Parikh RB, Teeple S, Navathe AS. Addressing bias in artificial intelligence in health care. JAMA. 2019;322(24):2377–8.
58. Yu K-H, Kohane IS. Framing the challenges of artificial intelligence in medicine. BMJ Qual Saf. 2018;28:238.
59. Ploug T, Holm S. The right to refuse diagnostics and treatment planning by artificial intelligence. Med Health Care Philos. 2020;23(1):107–14.
60. Karches KE. Against the iDoctor: why artificial intelligence should not replace physician judgment. Theor Med Bioeth. 2018;39(2):91–110.
61. Heidegger M. The question concerning technology, vol. 99; 1954. p. 113.
62. Descartes R. Meditations on first philosophy. In: Seven masterpieces of philosophy. London: Routledge; 2016. p. 63–108.
63. Dreyfus HL. On the internet. London: Routledge; 2008.
64. Carel H. Illness: the cry of the flesh. London: Routledge; 2018.
65. Merleau-Ponty M, Smith C. Phenomenology of perception. London: Routledge; 1962.
66. Topol E. Deep medicine: how artificial intelligence can make healthcare human again. London: Hachette UK; 2019.

The Future of Economy in Healthcare

12

Marzieh Nosrati, Mohammad Hossein Sharifnia, Soroush Ahmadi, Negar Sarhangi, Mandana Hasanzad, and Shekoufeh Nikfar

What Will You Learn in This Chapter?
This chapter generally covers subjects, including the significance of using economic concepts in the field of health, the evolution of health costs in recent years, the effects of developing technology, and events on healthcare expenditures. Statistics demonstrate that healthcare expenses have increased at a remarkable rate in recent years. The health system's spending is predicted to rise in the upcoming years because of the promotion of expensive technologies and interventions. On the other hand, the use of artificial intelligence and machine learning techniques can help control and lower part of these expenses, including those associated with medication development, prevention, diagnosis, and treatment.

Rational and Importance
Economics plays a crucial role in healthcare systems, as it helps allocate limited resources efficiently. Recent increases in healthcare costs have exceeded economic development, raising concerns for governments. Factors such as advancements in computer and electronic sciences, medical devices, and new treatments can influ-

M. Nosrati
Personalized Medicine Research Center, Endocrinology and Metabolism Clinical Sciences Institute, Tehran University of Medical Sciences, Tehran, Iran

Faculty of Pharmacy, Department of Pharmacoeconomics and Pharmaceutical Administration, Tehran University of Medical Sciences, Tehran, Iran

M. H. Sharifnia
Faculty of Pharmacy, Department of Pharmacoeconomics and Pharmaceutical Administration, Tehran University of Medical Sciences, Tehran, Iran

School of Chemistry, University of Tehran, Tehran, Iran

S. Ahmadi
Division of Pharmaceutical Outcomes and Policy, UNC Eshelman School of Pharmacy, Chapel Hill, NC, USA

N. Sarhangi · M. Hasanzad
Personalized Medicine Research Center, Endocrinology and Metabolism Clinical Sciences Institute, Tehran University of Medical Sciences, Tehran, Iran

S. Nikfar (✉)
Faculty of Pharmacy, Department of Pharmacoeconomics and Pharmaceutical Administration, Tehran University of Medical Sciences, Tehran, Iran

Endocrinology and Metabolism Clinical Sciences Institute, Tehran University of Medical Sciences, Tehran, Iran

M. Hasanzad (ed.), *A Glimpse at Medicine in the Future*,
https://doi.org/10.1007/978-981-97-4974-4_12

ence health expenditures. If policymakers consider this issue and pay attention to how health costs will change in the upcoming years and the causes affecting these changes, they can make far more intelligent decisions about which way to allocate the budgets.

12.1 An Introduction to the Role of Economics in the Healthcare System

Considering the concept of economics in the field of health care involves several aspects. Generally, the conflict between unlimited needs and scarce resources is one of the major issues facing healthcare systems. Therefore, limited resources have to be efficiently utilized to meet needs in healthcare systems, and economics always has a remarkably important role in decision-making regarding the allocation of limited resources. Moreover, spending on healthcare generally improves society's health status and increases economic growth. However, in recent years, the rate of increase in health system costs has exceeded the rate of increase in economic development, raising concerns. For instance, the OECD's yearly rise in health spending from 2000 to 2015 was 3.0%, whereas GDP growth over the same period was 2.3%. Since preserving financial stability is one of the top priorities for governments, forecasting changes in healthcare expenditures over the next few years can be extremely valuable in defining goals and developing strategies for the future [1]. Numerous factors, some of which will be covered in this chapter, might influence both the amount and the direction of alterations in health expenditures. For example, the tremendous development of computer and electronic sciences, massive inventions in the area of medical and biomedical devices, and novel medical and pharmaceutical interventions which have been introduced in the past few decades, result in higher expenses. Besides, the remarkable advancements of digital, electronic, and intelligent computerized technologies could lead to dramatic cost-minimizations of several ambitious treatment approaches such as precision medicine, gene therapy, cell therapy, and any other targeted therapeutical interventions that once were categorized as unattainable and unaffordable, but now they are possessing a considerable position with more purposeful prospects, by playing a more significant role in future's medical interventions.

Economics has always played an important role in making medical decisions to allocate resources more efficiently. Regarding the extremely expensive newly introduced medical technologies, which possess promising therapeutic effects on treating or even on completely curing and remedying once untreatable diseases, the role of economics in making decisions about resource allocation to these interventions has become more important than before.

This chapter lists some factors or events, new emerging interventions, or technologies that may have a significant impact on alterations to health system spending and to address how effective they might be on the costs in the future years.

12.2 Some Factors and Events Affecting the Future of Healthcare Costs

12.2.1 Climate Change

Climate changes, including rising sea levels, extreme weather events, droughts, flooding, and wildfires, are causing health issues like infectious diseases, respiratory conditions, heat-related morbidity, and undernutrition [2–5]. A study, which is an overview of the systematic review, lists all the health impacts resulting from climate change. Climate impact is primarily associated with changes in meteorological factors (such as high temperature, heat, sunlight intensity, cold, and humidity), extreme weather events (flooding), air quality (e.g., wildfire smoke exposure), and others (environmental pollutants). These events can cause increasing occurrences of infectious diseases, diabetes, malignancies, respiratory (probably through air quality), cardiovascular, neurological diseases, mental disorders (e.g., distress, anxiety, depression), and birth defects, as well as other nutritional (presumably through crop productivity and food insecurity), skin, allergies, and renal complications [6]. Research findings indicate that particularly in low- and middle-income nations there is a relationship between overall climate change and increasing disability-adjusted life years (DALY), a measure of the burden of disease [7]. Furthermore, as the prevalence of these illnesses rises, there may be a greater need for the services provided by the health system, which would raise expenses [6]. Immediate health sector reaction costs (extra medication and treatment expenses), as well as other expenditures such as sickness, injuries, fatalities, and missed work time, are examples of potential health effects that might have an impact on the health system's expenses [8].

12.2.2 Aging

Today, the average lifespan is increasing due to improvements in social and economic conditions, advancements in medical technology, and rising living standards [9]. Economic growth drives up the demand for medical services, consequently raising healthcare costs and extending life expectancy [10]. As a result, the elderly population is expanding, leading societies to confront aging demographics. An aging society is typically characterized as having either 10% or 7% of its population aged 65 years or older, or 60 years or older, respectively, based on the total population [9]. Research indicates that healthcare expenses begin to rise around the age of 50 and continue escalating until around age 90, after which they start to gradually decrease [11]. A significant challenge to the fiscal sustainability of public finances globally is the rising proportion of the population aged 65 years or older. This demographic shift has driven up healthcare spending worldwide due to heightened demand for medical and long-term care services. The increasing numbers of older adults have amplified the need for long-term healthcare services, including expensive technologies and hospitalizations. Additionally, there's been a surge in the prevalence of chronic diseases like diabetes, cancer, chronic renal disease, and cardiovascular conditions, along with their associated complications [10]. These factors have put the sustainability of healthcare systems at risk due to the increasing healthcare expenditures [9, 10].

12.2.3 Pandemic

The occurrence of a pandemic is associated with a significant effect on the pattern of changes in health costs. In the last pandemic, the resulted health and economic difficulties became worse in 2020 when governments around the world increased health spending to a record high of US $9 trillion (10.3% of global GDP) [12]. Between 2019 and 2021, spending per capita at all provider types grew, with preventive care providers—such as public health institutions and disease control agencies—growing at the greatest rate. COVID-19's effect on Investment in Health Capital can be listed as:

- A total of 64 nations had an increase in health capital investment, which is equal to 5.2% of current health spending or 0.4% of GDP.
- Countries with low and lower-middle incomes had the fastest investment growth.
- Approximately 66% of health providers in countries with high income and slightly over 50% in low- and middle-income countries received funding from hospitals.
- The purchase of devices and machinery increased dramatically in low-income nations.
- More than 75% of capital investments in health care in high- and middle-income nations are supported by the government [13].

In general, the COVID-19 pandemic affected healthcare expenditures in various aspects:

- COVID-19 caused a spike in inpatient admission to hospitals and other healthcare institutions, which increased the price of staffing, medical supplies, and equipment used in patient care [14].
- Healthcare costs grew as a result of the rapid growth in demand for personal protective equipment (PPE), which led to shortages and price increases [15].
- The pandemic accelerated the adoption of digital health and telemedicine solutions,

requiring investments in platforms [16]. While these technologies can offer cost-saving benefits in the long run, initial investments and setup costs contributed to increased healthcare spending during the pandemic.
- Significant resources were directed towards the research and development of COVID-19 vaccines, treatments, and therapeutics, contributing to healthcare spending [17].
- Disruptions in non-COVID-19 care, such as elective surgeries [18] and routine clinical practices or important visits [19], led to revenue losses, straining healthcare budgets.

In summary, the pandemic (like the last experience, COVID-19) has had multifaceted effects on healthcare expenditure, with increased costs in certain areas offset by reductions in others. The long-term financial implications and the sustainability of healthcare funding models remain topics of discussion and concern for policymakers, healthcare providers, and stakeholders globally.

12.3 New Emerging Interventions

12.3.1 Regenerative Medicine

Regenerative medicine has promise for normalizing congenital abnormalities and healing or replacing tissues and organs damaged by illness, trauma, or aging [20]. It has various kinds of interventions, such as stem cell treatments, cartilage regeneration, Platelet-Rich Plasma (PRP), and prolotherapy. The rise in chronic and degenerative diseases, which put financial strain on healthcare providers, along with developments in emerging technologies like nanotechnology, bioengineering, and stem cell therapy, has created a demand for regenerative medicine products [21].

Although there may be therapeutic uses for regenerative medicine, product development and clinical trials will require an immense amount of funding. Scientists and startups need to show the technology is profitable to persuade governments and large pharmaceutical companies to finance it [22].

12.3.2 Gene Therapy

Gene therapy has the potential to revolutionize healthcare economically by reducing long-term treatment costs. While initial expenses for gene therapy are high, the possibility of curing or significantly mitigating genetic diseases can lead to substantial savings in chronic healthcare expenses. Additionally, it may improve overall productivity as individuals benefit from better health, potentially reducing the economic burden of illness. This approach offers numerous benefits, resulting in significant economic impacts in healthcare, as below:

1. Reducing the burden of chronic diseases and rare diseases.
2. Improved productivity and quality of life.
3. Saving costs associated with less effective or invasive methods.
4. Economic opportunities in biotechnology industry.
5. Shift from palliative treatments toward more effective and curative ones.

Although these therapies have incredible benefits, challenges such as affordability, accessibility, and long-term effects on the healthcare system still need to be considered [23–27].

Some of the examples of gene therapy addressed these two aspects (advantages and disadvantages) of applying gene therapy in practice, such as:

- Gene therapy offers long-term, permanent treatment for hemophilia, with initial costs potentially reducing over time. While frequent clotting factor infusions are part of standard treatment, they require ongoing expenses [28, 29].
- Gene therapy targets specific genetic abnormalities in malignancies, offering a focused, potentially curative treatment with fewer radiation or chemotherapy sessions, saving money over time compared to conventional treatments [30].
- Emerging gene therapies focus on the genetic cause of cystic fibrosis, offering a more potent

and potentially curative treatment than traditional approaches [31, 32].

- Gene therapy for diabetes, despite being in its early stages, could potentially increase insulin production, potentially reducing insulin injections and drug costs, while insulin therapy, oral drugs, and lifestyle changes as part of the standard diabetes treatment are lifelong conditions [33].

Gene therapy success and economic impact depend on factors like efficacy, safety, and healthcare system issues. It has the potential to yield substantial advantages, as seen in the cases mentioned above. Nevertheless, the main barrier is its remarkably expensive cost. Therefore, it is important to weigh every advantage and disadvantage of implementing this novel intervention in healthcare by applying the principles of health economics to allocate resources efficiently.

It's important to consider that the economic impact of gene therapies extends beyond their initial costs. Factors such as the potential for a cure, reductions in other healthcare costs, and improvements in patients' quality of life contribute to the overall economic assessment. Ongoing research and real-world evidence are crucial for a more comprehensive understanding of the economic implications of gene therapies in different therapeutic areas.

12.3.3 Precision Medicine

Personalized therapy, which is sometimes referred to as precision medicine or personalized medicine, is an approach to providing medical care to each patient according to their particular traits, including genetics, environment, lifestyle, and biomarkers. By providing more effective and safer interventions, this tailored strategy seeks to enhance patient outcomes [34]. Precision medicine can have a variety of financial effects in healthcare systems. It frequently involves more precise and individualized interventions for prediction, prevention, and treatment, which may save total healthcare expenditures by avoiding unnecessary procedures and adverse consequences. It may also result in better patient outcomes and lower long-term healthcare costs. Research indicates that the application of precision medicine has increased in the past few years. The findings of the study by GenomWeb[1] indicate that not only are most organizations across all sectors implementing or preparing to implement precision medicine programs, but they are also eager to expand their use to cover a wider range of disease areas [35]. Moreover, in 2022, 34% of all newly approved medicines by the FDA are precision medicine interventions [36]. However, it is noticeable that there may be significant up-front expenses associated with the creation and use of these kinds of therapies. Considering all these points, economics will be having a major role in determining the amount of budget and resource allocation to these interventions for more routine use in practice.

Precision medicine includes targeted therapies, which are primarily used in the oncology area, as one of its key subcategories. From the standpoint of health economics, there are some points about this sort of intervention, such as:

- Targeted therapies may have higher drug costs but may reduce expenses related to adverse effects and hospitalizations, comparing directly with current standard treatments.
- The widespread utilization of these therapies is followed by lengthy, costly, and probable false positive and negative answers, in addition to the high expenses associated with targeted treatments.
- In economic evaluation studies of targeted therapies, it needs to be assessed the long-term outcomes of targeted medicines, considering their potential for extended treatment responses, reduced disease recurrence, and

[1] GenomeWeb is an Internet news source founded in 1997 to cover economic and scientific news regarding genomics.

lower healthcare expenses compared to current standard treatments.
- Investigating the effects of targeted medicines on patients' capacity for work and overall productivity is crucial [37–41].

12.3.4 Towards Value-Based Healthcare

The enormous advantages of emerging medical techniques, including regenerative medicine, gene therapy, and precision medicine, will drive up demand for their utilization in future decades. These interventions appear to have the potential to significantly raise the expenditures of the healthcare system in the upcoming years, given their high cost. However, a number of these interventions' advantages—many of which were covered in-depth above—will also have a beneficial economic effect on the healthcare system. Fortunately adopting a value-based approach to decision-making in the health system may influence how policymakers allocate resources for these therapies and determine how widely they are used regularly in practice.

As health systems are increasingly encouraged to embrace a value-based care model, a paradigm shift is observed within the health economics and health services research to incorporate patient perspectives in policy, as well as clinical decision-making. Value-based care puts the patient at the center of all efforts and seeks to optimize outcomes from the patient's perspective while containing costs and enhancing efficiency [42]. Achieving these goals requires the intentional design of the incentive system to motivate higher-value care. The development of novel value-based reimbursement schemes, such as the design of Accountable care organizations (ACOs) by the Centers for Medicare & Medicaid Services (CMS) to provide high-quality medical care to Medicare patients, are an example of this transition [43]. Moreover, research is evolving on measuring and improving outcomes that matter most to individuals and fostering a more patient-centered and coordinated care approach. Enhancing the overall patient experience and quality of care, while promoting the efficient use of resources within the healthcare ecosystems, are the major goals. The emerging care delivery models emphasize a collaborative team approach to patient care and the exchange of patient information, facilitating coordinated care and streamlined measurement of outcomes. While patient-reported outcomes are not a new concept, the health care systems are becoming more efficient at their real-time measurement and integration with other health data by harnessing the power of technology. In addition to using these data to support shared clinical decision-making, aligned with patient's preferences, this will also impact the evaluation of provider performance and comparative effectiveness research in the future [44]. For instance, in an effort to support the research and provide access to the information that can enable clinicians and patients to make better-informed choices and accelerate the care transformation, the Affordable Care Act (ACA) created the Patient-Centered Outcomes Research Institute (PCORI). PCORI is committed to producing information that can assist healthcare professionals, caregivers, and patients in making clinical decisions and is collaborating with the Agency for Healthcare Research and Quality (AHRQ) to distribute this information [45]. Moreover, to broaden and standardize the concept of value, the Special Task Force by the International Society for Pharmacoeconomics and Outcomes Research (ISPOR) defined a series of novel elements to be considered in value assessments of health care technologies. In addition to the quality-adjusted life-years (QALYs) and costs that have been routinely evaluated in the current Health Technology Assessment (HTA) practices, they have identified novel elements such as productivity, adherence-improving, reduction in uncertainty, fear of contagion, insurance value, severity of disease, value of hope, real option value, equity, and scientific spillovers to be considered [46]. More research on how to systematically measure and incorporate these elements in policy and regulatory decision-making will be essential in the future. Despite the recent efforts and advancements, the transition to value-based healthcare has proven to be more challenging than initially anticipated, and much of the hard work lies ahead of us.

12.4 New Technologies

12.4.1 Artificial Intelligence

Artificial intelligence (AI) is one of the recently introduced computerized tools that has shown some promising effects on facilitating processes. One of the most challenging and expensive issues facing the healthcare system is how to analyze large-scale data [47, 48]. Artificial intelligence and machine learning are tools that facilitate big data analysis and provide an opportunity to significantly improve outcomes in the prevention, diagnosis, and treatment of diseases [34]. Gathering suitable and accurate clinical data is the most pivotal element of making proper clinical decisions. The process of decision-making in clinical situations is extremely sensitive and is one of the key role complexities that exist in the healthcare systems. Without proper clinical evaluations, diagnoses, and observations, it could not be expected to achieve accurate clinical data, and certainly, the eventual medical interventions' purposes will not be achieved. AI can facilitate both the data-driven processes of analyzing clinical situations and evaluating clinical observations and diagnoses, as well as decision-making processes based on the obtained clinical data [49, 50].

- *AI assistance in disease prevention*: The combination of genotype analysis and AI creates a potential for prediction of the phenotypes, which can be effective in adopting the required actions to prevent diseases, alleviate the economic burden of diseases, and manage healthcare costs.

- By AI, machine learning, and other technologies, it will be possible to create predictive models that predict future events and lead to better patient outcomes and lower costs. One area where predictive analytics can be helpful is in identifying patients who may be at risk of developing chronic conditions like endocrine or cardiac diseases. Predictive models created by data analysis can identify patients who may be at higher risk of developing these conditions and target interventions to prevent or treat them.
- *AI assistance in disease diagnosis*: Despite all of the advances in science, accurately diagnosing diseases remains a global challenge. AI diagnostic technologies have the potential to save time and money and enhance accuracy over traditional methods [34]. AI has proven its capabilities in diagnosing multiple diseases and forecasting the possible future of patients' clinical conditions. AI algorithms, including convolutional neural networks (CNNs), as the most prevalent deep learning and machine learning architecture typologies, deep autoencoders, and generative adversarial networks that were proven to have a more suitable function for making clinical decisions and conducting unsupervised learning considering unlabeled data, have demonstrated their ability to analyze medical images such as CT scans, MRIs, and PET scans to detect different diseases [51]. AI has shown its medical images' analyzing potential in detecting tuberculosis, stroke, and cerebrovascular disease; also, it has demonstrated a promising aid in evaluating clinical input data such as blood pressure, heart rate, demographics, and other related health function parameters (e.g., ejection fraction, pulmonary capillary wedge pressure, end-systolic and diastolic volumes, etc.) to diagnose hypertension, cardiovascular, and other chronic diseases [52, 53].
- *AI assistance in disease treatment*: AI has grown into a useful tool for improving the application of precision medicine interventions, using its capacity to evaluate large, complicated datasets, predict outcomes, and enhance therapeutic approaches. Moreover, it has an enormous effect on dose optimization and adverse medication event prediction, which improves patient safety and treatment outcomes. Besides, due to the rapid healthcare-related technological advancements, many

possible options exist for curing and treating different patients, but electing the best plausible treatment approach for each patient remarkably relies on multiscale clinical data analysis that AI can facilitate a more rational selection of these medical intervention approaches [34].

Besides, the advancement of digital technology and social digital literacy has created further challenges for the healthcare system. For instance, social media, which is called "an era of fake news," brought many problems for effective medical administrations due to the creation of fear about the side effects or the probability of ineffectiveness of some medical interventions that have led to noncompliance or even insubordination of patients to comply with the provided medical instructions for treating their diseases by healthcare experts [47, 48]. Patient education is crucial for healthcare delivery, with AI enhancing this by providing personalized, interactive information and guidance to patients and their caregivers [34].

12.5 Conclusion

The future of healthcare expenditure is influenced by various factors, including demographic shifts and an aging population, technological advancements, policy changes, and global health challenges.

Aging will likely cause an increased demand for healthcare services, especially for chronic and age-related conditions, leading to higher healthcare expenditures due to the need for specialized care and long-term treatments. In addition, new emerging medical interventions, such as regenerative medicine, precision medicine, gene editing, and AI-driven diagnostics, have the potential to improve health outcomes but may also contribute to increased costs as new technologies often come with higher price tags. Also, ongoing and emerging global health challenges, such as pandemics and climate-related health risks, may require substantial investments in preparedness, research, and healthcare infrastructure, impacting future healthcare budgets. However, including shifts towards universal healthcare coverage, value-based care models, and alternative payment models, can influence healthcare expenditure by altering how services are delivered, funded, and reimbursed.

Balancing the need for quality healthcare services with financial sustainability will be a key challenge for healthcare systems globally. Innovations in healthcare delivery, financing mechanisms, and cost-effective interventions will be essential to manage and optimize healthcare expenditure.

In summary, the future of healthcare expenditure is shaped by a complex interplay of demographic, technological, policy, and global health factors. Proactive planning, innovation, and collaboration across sectors will be crucial to address the evolving challenges and opportunities in healthcare financing and delivery.

References

1. https://www.oecdlibrary.org//sites/3d1e710cen/index.html?itemId=/content/component/3d1e710c-en#.
2. Watts N, Amann M, Arnell N, Ayeb-Karlsson S, Beagley J, Belesova K, et al. The 2020 report of the Lancet Countdown on health and climate change: responding to converging crises. Lancet. 2021;397(10269):129–70.
3. Hsiang SM, Burke M. Climate, conflict, and social stability: what does the evidence say? Clim Chang. 2014;123:39–55.
4. Hsiang SM, Burke M, Miguel E. Quantifying the influence of climate on human conflict. Science. 2013;341(6151):1235367.
5. Patz JA, Frumkin H, Holloway T, Vimont DJ, Haines A. Climate change: challenges and opportunities for global health. JAMA. 2014;312(15):1565–80.
6. Rocque RJ, Beaudoin C, Ndjaboue R, Cameron L, Poirier-Bergeron L, Poulin-Rheault R-A, et al. Health effects of climate change: an overview of systematic reviews. BMJ Open. 2021;11(6):e046333.
7. Zhang Y, Bi P, Hiller JE. Climate change and disability–adjusted life years. J Environ Health. 2007;70(3):32–8.
8. Ebi KL, Hess JJ, Watkiss P. Health risks and costs of climate variability and change. In: Disease control priorities. Washington, DC: World Bank; 2017. p. 7.
9. Li L, Du T, Hu Y. The effect of population aging on healthcare expenditure from a healthcare demand perspective among different age groups: evidence from

Beijing City in the People's Republic of China. Risk Manag Healthc Policy. 2020;13:1403–12.
10. Lopreite M, Zhu Z. The effects of ageing population on health expenditure and economic growth in China: a Bayesian-VAR approach. Soc Sci Med. 2020;265:113513.
11. Cylus J, Roubal T, Ong P, Barber S. The economics of healthy and active ageing series. In: Sustainable health financing with an ageing population: implications of different revenue raising mechanisms and policy options. Brussels: European Observatory on Health Systems and Policies and the WHO Centre for Health Development; 2019.
12. https://www.who.int/publications-detail-redirect/9789240064911.
13. https://reliefweb.int/report/world/global-spending-health-coping-pandemic.
14. Emanuel EJ, Persad G, Upshur R, Thome B, Parker M, Glickman A, et al. Fair allocation of scarce medical resources in the time of Covid-19. Mass Med Soc. 2020;382:2049–55.
15. Ranney ML, Griffeth V, Jha AK. Critical supply shortages—the need for ventilators and personal protective equipment during the Covid-19 pandemic. N Engl J Med. 2020;382(18):e41.
16. Wong MYZ, Gunasekeran DV, Nusinovici S, Sabanayagam C, Yeo KK, Cheng C-Y, et al. Telehealth demand trends during the COVID-19 pandemic in the top 50 most affected countries: infodemiological evaluation. JMIR Public Health Surveill. 2021;7(2):e24445.
17. Wouters OJ, Shadlen KC, Salcher-Konrad M, Pollard AJ, Larson HJ, Teerawattananon Y, et al. Challenges in ensuring global access to COVID-19 vaccines: production, affordability, allocation, and deployment. Lancet. 2021;397(10278):1023–34.
18. Mehta A, Awuah WA, Ng JC, Kundu M, Yarlagadda R, Sen M, et al. Elective surgeries during and after the COVID-19 pandemic: case burden and physician shortage concerns. Ann Med Surg. 2022;81:104395.
19. Barach P, Fisher SD, Adams MJ, Burstein GR, Brophy PD, Kuo DZ, et al. Disruption of healthcare: will the COVID pandemic worsen non-COVID outcomes and disease outbreaks? Prog Pediatr Cardiol. 2020;59:101254.
20. Mao AS, Mooney DJ. Regenerative medicine: current therapies and future directions. Proc Natl Acad Sci. 2015;112(47):14452–9.
21. Jessop ZM, Al-Sabah A, Francis WR, Whitaker IS. Transforming healthcare through regenerative medicine. BMC Med. 2016;14:1–6.
22. Mason C, Dunnill P. The strong financial case for regenerative medicine and the regen industry. Regen Med. 2008;3:351.
23. Ten Ham RM, Klungel OH, Leufkens HG, Frederix GW. A review of methodological considerations for economic evaluations of gene therapies and their application in literature. Value Health. 2020;23(9):1268–80.
24. Qiu T, Pochopien M, Liang S, Saal G, Paterak E, Janik J, et al. Gene therapy evidence generation and economic analysis: pragmatic considerations to facilitate fit-for-purpose health technology assessment. Front Public Health. 2022;10:773629.
25. Garrison LP, Jackson T, Paul D, Kenston M. Value-based pricing for emerging gene therapies: the economic case for a higher cost-effectiveness threshold. J Manag Care Spec Pharm. 2019;25(7):793–9.
26. Wong CH, Li D, Wang N, Gruber J, Conti RM, Lo AW. Estimating the financial impact of gene therapy in the US. Cambridge: National Bureau of Economic Research; 2021.
27. Quinn C, Young C, Thomas J, Trusheim M, Group MNFW. Estimating the clinical pipeline of cell and gene therapies and their potential economic impact on the US healthcare system. Value Health. 2019;22(6):621–6.
28. Jayasundera KT, Abuzaitoun RO, Lacy GD, Abalem MF, Saltzman GM, Ciulla TA, et al. Challenges of cost-effectiveness analyses of novel therapeutics for inherited retinal diseases. Am J Ophthalmol. 2022;235:90–7.
29. Bolous NS, Chen Y, Wang H, Davidoff AM, Devidas M, Jacobs TW, et al. The cost-effectiveness of gene therapy for severe hemophilia B: a microsimulation study from the United States perspective. Blood. 2021;138(18):1677–90.
30. Lyman GH, Nguyen A, Snyder S, Gitlin M, Chung KC. Economic evaluation of chimeric antigen receptor T-cell therapy by site of care among patients with relapsed or refractory large B-cell lymphoma. JAMA Netw Open. 2020;3(4):e202072.
31. Pranke I, Golec A, Hinzpeter A, Edelman A, Sermet-Gaudelus I. Emerging therapeutic approaches for cystic fibrosis. From gene editing to personalized medicine. Front Pharmacol. 2019;10:121.
32. Miah KM, Hyde SC, Gill DR. Emerging gene therapies for cystic fibrosis. Expert Rev Respir Med. 2019;13(8):709–25.
33. Tan SY, Wong JLM, Sim YJ, Wong SS, Elhassan SAM, Tan SH, et al. Type 1 and 2 diabetes mellitus: a review on current treatment approach and gene therapy as potential intervention. Diabetes Metab Syndr Clin Res Rev. 2019;13(1):364–72.
34. Alowais SA, Alghamdi SS, Alsuhebany N, Alqahtani T, Alshaya AI, Almohareb SN, et al. Revolutionizing healthcare: the role of artificial intelligence in clinical practice. BMC Med Educ. 2023;23(1):689.
35. https://www.oracle.com/a/ocom/docs/industries/healthcare/research-report-genomeweb-trends-precision-medicine.pdf.
36. https://www.personalizedmedicinecoalition.org/.
37. Ye F, Dewanjee S, Li Y, Jha NK, Chen Z-S, Kumar A, et al. Advancements in clinical aspects of targeted therapy and immunotherapy in breast cancer. Mol Cancer. 2023;22(1):105.
38. Piawah S, Venook AP. Targeted therapy for colorectal cancer metastases: a review of current methods of molecularly targeted therapy and the use of tumor

biomarkers in the treatment of metastatic colorectal cancer. Cancer. 2019;125(23):4139–47.
39. Ward ZJ, Scott AM, Hricak H, Atun R. Global costs, health benefits, and economic benefits of scaling up treatment and imaging modalities for survival of 11 cancers: a simulation-based analysis. Lancet Oncol. 2021;22(3):341–50.
40. Seo MK, Cairns J. How are we evaluating the cost-effectiveness of companion biomarkers for targeted cancer therapies? A systematic. BMC Cancer. 2021;21:980.
41. Tarride J-E, Cheung M, Hanna TP, Cipriano LE, Regier DA, Hey SP, et al. Platform, basket, and umbrella trial designs: issues around health technology assessment of novel therapeutics. Can J Health Technol. 2022;2(7).
42. Porter ME. What is value in health care. N Engl J Med. 2010;363(26):2477–81.
43. NEJM Catalyst. What is value-based healthcare? NEJM Catal. 2017;3(1).
44. Baumhauer JF, Bozic KJ. Value-based healthcare: patient-reported outcomes in clinical decision making. Clin Orthop Relat Res. 2016;474:1375–8.
45. Burwell SM. Setting value-based payment goals—HHS efforts to improve US health care. N Engl J Med. 2015;372(10):897–9.
46. Lakdawalla DN, Doshi JA, Garrison LP Jr, Phelps CE, Basu A, Danzon PM. Defining elements of value in health care—a health economics approach: an ISPOR Special Task Force report [3]. Value Health. 2018;21(2):131–9.
47. Yang YC, Islam SU, Noor A, Khan S, Afsar W, Nazir S. Influential usage of big data and artificial intelligence in healthcare. Comput Math Methods Med. 2021;2021:5812499.
48. Rallapalli S, Gondkar R, Ketavarapu UPK. Impact of processing and analyzing healthcare big data on cloud computing environment by implementing hadoop cluster. Procedia Comput Sci. 2016;85:16–22.
49. Goodfellow I, Pouget-Abadie J, Mirza M, Xu B, Warde-Farley D, Ozair S, et al. Advances in neural information processing systems, vol. 27. Red Hook: Curran Associates, Inc; 2014. p. 2672–80.
50. Kingma DP, Welling M. Auto-encoding variational bayes. arXiv preprint arXiv:13126114; 2013.
51. Litjens G, Kooi T, Bejnordi BE, Setio AAA, Ciompi F, Ghafoorian M, et al. A survey on deep learning in medical image analysis. Med Image Anal. 2017;42:60–88.
52. Kumar Y, Koul A, Singla R, Ijaz MF. Artificial intelligence in disease diagnosis: a systematic literature review, synthesizing framework and future research agenda. J Ambient Intell Humaniz Comput. 2022;14:1–28.
53. Shimron E, Perlman O. AI in MRI: computational frameworks for a faster, optimized, and automated imaging workflow. Bioengineering. 2023;10:492.

GPSR Compliance

The European Union's (EU) General Product Safety Regulation (GPSR) is a set of rules that requires consumer products to be safe and our obligations to ensure this.

If you have any concerns about our products, you can contact us on ProductSafety@springernature.com

In case Publisher is established outside the EU, the EU authorized representative is:

Springer Nature Customer Service Center GmbH
Europaplatz 3
69115 Heidelberg, Germany

Batch number: 10732801

Printed by Printforce, the Netherlands